United States Army in Vietnam

Public Affairs: The Military and the Media, 1962–1968

by

William M. Hammond

Center of Military History
United States Army
Washington, D.C., 1990

Library of Congress Cataloging-in-Publication Data

Hammond, William M.
Public affairs.

(United States Army in Vietnam)
"GPO PO No. 80001."
Bibliography: p.
Supt. of Docs. no.: D 114.7/3:P96
1. Vietnamese Conflict, 1961-1975—Journalists. 2. Armed Forces and mass media—United States—History. I. Title. II. Series.
DS559.46.H36 1988 959.704'38 88-931

First Printed 1988—CMH Pub 91-13-1

For sale by the Superintendent of Documents, U.S. Government Printing Office Washington, D.C. 20402

United States Army in Vietnam

David F. Trask, General Editor

Advisory Committee

(As of 8 July 1987)

U.S. Army Center of Military History

. . . to Those Who Served

Foreword

The U.S. Army in Vietnam series documents the Army's role in the Vietnam War. Most of the studies in the series deal with combat operations, staff relations, or with technical aspects of the war: logistics, engineering, and communications. A few depart from that format, taking their direction from the unique nature of the conflict and the circumstances that came to surround it. This is such a book.

The Vietnam War was the first in modern history fought without the filtering that Army field press censorship had provided during World Wars I and II and the Korean War. As a result, the American news media assumed an unprecedented role in describing and defining the nature of the conflict for the American public and the Congress. Official preoccupation with the public reaction to news, both good and bad, in turn became an important influence on the policies governing what Army commanders could and could not do in the field.

This book examines the tensions and controversies that developed as the war lengthened and the news media went about their traditional tasks. The first of two volumes on the subject, it draws upon previously unavailable Army and Defense Department records to interpret the role the press played during the war. It also sheds new light on official policies designed to govern relations between the military and the media in Vietnam.

The story has been difficult to write. Many of the pertinent documents were lost in the chaos that accompanied the fall of Saigon; others have disappeared with the passage of time. The author has nevertheless managed to reconstruct much of what happened, to the benefit of future generations of American soldiers and newsmen who regrettably may again find themselves involved in battle.

Washington, D.C.
15 August 1987

WILLIAM A. STOFFT
Brigadier General, USA
Chief of Military History

The Author

William M. Hammond is a graduate of the Catholic University of America, where he received the S.T.B., M.A., and Ph.D. degrees. He has taught American history at the University of Maryland Baltimore County and political science at Trinity College in Washington, D.C. He is the author of the Army's history of the selection and interment of the Vietnam Unknown Soldier, *The Unknown Serviceman of the Vietnam Era*; several chapters in *The Vietnam War* (Crown Publishers, Inc.); and numerous shorter articles and publications. He is currently writing the companion volume to this study, Public Affairs: The Military and the Media, 1968–1973.

Preface

As the war in South Vietnam developed, a belief grew in official circles that the attitude of the American public would play a major part in determining whether the United States would achieve its goals in that conflict. Reasoning that the news media had a profound influence on public opinion, civilian officials assumed a larger role than ever before in the formulation of military policies to manage the press at the scene of the fighting. In the process they affected not only the handling of the news media in Vietnam but in some measure also the conduct of the war.

This study examines the evolution of the U.S. government's public affairs policies in Vietnam between 1962 and 1968. Adopting a broad viewpoint in order to depict the many influences—civilian and military, political and diplomatic—that bore upon the conduct of public affairs, the work describes the tensions that developed between the institutions of the press and the military as the war grew and as each served its separate ends. It observes events from the perspective of the Military Assistance Command's Office of Information in Saigon, which carried much of the burden of press relations, but necessarily considers as well the role of the White House, the State and Defense Departments, and the U.S. embassy in Saigon in the creation of information policy. By drawing together many disparate strands, the book seeks to delineate some of the issues and problems that can confront an open society whenever it wages war.

Many people contributed to the successful completion of this book. Although I cannot mention all here, a number deserve special notice. Former Chiefs of Military History Brig. Gens. James Collins and Douglas Kinnard and Chief Historians Maurice Matloff and David Trask approved the concept of the book and offered continual advice and support. Ann David, Douglas Shoemaker, and James Broussard contributed valuable research assistance. Maj. Gen. Winant Sidle, Barry Zorthian, John Mueller, Lawrence Lichty, Peter Braestrup, Rodger Bankson, Charles MacDonald, Stanley Falk, John Schlight, Col. James Ransone, Col. James Dunn, Lt. Col. Richard Perry, George MacGarrigle, Vincent Demma, Richard Hunt, Jeffrey Clarke, Joel Meyerson, and Ronald Spector all read portions of the draft, contributing important observations. The CMH librarian, Carol Anderson, gave valuable assistance, as did Harry Zubkoff and his staff at the Department of the Air Force's News Clipping and Analysis Service. Jack Shulimson of the Office of Marine Corps History and William Heimdahl of the Office of Air Force History also deserve special thanks, as does Gustenia Scott who typed the manuscript and inserted corrections and revisions.

Thanks, as well, should go to Arthur Hardyman, who designed the layout of the book; to Howell Brewer, Jr., who coordinated the collection of the photography; and to Linda Cajka, who researched and prepared the maps. Special recognition belongs to the editors—Catherine Heerin, Barbara Gilbert, and Diane Sedore Arms—who gave much more than duty required.

My wife, Lilla, and my children, Michael and Elizabeth, deserve a special mention for tolerating a husband and parent who seemed at times more attentive to his book than to them.

I alone am responsible for interpretations and conclusions, and for any errors that appear.

Washington, D.C.
15 August 1987

WILLIAM M. HAMMOND

Contents

Maps

Illustrations

Page

Illustrations courtesy of the following sources: p. 6, *Indochine;* pp. 26 and 276, *New York Times;* p. 27, *Newsweek;* pp. 37, 71, 88, 89, and 223, Indochina Archives, Berkeley; pp. 40, 153, 210, and 351, Wide World Photos; p. 143, Magnum Photos; p. 50, *Time* Magazine; p. 77, Mrs. Connie Shank; p. 83, Peter Bankson; pp. 108, 141, 144, 183, 235, and 323, Barry Zorthian; pp. 180, 288, 380, and 383, Lyndon Baines Johnson Library; p. 189, Lawrence W. Lichty; p. 239, Col. Robert L. Burke; pp. 254 and 267, UPI/Bettmann Newsphotos; p. 271, Camera Press Ltd.; p. 277, Harrison Salisbury, *New York Times;* and p. 358, Larry Burrows, *Life* Magazine, copyright 1969, *Time,* Inc. All other illustrations from the files of the Department of Defense and the U.S. Army Center of Military History.

Public Affairs: The Military and the Media, 1962–1968

Prologue

The governments of free nations first learned to bargain with the press in time of war in 1854, when a correspondent for the *London Times*, William Howard Russell, accompanied the British Army into combat in the Crimea. Russell soon discovered that disease was decimating the troops and that outright blundering was destroying any chance for victory. Supported by the *Times*, he aroused Britain's middle classes with descriptions of the horrors of the army's hospital at Sevastopol, the ill-advised charge of the Light Brigade, and the fumbling of the British command. The British establishment responded with charges that the reporter had ruined Britain's worldwide public image and accused him of betraying sensitive military information to the enemy. So damaging were his revelations, nevertheless, that the government of Britain's prime minister, Lord Aberdeen, fell in a parliamentary vote of no confidence.[1]

The unseating of a prime minister by a newspaperman was a lesson governments never forgot. In the years that followed, each time a war occurred the nations involved attempted either to enlist the cooperation of the press or to restrain it. The history of warfare in the nineteenth and early twentieth centuries thus became at times as much a history of governmental attempts to control the press as a chronicle of battles.

The first efforts were tentative. During the Indian mutiny of 1857, when the *Times* announced that Russell would once more accompany the army, the British government simply took pains to ensure that the reporter received the best of impressions about everything. The commander in the field, Sir Colin Campbell, even gave Russell access to official reports—"every information I have myself"—provided he refrained from mentioning what he knew in camp and used it only in his letters to the *Times*. The policy had the desired result. Although Russell condemned the pillaging of Lucknow by British troops and various atrocities against captured mutineers by individual British soldiers, he supported the army and never broke his pledge.[2]

Circumstances changed during the American Civil War, when a telegraph office or a railroad was nearly always within reach of a man with a good horse. During August 1861 the commander of the Army of the Potomac, Maj. Gen. George B.

[1] This section is based on Joseph J. Mathews, *Reporting the Wars* (Minneapolis: University of Minnesota Press, 1957), pp. 31–51, and William M. Hammond, *The Light of Controversy: Five Essays on the Rise of the War Correspondent* (Ann Arbor: University Microfilms, 1972).

[2] Quote from William Howard Russell, *My Diary in India* (London: Routledge, Warner, and Routledge, 1860), p. 184; *The Times of London: The History of The Times* (New York: Macmillan, 1947), 2: 316.

McClellan, who had been an official American observer with the British Army in the Crimea tried to work out an *entente cordiale* with the press similar to the one Russell had agreed to, but the arrangement broke down because of rivalry between various newspapers and disagreements over what information was fit to print. Attempting to control the transmission and dissemination of strategically important information, President Abraham Lincoln gave the military control of all telegraph lines and made censorship of the press a function of the War Department. The move was at best marginally successful. War correspondents released information of value to the enemy with such regularity that the commander of the Confederate Army of Northern Virginia, General Robert E. Lee, read northern newspapers assiduously throughout the war. He even came to know which reporters were the most accurate, commenting on one occasion that he liked the work of a particular correspondent for the *Philadelphia Inquirer* because the man "knew what he reported and reported what he knew."[3]

By the end of the nineteenth century, governments began to become sophisticated in their handling of the press in wartime. Although the United States failed to censor correspondents effectively during the war with Spain, the British eliminated most problems with the press during the Boer War by commissioning reporters as officers in the army and subjecting them to field regulations. After a period of confusion the British Army also centralized its censorship effort, for the first time clearly defining the categories of information of most value to the enemy. The Japanese carried the process one step further during their war with Russia in 1904. Welcoming reporters enthusiastically, they took pains to keep them occupied and entertained but procrastinated for months before taking them into the field. When the reporters at last visited the scene of battle, the Japanese kept them as far from the action as possible.

With the outbreak of World War I in 1914, the British and the French excluded newsmen from the battlefield, but their opponents were more openhanded. Recognizing an opportunity to influence world public opinion, German officials allowed reporters from neutral countries to visit their armies under escort and singled out prominent correspondents for especially lavish treatment. The British and French held the line for only a few months, relenting gradually when they realized that continued suppression of independent news from the front damaged civilian morale. The French allowed newsmen to visit field units under escort as early as December 1914. The British followed suit in May 1915, accrediting a number of reporters to cover their headquarters in France. Although the relationship between the press and military authorities remained turbulent, the press from that time on strengthened its right to cover the war. Everything newsmen wrote was censored, but by the time American forces arrived in France in

[3] Lee is quoted in Douglas Southall Freeman, *R. E. Lee: A Biography*, 4 vols. (New York: Charles Scribner's Sons, 1935), 4: 171. See also Robert W. Coakley, Paul J. Scheips, and Emma J. Portuondo, Antiwar and Antimilitary Activities in the United States, 1846–1954, OCMH (Office of the Chief of Military History, which is now the Center of Military History [CMH]) Study, 1970, pp. 25–26.

1917 some fifty reporters were regularly assigned to the British Army, which also played host to a steady stream of visiting correspondents.[4]

American commanders for the most part accepted the Allied scheme for controlling the press but allowed newsmen greater freedom to accompany troops in the field. A number of American reporters thus took up station with units of their choice, returning to headquarters only to have their reports censored and dispatched. At first U.S. commanders attempted to restrict the number of correspondents accredited to the Army to 31, but so many visiting reporters arrived—at one point over 411—that the system collapsed. Although accredited correspondents supposedly held privileged positions, they had constantly to compete for support with the visitors, some of whom stayed so long that they became known as divisional correspondents.[5]

World War I had a profound effect upon the way governments handled the press in future wars. A total war that massed not only armies but entire economies and peoples against one another, it made news a strategic commodity, an all-important means of buttressing civilian morale. Since they were expending huge amounts of national treasure and millions of lives—1,265,000 casualties on all sides at the Battle of the Somme alone—the governments involved could hardly afford to give their citizens the whole truth. Instead they softened the bad news by censoring the facts and striving constantly to cast the war in the context of a noble endeavor, "a war," as the Western Allies put it, "to end all wars." The press cooperated, yielding to censorship and concentrating on morale-building human interest stories.[6]

News remained a strategic commodity when World War II began in 1939. Although the French appeared to have forgotten the lessons of the earlier war, invoking immediate, drastic censorship, the Germans once more allowed correspondents from neutral countries to report more or less as they wished, making certain all the while that they saw nothing truly damaging to the German cause. The British practiced censorship but nevertheless allowed the press to report the Battle of Britain and the bombing of London with relative freedom. The result was a public relations coup that gave the widest possible publicity to German barbarity and British determination.

The ability to transmit information electronically introduced another element into the management of news during the war. Because all sides were able to broadcast their respective interpretations of events, none could insulate its people completely from the outside world or cut off all word of important military events. They also found it impossible to release news in a vacuum, without considering what the other side was saying. The Germans prescribed the death penalty for citizens caught listening to foreign broadcasts, yet at times more than 50 percent

[4] This section is based on Mathews, *Reporting the Wars*, pp. 155–216.

[5] Wilbur Forrest, *Behind the Front Page: Stories of Newspaper Stories in the Making* (New York: D. Appleton-Century Co., 1934), pp. 164–65.

[6] Mathews, *Reporting the Wars*, pp. 155, 175; Frederick Palmer, *With My Own Eyes: A Personal Story of Battle Years* (Indianapolis: Bobbs-Merrill Co., 1933), p. 476.

The French Censorship Office, Hanoi, 1954

of the German people tuned in to British and American programs.[7]

Throughout the war American and British correspondents cooperated with the military. The military, for its part, saw to it that reporters in the field obtained a good overall picture of what was happening, specifying only that newsmen submit their work for censorship. Reporters complained that the American public received news of the battles of Midway and the Coral Sea far too long after those events had occurred. Yet when Lt. Gen. George S. Patton, Jr., slapped a battle-fatigued soldier during the Sicilian campaign, they suppressed the story at General Dwight D. Eisenhower's request because it might have been useful to the enemy as propaganda. Word of the incident surfaced three months later, when columnist Drew Pearson in Washington learned of it, but even then Pearson submitted the story to local censors, who passed it on the grounds that killing an article merely for reasons of morale exceeded their authority.

The handling of the press during the Korean War differed significantly from that of earlier conflicts. Lacking facilities to censor news reports, the U.S. commander in Korea, General of the Army Douglas MacArthur, imposed a system of voluntary guidelines similar to the one offered by General McClellan during the Civil War. Hampered by fierce competition among reporters and by a failure clearly to specify what news was of value to the enemy, MacArthur's system broke down much as had McClellan's—so much so that breaches of security by the press became an almost daily occurrence. The revelations at first made little difference. North Korean troops, in retreat after the Inchon landing, were unable to take advantage of the information. The situation became more desperate in November 1950, when Communist China entered the war, forcing American troops onto the defensive. Unable to tolerate further security violations, MacArthur imposed censorship.[8]

Although censorship reduced the number of security violations, it failed to eliminate them entirely because members of the press disposed to violating the rules could still report freely when they traveled to Tokyo and the United States. On 18 June 1951, for example, *Newsweek* published a map detailing the order of

[7] Mathews, *Reporting the Wars*, p. 214.

[8] This section is based on B. C. Mossman, Command and Press Relationships in the Korean Conflict, OCMH Study [1967], CMH files.

battle for the entire U.S. Eighth Army. In order to score against the competition, a few reporters also collaborated with the correspondent of the Paris Communist newspaper *Le Soir*, Wilfred Burchett, to receive from behind enemy lines and to publish carefully screened photographs of smiling and well-fed American prisoners of war—in effect providing major international outlets for enemy propaganda. Military information officers, for their part, several times provoked the press by withholding legitimate news. When the inmates rioted at a United Nations prisoner-of-war facility, the U.S. Army withheld all word of the event lest it become an issue in armistice negotiations. American officials also held back when enemy prisoners seized the commander of the Koje-do prisoner-of-war camp, Brig. Gen. Francis T. Dodd, in May 1952. In both cases word finally surfaced in the form of damaging newspaper exposes.

Ho Chi Minh

As the war in Korea continued, the American news media also followed events in Indochina, reporting the French struggle against Ho Chi Minh and the Viet Minh, the battle of Dien Bien Phu, and the rise of Ngo Dinh Diem. Prominent among the newsmen present during those years were Robert Shaplen and Harold Isaacs of *Newsweek*, both of whom had arrived during the late 1940s. By the 1950s Bernard Fall was in Indochina, conducting academic research but also submitting articles to such magazines as the *Nation* and the *Far Eastern Survey*. Also reporting were Larry Allen and Forest Edwards of the Associated Press, James Robinson of NBC News, and the Australian freelancer Denis Warner. Because they covered all of Southeast Asia most of those reporters spent limited amounts of time in Vietnam, traveling to the country when significant news broke and departing shortly after the story was told. Only a few remained long enough to gain any expertise.[9]

Official agencies dealing with the press in Vietnam during the early years of the war had little control over reporting. The French instituted censorship and a system of press camps where newsmen received food, shelter, and official briefings, but reporters had only to leave the country to say what they wanted. As a result, news of the war originating in Vietnam was heavily censored at the source, while stories on the same subject dispatched from Hong Kong and Singapore flowed freely. The United States handled the press much more easily than

[9] "Camp de Presse, Hanoi," *Indochine*, no. 31 (July 1954), pp. 20–24.

the French, meeting the few problems that developed as they occurred. Since most of the articles on South Vietnam appearing in American journals concentrated on the Communist menace and portrayed the president of South Vietnam, Ngo Dinh Diem, as one of Asia's ablest leaders, a more formal policy seemed unnecessary.[10]

An example of the way the U.S. mission in Saigon handled the press occurred in November 1959, when Jim Lucas of the Scripps-Howard syndicate arrived in Saigon to examine the American aid program. Several months earlier Scripps-Howard had sparked a congressional investigation by publishing allegations by correspondent Albert Colegrove to the effect that the U.S. mission was squandering millions of dollars on high living and unnecessary projects. Although Lucas had been a celebrated Marine combat correspondent during World War II and had won a Pulitzer Prize, his arrival in South Vietnam was greeted with considerable misgiving. The commander of the U.S. Military Assistance Advisory Group, Lt. Gen. Samuel T. Williams, received instructions from the American embassy to report carefully on all his conversations with the reporter. So too did other members of the U.S. mission. To eliminate all surprises, the mission also monitored the reporter's dispatches, receiving copies from the Saigon cable office through South Vietnamese officials shortly after they were transmitted and carefully marking each with a Confidential security classification. For the rest, Lucas received every consideration. He stayed at Williams' home his first night in Saigon, accompanied the general on trips into the countryside, and wrote what he wanted. Not so Colegrove. He was denied a visa to reenter the country by the South Vietnamese government.[11]

If Williams was under pressure, so was Lucas. His employers expected him to support Colegrove's allegations, but he distrusted his colleague's sources. In the end he cabled his editors that he would report what he saw fit. He nevertheless made certain that he balanced his praise for successful American projects with careful attention to Colegrove's original allegations. Lucas reported that some members of the U.S. mission in Saigon indeed lived in expensive rented housing and that the attempt to build a radio station for the Diem regime had been seriously mismanaged. Deeply concerned about Diem's violations of South Vietnamese civil rights, he also passed along charges by one of Diem's political opponents, Dr. Phan Quang Dan, that the regime was violating its own laws and attempting to set up a political dictatorship.[12]

Although allegations such as those of Lucas and Colegrove received wide play in the American news media, the war in South Vietnam attained little promi-

[10] E. K. Lindley, "Ally Worth Having: South Vietnam," *Newsweek*, 29 Jun 59, p. 53; "Revolt at Dawn," *Time*, 21 Nov 60, p. 76; "The Coming Showdown in South Vietnam," *Reader's Digest*, Nov 61, p. 257.

[11] Memo, Lt Gen Samuel T. Williams for Ambassador, 4 Nov 59, sub: Contacts With Mr. Jim Lucas, file 93, Samuel T. Williams Papers, CMH. File 93 also contains copies of Lucas' dispatches, often accompanied by the final newspaper version.

[12] Memo, Arthur Z. Gardiner, Director, U.S. Operations Mission, Saigon, 5 Nov 59, sub: Memo of Conversation With Jim Lucas; Memo for the Record (MFR), MAAG J5-CH, 10 Nov 59, sub: Lucas Interview; Jim Lucas, "Vietnam Aid Gets Spotlight," *New York World-Telegraph*, 23 Nov 59; Lucas,

nence in American newspapers until the end of 1960, when an attempted coup against Diem in which four hundred civilians were killed prompted the *New York Times* to station in Saigon a permanent correspondent, the Pulitzer Prize–winning reporter Homer Bigart. Joining Malcolm Browne of the Associated Press, Ray Herndon of United Press International, Nicholas Turner of Reuters, Pierre Chauvet of Agence France Presse, and part-time reporters James Wilde of *Time* and Francois Sully of *Newsweek*, Bigart became the first of the flood of reporters who arrived in South Vietnam as the United States and North Vietnam escalated the war.

"Hanging Sam's Men Advise on Communists, Tigers," unattributed newspaper clipping, 14 Dec 59; Lucas, "Landlords in Saigon Find U.S. $2 Million Customer," *Fort Worth Press*, 14 Nov 59; Msg, USINFO Washington to USIS Saigon, 5 Dec 59, sub: Article by Lucas, *Washington News*. All in file 93, Williams Papers.

1

Taking Sides

The claim that South Vietnam was fighting Communist aggression on its own with the United States providing only advice and support shaped U.S. relations with American newsmen in Saigon from the very beginning of the U.S. involvement in South Vietnam. Seeking to reinforce that contention, American policy makers sought to emphasize the role of the South Vietnamese government in the release of news to the press. Although U.S. official spokesmen might brief newsmen on the activities of Americans in South Vietnam, they followed South Vietnamese press guidance on all matters involving the country itself.[1]

The approach seemed appropriate. U.S. policy sought to strengthen South Vietnam by fostering the confidence and self-reliance of the country's leaders, an end easily frustrated if Americans began assuming functions proper to South Vietnamese officialdom. The president of South Vietnam, Ngo Dinh Diem, was also sensitive to any infringement on his nation's prerogatives. American policy makers believed that he would resent any attempt by American diplomats to become the source of news for South Vietnam and that he might even retaliate by curtailing the flow of information between his government and the U.S. embassy, a development almost certain to hamper the effort against the Communist insurgency in the countryside.[2]

The Origins of Controversy, 1961–1963

Information officers at the embassy were caught between their government's concern for South Vietnamese sovereignty and the desire of American newsmen

[1] Msg, Saigon 726 to State, DAIN 14863, 5 Feb 63, U.S. Department of State, Foreign Affairs Information Management, Bureau of Intelligence and Research (FAIM/IR), Record Group (RG) 951. Unless otherwise indicated, State Department records cited in this chapter may be found in RG 951. Messages with a DAIN number are in the Army's Staff Communications Center files, which are presently housed at the Army War College, Carlisle, Pa.

[2] Msg, Saigon 726 to State, DAIN 14863, 5 Feb 63; Msg, State 1006 to Saigon, 21 Feb 62, FAIM/IR; Msg, Saigon 656 to State, DAIN 5969, 8 Jan 63.

covering the war to learn as much as possible. They sometimes briefed reporters in private about South Vietnamese military operations—especially when the Diem regime's news releases were less than candid—but their efforts were at best partially successful. When the government of South Vietnam ordered news of a military operation suppressed, U.S. Army information officers had no choice but to respond to queries from reporters with "I have been ordered by the Vietnamese Joint General Staff not to talk to you about this subject." No amount of explaining afterward could totally erase the newsmen's impression that the United States was somehow cooperating with the South Vietnamese to inhibit the flow of news.[3]

In fact, American policy makers were inclined to release information about the war because it fit their concept of good public relations. During a conference in Honolulu in January 1962 Secretary of Defense Robert S. McNamara declared that pessimistic anti-Diem reports in American newspapers were inimical to U.S. interests in South Vietnam and that adverse reporting hurt "our case with the public, with congress, and with our own officials." Instead of demanding restrictions on the press, McNamara advocated a policy of greater openness. The U.S. mission in Saigon, he said, should respond to the complaints of newsmen by declassifying as much information as it could.[4]

U.S. Ambassador to South Vietnam Frederick E. Nolting, Jr., replied with a practical objection. The South Vietnamese must receive credit for winning the war, he told McNamara; otherwise, they might begin to consider the conflict an American endeavor and reduce their efforts. "This must continue to be their fight. The U.S. must keep in the background."

Because he believed that good relations with the press were essential, McNamara instructed military public affairs officers in Saigon to declassify information whenever possible. He nevertheless accepted Nolting's argument. When he issued the Recommendations on Actions To Be Taken stemming from the conference he thus said nothing about opening military operations to the press. Instead, U.S. military commanders in charge of the war were to declassify "within their judgment." Since basic policy deferred to the South Vietnamese, who wanted information cut off rather than opened, that judgment continued to be closely constrained.

In addition, a tangle of military, diplomatic, and political concerns argued against a policy of open information. Military security, first of all, demanded secrecy. An open society with little authority over the press except in time of declared war, the United States faced an enemy who had complete control over every word published in areas under his domination. While U.S. intelligence analysts had to work hard to get anything more than propaganda from Communist periodicals, the Communists had only to read the American press to learn

[3] Msg, Saigon 726 to State, DAIN 14863, 5 Feb 63; Msg, Saigon 656 to State, DAIN 5969, 8 Jan 63; David Halberstam, "Curbs in Vietnam Irk U.S. Officers: Americans Under Orders To Withhold News," *New York Times*, 22 Nov 62; "Salinger Tells How Kennedy Tried To Hide Vietnam Build-Up," *U.S. News & World Report*, 12 Sep 66, p. 103.

[4] This section is based on HQ, CINCPAC, Record of the Secretary of Defense Conference at Honolulu, 15 January 1962 (hereafter cited as Honolulu Conference), pp. 49-50.

important details of what they wanted to know. A tightening of restrictions governing the access of newsmen to events and an appeal to the patriotism of the press to foster restraint thus seemed in order. It might save American lives.[5]

The U.S. decision in 1961 to bolster the sagging Diem regime by taking a more active role in the war provided what appeared to be a second argument for secrecy. Because the decision violated the 1954 Geneva Agreements, which had limited foreign intervention in South Vietnam, it made the United States vulnerable to Communist propaganda. For although American diplomats had never signed the agreements and the Communists had broken them for years, there seemed little doubt that the enemy would use every press release and news conference available under a policy of open information to document the allegation that the United States was the aggressor in South Vietnam.[6]

A concentration upon secrecy would have a third, even more important effect. By limiting the American public's knowledge of what was happening in South Vietnam, it would help to defuse any adverse domestic reaction to U.S. risk-taking in Southeast Asia. (*Map 1*) Recent events had inspired that line of thought. In a January 1961 speech at the 22d Communist Party Congress, Soviet Premier Nikita Khrushchev had advocated wars of national liberation. The attempt by an American-sponsored expeditionary force to invade Cuba at the Bay of Pigs had failed, and the neutralization of Laos in late 1961 had ended any hope that the United States might seal South Vietnam's borders to further Communist aggression. All had made American leaders alert to the danger of irresolute responses to Communist initiatives. The American people, however, while well disposed toward Diem, seemed little interested in a foreign war. If enthusiasm for the conflict in South Vietnam began to fade because of negative reporting in the press, the American effort to defeat Communist aggression in Southeast Asia would also begin to slip and might even fail for lack of support. A low profile, achieved through restraints on the press at the scene of conflict and designed to sustain the American public's support for the war, seemed a safer course.[7]

[5] John Mecklin, *Mission in Torment: An Intimate Account of the U.S. Role in Vietnam* (Garden City, N.Y.: Doubleday, 1965), pp. 105–06; Roger Hilsman, *To Move a Nation* (Garden City, N.Y.: Doubleday, 1967), pp. 508–45.

[6] This section is broadly based on the following sources and authorities: Mecklin, *Mission in Torment*, pp. 105–06; *The Senator Gravel Edition of the Pentagon Papers: The Defense Department History of United States Decisionmaking on Vietnam* (hereafter cited as *Pentagon Papers*), 4 vols. (Boston: Beacon Press, 1971), 2: 102–227; Hilsman, *To Move a Nation*, pp. 421–22; U.S. Army, Pacific, History of the U.S. Army Build-up and Operations in the Republic of Vietnam, 1 January 1961 to 31 January 1962, p. 32, CMH files; MFR, U.S. Military Assistance Command, Vietnam, Office of Information (MACOI), 9 May 64, sub: Investigation of Captain Shank's Allegations in Letters Home, 69A702 2/15, Washington National Records Center (WNRC); Memo, Col B. L. Baker, USAF, for Arthur Sylvester, ASD (PA), 12 Mar 64, sub: Restrictions on Release of Information in RVN, Directorate of Defense Information (DDI), News from Vietnam (56) file; Msg, State 1574 to Saigon, 24 Jun 61, and Msg, State 796 to Saigon, 20 Dec 61, both in FAIM/IR.

[7] Mecklin, *Mission in Torment*, pp. 105–06; Pierre Salinger, *With Kennedy* (Garden City, N.Y.: Doubleday, 1966), p. 134; Hilsman, *To Move a Nation*, pp. 134–35, 150, 349; Louis Harris, *The Anguish of Change* (New York: W. W. Norton, 1973), pp. 53–54. Examples of official concern for public opinion may be found in the *Pentagon Papers*, 2:113, 120, and 3:559; Msg, Saigon 726 to State, DAIN 14863, 5 Feb 63; Honolulu Conference, 15 Jan 62, p. 48.

NORTH VIETNAM
CHINA
BURMA
Dien Bien Phu
HANOI
Haiphong
Red R.
Mekong R.
LAOS
GULF OF TONKIN
HAINAN
Vinh
VIENTIANE
17th Parallel
Hue
Da Nang
THAILAND
Mekong R.
SOUTH VIETNAM
Pleiku
Qui Nhon
BANGKOK
CAMBODIA
Tonle Sap
Ban Me Thuot
Nha Trang
Snuol
Cam Ranh
GULF OF THAILAND
PHNOM PENH
Krek
Tay Ninh
QL-13
QL-1
Bien Hoa
Phan Thiet
Sihanoukville
SAIGON
My Tho
Can Tho
Ca Mau
QL-4
SOUTH CHINA SEA
SOUTHEAST ASIA
1962
0 100 MILES
0 100 KILOMETERS

MAP 1

On 21 February 1962, the U.S. Information Agency and the State and Defense Departments solidified their press policy in a message to the U.S. mission in Saigon. Widely known as Cable 1006, the directive stressed the need to reinforce the idea that the war was essentially a South Vietnamese affair. Although American newsmen would always tend to concentrate on the activities of Americans, "it is not . . . in our interest . . . to have stories indicating that Americans are leading and directing combat missions against the Viet Cong." While news stories critical of the South Vietnamese would likewise always exist, newsmen were to be made to understand that "frivolous, thoughtless criticism" of the South Vietnamese government made cooperation with the Diem regime difficult to achieve. To prevent that problem, correspondents were never to go along on military operations that might generate unfavorable news stories that the United States wanted to avoid.[8]

The cable's State Department authors justified the directive as an attempt to give local U.S. officials in Saigon more flexibility in dealing with newsmen, a major effort to achieve "maximum cooperation" with the press. While the directive recognized the right of American reporters to cover the war and ordered the U.S. ambassador to keep the press informed to an extent compatible with military security, it also stressed the need for officials to operate without the interference of newsmen. In that way, far from opening information, the cable prompted the U.S. mission in Saigon to persist in the practice of excessive classification to a degree that denied newsmen access to whole segments of the war.[9]

The Saigon correspondents perceived the hardening. Shortly after Cable 1006 arrived in Saigon, they began to complain in print of difficulties in getting information from embassy officers. Homer Bigart wrote that "American officials who 'leak' stories unflattering to the Saigon government . . . are tracked down and muzzled." He added that "correspondents who send gloomy dispatches are apt to be upbraided for lack of patriotism." David Halberstam, another *Times* correspondent, concurred. "United States military officers here have expressed some concern," he said, "because they feel they are being muzzled by the South Vietnamese government with the support of the United States. . . . American officers serving in the field and flying helicopters believe that Americans at home have too little knowledge and understanding of what is going on in Vietnam."[10]

The Assistant Secretary of State for Far Eastern Affairs, Roger Hilsman, later disputed the newsmen's contentions. The commander of the U.S. Military Assistance Command, Vietnam (MACV), General Paul D. Harkins, had issued a memorandum during November 1962 in which he advised U.S. officers in South

[8] Msg, State 1006 to Saigon, 21 Feb 62. Cable 1006 is elaborated on in U.S. Congress, House, Committee on Government Operations, Subcommittee on Foreign Operations and Government Information, *United States Information Problems in Vietnam*, 88th Cong., 1st sess., 1 October 1963, H. Rpt. 797 (hereafter cited as Moss Report), p. 3. See also Mecklin, *Mission in Torment*, pp. 111-19; Thomas C. Sorensen, *The Word War: The Story of American Propaganda* (New York: Harper & Row, 1968), p. 191.

[9] Msg, State 1006 to Saigon, 21 Feb 62; Moss Report, p. 12; Mecklin, *Mission in Torment*, p. 115.

[10] Homer Bigart, "Saigon's Regime Rejects Pressures for Reforms," *New York Times*, 3 Jun 62; Halberstam, "Curbs in Vietnam Irk U.S. Officers."

Vietnam to be "sincere and truthful" in their dealings with the press and never to "use security as an excuse" for failing to discuss unclassified matters. Hilsman said that Harkins' guidance reflected the true direction of official relations with the press in South Vietnam.[11]

In fact, whatever Harkins' affirmations, the U.S. mission in Saigon, with the approval of the State and Defense Departments, tended to define military security and the national interest narrowly and to treat the American involvement in Southeast Asia as if it were a clandestine operation. South Vietnam was far too open an environment, and the newsmen resident there far too freewheeling, for that approach. As a result, military secrets became known daily to the press, and the credibility of the U.S. mission in South Vietnam declined with each new revelation.[12]

Policy makers in Washington, for example, refused to allow information officers to mention the use of napalm in releases to the press lest the Communists make propaganda of it, but newsmen went into the field, observed napalm exploding, and recorded its effects with their cameras. One photograph even appeared on the cover of *Life* magazine in early 1962. The restriction thus accomplished little beyond lowering the confidence of newsmen in the candor of American diplomats and military officers.[13]

The same was true of a policy that limited information on the use of armed helicopters against the Viet Cong. On 25 July 1963, Peter Arnett of the Associated Press filed a story claiming that the United States had altered its rules of engagement in South Vietnam to permit helicopters to take offensive action against the enemy. The change had indeed occurred because the morale of U.S. helicopter crews demanded that they be allowed to fire at the enemy before he could fire at them. Despite that justification the United States refused to admit that Americans were taking a more active role in combat. It avowed instead that the American mission in South Vietnam was logistical, technical, and advisory and that U.S. helicopters fired only when fired upon. The chief of the Army Section of the U.S. Military Assistance Advisory Group tried to explain the change by telling newsmen that helicopter crews laid down only "suppressive fire" to keep the enemy at bay while they delivered their loads. He maintained that such defensive firing differed significantly from offensive artillery barrages and fixed-wing aircraft strikes. Reporters had nevertheless been in the field and had seen the helicopters firing. Rejecting hairsplitting distinctions, they lost more confidence in the embassy.[14]

Although the effects of the policies on napalm and helicopters were detrimental, nothing hurt the U.S. mission's credibility more than the practice of concealing the American role in the air war. American officials in both Saigon and

[11] Msg, Saigon 327 to State, 19 Dec 62, FAIM/IR; Moss Report, p. 5.

[12] Moss Report, p. 5.

[13] Memo, Baker for Sylvester, 12 Mar 64, sub: Restrictions on Release of Information in RVN; Mecklin, *Mission in Torment*, p. 115.

[14] Memo, Baker for Sylvester, 12 Mar 64, sub: Restrictions on Release of Information in RVN.

Washington saw no reason to tell the press that American pilots were flying combat missions for the South Vietnamese Air Force. Public affirmation of that fact might have harmed South Vietnamese morale while playing into the hands of Communist propagandists. Thus, American newsmen were denied permission to visit the South Vietnamese air base at Bien Hoa, near Saigon, where many U.S. airmen lived. The commander of the Pacific Air Forces, General Emmet O'Donnell, told reporters that American combat pilots were in South Vietnam only to train South Vietnamese airmen, not to fight the war themselves. While those pilots often accompanied their counterparts into battle, they did so only to advise their trainees in a practical context.[15]

The facts were different. Although South Vietnamese pilots flew smaller aircraft, few were qualified to fly the A-26 bomber and still fewer capable of conducting a combat mission in one. The so-called trainee was usually a low-ranking enlisted man who sat to one side while the Americans did the work.[16]

As air sorties numbering over a thousand per month by early 1963 began to produce U.S. casualties, there was no chance that reporters would miss what was going on. Noting the crash of a South Vietnamese A-26 bomber 260 miles north of Saigon, for example, the Associated Press reported on 9 April 1963 that a U.S. pilot and copilot had been aboard along with a South Vietnamese observer and that American pilots flew bombers belonging to the South Vietnamese Air Force because South Vietnam had too few trained pilots. The attempt to mislead newsmen about the extent of American involvement in the air war thus forced information officers to take a position that once more hurt their credibility.[17]

Official Optimism, 1962

Official disclaimers, the narrow definition of military security, and the need to get along with the Diem regime were only the beginnings of the American mission's problem with newsmen. Also important was the tendency of both the U.S. mission in Saigon and those agencies in Washington concerned with the war to state in public that everything was going well when the personal experience of newsmen at the scene suggested the opposite.

There seemed to be good reasons for this official stance. If American diplomats and military advisers were to acknowledge publicly that South Vietnam was faltering, the American people and their congressional representatives might question whether further U.S. aid was appropriate. Diem would become even more difficult to deal with, and the Communists would undoubtedly broadcast the affir-

[15] Ibid. O'Donnell is paraphrased in [AP], "U.S. Role Redefined," *New York Times*, 26 Feb 63. See also Mecklin, *Mission in Torment*, p. 115.

[16] Memo, Baker for Sylvester, 12 Mar 64, sub: Restrictions on Release of Information in RVN.

[17] Msg, Saigon 749 to State, 13 Feb 63, FAIM/IR; "Vietnamese B-26 Crashes While on Strafing Mission," *New York Times*, 9 Apr 63.

mation as an American admission of their own success. It thus seemed better, when the South Vietnamese were defeated in battle, to attempt to diminish the importance of the event by steering the press toward progress in other areas of the war.[18]

The Saigon correspondents, for their part, usually refused to go along. Since they had private sources of information, they knew most of what was happening in South Vietnam and viewed U.S. attempts to underscore South Vietnamese success as just one more proof of the U.S. mission's desire to placate Diem. More than diplomacy, however, was involved in the official optimism that emanated from Saigon and circulated in Washington. Senior American officials honestly believed that the South Vietnamese government's prospects were improving, and they could back their conclusion with convincing quantitative proof.[19]

The U.S. buildup after 1961, for example, seemed to have strengthened the South Vietnamese armed forces. By the end of 1962 an augmented corps of American advisers had centralized South Vietnamese logistical functions, improved intelligence reporting, and restructured the country's system of military training. In 1961 no reliable military communications network had existed in South Vietnam. By September 1962 all of South Vietnam's major military units were linked by telephone and all radio frequencies were standardized.[20]

These improvements seemed to show their worth in battle. By early 1963, U.S. officials claimed, South Vietnamese commanders had learned to use American-supplied M113 armored personnel carriers and had begun to operate with the help of American helicopters in such formerly inviolate enemy strongholds as War Zone D northeast of Saigon and the U Minh Forest in the Mekong Delta. Unfamiliar with the tactics the new equipment allowed and terrified of the helicopters, the enemy suffered a series of defeats and seemed increasingly on the defensive.[21]

American officials could also cite progress in the pacification program, the government's effort to win the peasantry to its side. During 1962 and 1963, they told newsmen, the South Vietnamese Army had cleared numerous "hard core" Viet Cong regions, moving the residents to fortified hamlets where psychological warfare groups, specialists in civic action, and first-aid teams could demonstrate directly the government's concern for its people. So successful had the

[18] Interv with Col Greene, Secretary of the MACV Joint Staff, 1963–64, 6 Jun 65, p. 17, CMH files; Msg, State 1006 to Saigon, 21 Feb 62; Hilsman, *To Move a Nation*, p. 441; Phil G. Goulding, *Confirm or Deny: Informing the People on National Security* (New York: Harper & Row, 1970), p. 20.

[19] Msg, Saigon 726 to State, DAIN 14863, 5 Feb 63.

[20] HQ, MACV Advisory Group, Final Report of Daniel B. Porter, Jr., Colonel, USA, Senior Advisor of III and IV Corps, 13 December 1962–13 February 1963, 13 Feb 63, and Memo, Gen Paul D. Harkins for President Ngo Dinh Diem, 15 May 63, both in CMH files; Brigadier General James Lawton Collins, Jr., *The Development and Training of the South Vietnamese Army, 1950–1972*, Vietnam Studies (Washington, D.C.: Government Printing Office, 1975), pp. 20–29.

[21] Ltr, Col Daniel B. Porter to Thomas Lewis, 3 Nov 71, quoted extensively in Thomas Lewis, "The Year of the Hare" (M.A. diss., George Washington University, 1972), p. 56. See also Military Assistance Command, Vietnam, Translation of the Viet Cong After Action Report on the Battle of Ap Bac, 2 Jan 63, IR 2903011563, 1 May 63, CMH files.

Ambassador Nolting Entertains Diem, 1962

program become, official spokesmen said, that by 1963 South Vietnam seemed on the verge of isolating the enemy from the people. They buttressed their claim with statistics. Total Viet Cong-initiated incidents in the Mekong Delta had dropped from 3,338 between January and June 1962 to 2,769 between July 1962 and January 1963. During the same period the percentage of the population of South Vietnam's Mekong Delta under enemy control had declined eight percentage points.[22]

The United States knew that much was wrong in South Vietnam but believed that this positive side of the war should be told, to bolster the confidence of South Vietnam's leaders, to spur further accomplishment, and to counter the negative reporting of the press. While they noted that the South Vietnamese and the Communists had fought to a draw and that the war would go on for many years, official spokesmen rarely failed to mention in their statements the growing "effectiveness" of U.S. aid and the fact that the South Vietnamese were gaining.[23]

Examples of the kind of rhetoric that resulted were abundant throughout July,

[22] Msg, Saigon 261 to State, DAIN 85011, 19 Aug 63; see also Porter's comments in his letter to Lewis, "The Year of the Hare," pp. 56-57.

[23] Mecklin, *Mission in Torment*, p. 117; Tad Szulc, "Vietnam Conflict Seen at Impasse," *New York Times*, 22 Aug 62; "McNamara Lauds Gains in Vietnam," *New York Times*, 7 Jul 62.

August, and September of 1962. On 7 July Secretary of Defense McNamara told a Pentagon press conference that U.S. aid to South Vietnam had been a positive influence over events in Southeast Asia while Communist effectiveness had declined. The South Vietnamese Army had improved its "kill ratios," forcing the Communists to stage fewer incidents and to attack with fewer men. Returning on 25 July from a conference in Hawaii, McNamara amplified his remarks. Although the war could last for years, the South Vietnamese were "beginning to hit the Viet Cong insurgents where it hurts most—in winning the people to the side of the government." Not only was American military assistance succeeding, the practice of collecting the country's peasants in fortified villages away from enemy influence, the strategic hamlet program, had provided the common people with protection while giving them "the opportunity to learn basic democratic practices in electing their own village leaders by secret ballot."[24]

Roger Hilsman echoed McNamara. In a widely reported September interview he predicted that the campaign to provide security for South Vietnam's countryside through military and social programs might take seven years but that there was reason for "guarded optimism" over its outcome. Vigorous U.S. logistical support had given "new confidence" to South Vietnam's armed forces and had sparked a number of gratifying victories. In the process more than two thousand hamlets had become fortified villages closely identified with the Diem government through medical, economic, and educational assistance. Meanwhile, Hilsman said, the Viet Cong defection rate had risen, the number of enemy recruits had fallen off, and portions of South Vietnam closed to outsiders just a few months before had become open and safe.[25]

More hesitant than official spokesmen, American newspapermen in South Vietnam at first shared this optimism. While some—Homer Bigart, for example, and *Newsweek* stringer Francois Sully—remained implacable opponents of the Diem regime, others, such as Robert Trumbull and David Halberstam of the *New York Times*, were impressed by the enlarged U.S. commitment to South Vietnam. In a 7 July 1962 story entitled "Vietnamese Rout Red Unit," Trumbull featured a South Vietnamese helicopter operation that had surprised and put to flight a group of Communist guerrillas on the outskirts of Saigon. He called the action "a notable victory in the kind of widespread, small-scale warfare being fought in Vietnam." Halberstam was also optimistic. Referring to a South Vietnamese victory at An Hu in the Plain of Reeds, an area west of Saigon near the Cambodian border, the reporter observed that 153 Viet Cong had perished because American logistical support, especially helicopters, had allowed South Vietnamese units to seek out the enemy on his own ground. In later reports Halberstam reinforced official contentions that U.S. assistance was paying off with his repeated assertions that the South Vietnamese were keeping the enemy off balance and that

[24] "McNamara Lauds Gains in Vietnam"; [AP], "McNamara Hails Gains in Vietnam," *New York Times*, 25 Jul 62.

[25] "U.S. Aide Guardedly Optimistic," *New York Times*, 19 Sep 62; Hanson W. Baldwin, "Cautious Optimism Voiced by U.S. in War on Reds," *New York Times*, 28 Sep 62.

the Viet Cong were in awe of the helicopter. So favorable was his work that the State Department sent a letter to his employers at the *Times* commending him for his accuracy and fairness.[26]

Although the infusion of American men and materiel had indeed thrown the Viet Cong off stride, the gain was at best temporary. The enemy learned to cope with helicopters and armored personnel carriers and soon regained the initiative. As South Vietnamese fortunes declined, the Saigon correspondents began once more to criticize, citing the inability of the South Vietnamese armed forces to deal with the enemy and the fact that, for all the promise of the strategic hamlets, the effort to win the peasantry to the side of the government remained much in doubt.

Their statements reflected a split that had developed within the U.S. mission in Saigon. The ambassador and his top military aides contended that the war was going well and that success in battle would ensure the adherence of South Vietnam's peasantry to the Diem regime. Many junior officers believed just the opposite. Critical of South Vietnamese efforts, they argued that Diem was losing the war and that the United States had to do more to win the people to the side of the government. Because they were rarely exposed to the ambassador and his aides but in frequent contact with lower-ranking U.S. advisers to South Vietnamese combat units, newsmen tended to reflect the more negative viewpoint in their reports.[27]

Shortly after McNamara delivered his cautiously optimistic 25 July report on the progress of the war, for example, Jacques Nevard of the *New York Times* expressed doubts that the war was going as well as everyone said. Drawing upon his own sources, he said he believed that the Diem regime had at best a "50-50" chance for survival and that many U.S. officers training South Vietnamese units considered the odds even slimmer. The reporter argued that most of the people he had interviewed disagreed with McNamara. While the secretary thought that the South Vietnamese were winning the people to the side of the government, the consensus among Americans serving in the field was that political apathy or even hostility on the part of large segments of the population continued to be the Communists' biggest asset. McNamara spoke of democratic practices in the strategic hamlets, but most Americans in contact with day-to-day operations suspected that many occupants of the hamlets had been rounded up and resettled forcibly and that the walls protecting the villages were designed as much to keep the residents in as to keep the Viet Cong out. Nevard concluded that although there was general agreement that the South Vietnamese armed forces, supported by 9,000 United States soldiers, airmen, and marines, had been fight-

[26] Robert Trumbull, "Vietnamese Rout Red Unit," *New York Times*, 7 Jul 62; David Halberstam, "South Vietnamese Inflict Major Defeat on Reds," *New York Times*, 19 Sep 62, and "Viet Cong Serves Tea and Weapons," *New York Times*, 12 Oct 62; Msg, State 562 to Saigon, 30 Nov 62, FAIM/IR.

[27] Memo, William H. Sullivan for Robert McNamara [Sep 63], sub: Divergent Attitudes in U.S. Official Community, p. 3, CMH files.

ing more efficiently than before, there was also agreement that the insurgents were becoming stronger all the time.[28]

Although less critical than Nevard, Halberstam soon began to display similar doubts. In a 21 October article on the background of the war he noted that despite a few successes on the part of the South Vietnamese, most of the optimism emanating from the U.S. government seemed unwarranted. "This is a war fought in the presence of a largely uncommitted or somewhat unfriendly peasantry," he said, "by a government that has yet to demonstrate much appeal to large elements of its own people. The enemy is lean and hungry, experienced in this type of warfare, patient in his campaign, endlessly self-critical, and above all, an enemy who has shown that he is willing to pay the price." There was considerably less optimism in the field than in Washington or Saigon. "The closer one gets to the actual contact level of this war, the further one gets from official optimism."[29]

If Nevard's and Halberstam's appraisals were pessimistic, they were still closer to the truth than McNamara's. The two reporters had based their conclusions on the practical, concrete testimony of American advisers at the scene of the action. McNamara, on the other hand, placed great store in statistics which, although useful as indicators of enemy activity, failed to grasp the basically political, human essence of the war. The loyalty of the South Vietnamese people to their leaders, the quality of local governments, and the success of attempts to change the opinions of the populace were the elements officials needed to measure, and they were not susceptible to measurement through the kill ratios, estimates of the percentage of the population pacified, and tallies of enemy-initiated incidents upon which McNamara relied.

Complicating the problem was the fact that McNamara's numbers were drawn mostly from South Vietnamese sources. The secretary's view of the war was thus shaped by the statements of South Vietnamese military commanders, who tended to report what they believed the Americans wanted—large numbers of sorties against the Viet Cong, heavy body counts, action—and to magnify their own achievements to gain approval.[30]

Most U.S. observers admitted a wide margin for latitude and doubt in the war's statistics, but that recognition only tempted American officials to inject their own interpretations into reports from the field in order to explain away deficiencies. During 1963, for example, junior civilian members of the country team began to argue through channels that South Vietnamese military operations had done little to improve acceptance of Diem by the peasantry. Senior officials within the

[28] Jacques Nevard, "Americans Voice Doubt on Vietnam," *New York Times*, 29 Jul 62.

[29] David Halberstam, "U.S. Deeply Involved in the Uncertain Struggle for Vietnam," *New York Times*, 21 Oct 62.

[30] Robert B. Rigg, CINCPAC Director of Intelligence, 1962–63, "The Asian Way," *Army* 20 (July 1970): 45–46. Memo, Lt Col John Cushman, Senior Adviser to the ARVN 21st Division, 1964, for Lt Col Robert Montague, Strategic Hamlets Adviser, 1964 [1964], sub: Performance Evaluation in Vietnam; Ltr, William P. Bundy, Assistant Secretary of State for Far Eastern Affairs, to David Ness, Deputy Chief of Mission, Saigon, 16 Jan 64, both in CMH files.

U.S. mission who disagreed either moderated the negative reports that passed across their desks or neglected to send them on to Washington.[31]

The same thing happened to the Saigon correspondents. Although reporters in the field disagreed more and more with U.S. policy, the opinions of their editors turned critical only gradually. When the newsmen began to object to the U.S. mission's restrictive policies and unrealistic assessments, those editors doubted their younger subordinates' statements and chose to moderate them before sending them to press.[32]

As the difficulty of working with Diem increased, second- and third-level embassy officers and military advisers, who had the most direct experience of the Diem regime's failings, began to disagree more and more vehemently with their superiors, who appear to have interpreted any criticism of U.S. progress in South Vietnam as a personal affront. Secretary of Defense McNamara commissioned William H. Sullivan, a Foreign Service officer of long experience with Southeast Asian affairs, to look into the problem. Sullivan reported during September 1963 that as doubts grew and emotional pressures built up, internecine warfare had flared between the two groups. Unable to obtain a sympathetic hearing, the dissenters began leaking classified information to the press in hopes that publication of their point of view would attract the notice of Washington agencies and bring change. Although many American editors cut the stories that resulted, some, especially the editors of *Newsweek* and the *New York Times*, printed enough to alarm the governments of both the United States and South Vietnam.[33]

U.S. policy makers were ambivalent in their response. Aware that the practice of leaking derogatory information to the press would identify them too closely with criticisms of Diem and damage the American ability to work with the South Vietnamese government, they attempted to restrain official dissenters in private. Yet they were unwilling to tamper openly with the institution of the press and opposed any form of direct censorship of news dispatches. In general, they considered outright suppression of the Saigon correspondents a mistake and South Vietnamese success the best antidote to a bad press.[34]

Diem shared neither the Americans' fears nor their principles. He found it difficult to understand how the American press could publish anything derogatory about an ally in mortal combat with a mutual enemy and expected the same broad support from American newsmen that he received from the U.S. government. When he found that support wanting and observed that reporters actually used leaked information to attack his policies, he reacted with indignation. Every critical newspaper dispatch became a sign of the reporters' sympathy for the Viet

[31] Memo, Sullivan for McNamara [Sep 63], sub: Divergent Attitudes in U.S. Official Community, p. 3.

[32] Msg, Saigon 726 to State, DAIN 14863, 5 Feb 63; David Halberstam, *The Making of a Quagmire* (New York: Random House, 1964), pp. 266–69.

[33] Memo, Sullivan for McNamara [Sep 63], sub: Divergent Attitudes in U.S. Official Community, p. 2; Mecklin, *Mission in Torment*, pp. 61, 105, 118.

[34] Ltr, Frederick B. Nolting to Chalmers Wood, Director of the Dept of State Vietnam Working Group, 1 Nov 62, FAIM/IR; see also Msg, Saigon 252 to State, 27 Aug 63, FAIM/IR.

Cong, and that sympathy, with its treasonous implications, made the newsmen seem "worse than the communists."[35]

Confrontation: The Bigart and Sully Cases

Heightened by the U.S. mission's own ambivalence, the tensions growing between Diem and the press undermined official relations with newsmen from early 1962 onward. For at the very moment when the United States was attempting to put as good a face on the war as possible, the Diem regime was moving to stifle its most outspoken critics in the press, Homer Bigart and Francois Sully. The controversy that resulted crystallized the prejudices of the Saigon correspondents.

The more experienced of the two reporters, Bigart had few illusions about the war. Doubting that a regime as weak as Diem's could overcome an enemy as resourceful as the Viet Cong, he wrote his conclusions with a vigor that galled the South Vietnamese. A "stringer" (occasional reporter) for *Newsweek*, Sully lacked Bigart's experience as a newsman but compensated with an acid pen. Since his editors at *Newsweek* were themselves critical of Diem, he attracted the president's antipathy because of what he wrote but also simply because he was *Newsweek*'s representative.[36]

The hostility surfaced in March 1962, when Diem, emboldened by several military successes on the battlefield and by American official optimism, summarily ordered both newsmen deported. Although hardly satisfied with the work of Bigart and Sully, the U.S. mission in Saigon had little choice but to support the two reporters. As State Department officials observed, the expulsion of correspondents representing periodicals as influential as the *New York Times* and *Newsweek* would make it much more difficult for the United States to maintain American public and congressional support for the war and greatly complicate U.S. relations with South Vietnam.[37]

The task of negotiating with the South Vietnamese fell to the U.S. mission's Charge d'Affaires, William Trueheart, who on 23 March won a three-day extension for Bigart but nothing for Sully. Although Diem was willing to bargain, he clearly intended to assert his independence. Nolting then intervened, winning recision of the expulsion order against Bigart but again nothing for Sully. Diem was adamant. Sully, he said, had for years maligned the Ngo family with impunity, undermining the loyalty of South Vietnam's people to their government and contributing to an unsuccessful coup in 1962. The reporter's presence in South

[35] Msg, Saigon 1164 to State, 10 Dec 60, and Msg, Saigon 258 to State, 17 Sep 60, both in FAIM/IR; Msg, Saigon 726 to State, DAIN 14863, 5 Feb 63.

[36] Mecklin, *Mission in Torment*, pp. 129–30.

[37] Msg, State 1131 to Saigon, 23 Mar 62, FAIM/IR.

Vietnam constituted both a continuing embarrassment and a threat to the existence of the state.[38]

Nolting responded that the success of joint U.S.-South Vietnamese helicopter operations had at last brought recognition in the American press for the Diem regime's ability to win and that the expulsion of the two correspondents would jeopardize the trend. Moved by the argument, Diem agreed to countermand the order against Sully but vowed that neither reporter would remain in South Vietnam after his current visa expired. To enforce his own opinion of what good press relations should be, Diem then cut off many of Sully's South Vietnamese sources and excluded the reporter from government-sponsored trips for the press into the field.[39]

Considering an attack on one a threat to all, the Saigon correspondents also refused to yield. While *Newsweek*, according to Sully, did attempt to balance its critical tone with more positive stories and the *Times* rotated Bigart home in July, press support for South Vietnam remained less than the total commitment Diem seemed to expect. Reporters continued to qualify their stories of South Vietnamese victories with statements that cautioned against overoptimism, and occasional critical commentaries on the war continued to appear.[40]

The situation worsened during July and August. In July Bigart wrote a wrap-up of his tour in South Vietnam for the *Times*. Blaming the war's lack of progress upon Diem's inability to win the loyalty of the South Vietnamese people, the reporter called for a complete reassessment of the American aid program to South Vietnam. He then made what seemed at the time an exceedingly pessimistic prediction. If the United States failed to make the Diem regime reform, he wrote, it would face two equally undesirable alternatives. It would have either to replace Diem with some sort of military junta or to commit American troops to the war to bolster South Vietnam's sagging prospects.[41]

Sully followed shortly thereafter with an unpleasant article of his own. In it the reporter admitted the tactical strides the United States had made in South Vietnam but countered them with a military Gresham's law in which bad policies drove out good. Quoting Bernard Fall, by then a well-known scholar, he argued that the war in Southeast Asia was more political than military, that American advisers had trained the South Vietnamese to wage a purely conventional war when unconventional methods were needed, and that the U.S. Marine Corps might fly helicopters for the South Vietnamese but could never give peasant soldiers an ideology worth dying for. He cited as evidence an interview he had held with an anonymous South Vietnamese general who had attributed most of the country's ills to the Diem regime. The military chain of command was defunct in South Vietnam, the general had said, because Diem held all the strings. The war in the central provinces was likewise going badly because the people refused

[38] Msg, Saigon 1231 to State, 27 Mar 62, and Msg, Saigon 1215 to State, 23 Mar 62, both in FAIM/IR.
[39] Msg, Saigon 1380 to State, 30 Apr 62, FAIM/IR.
[40] Ibid.
[41] Homer Bigart, "Vietnam Victory Remote," *New York Times*, 25 Jul 62.

Homer Bigart

to trust the government and aided the enemy instead.[42]

Sully's comments were little different from what was being reported elsewhere in the press, but *Newsweek* drew them to Diem's attention by illustrating the article with a picture of a group of South Vietnamese militia women under the command of Diem's sister-in-law, Madame Ngo Dinh Nhu. The accompanying caption read: "Female militia in Saigon: the enemy has more drive and enthusiasm."[43]

By the time his article appeared, Bigart was out of the country and beyond Diem's reach; but Sully lacked that advantage. A resident of South Vietnam, he was an easy target for all the antipathy Diem had built up against the press. Upon the appearance of his article, therefore, Madame Nhu immediately responded with an open letter expressing "profound indignation" at the reporter's disrespect for South Vietnamese womanhood. The *Times of Vietnam* and other state-controlled Saigon newspapers followed with a series of bold-faced attacks against Sully and his employers. The newsman was accused of being an opium smuggler, a Viet Cong spy, and a patron of sex orgies. The police put him under surveillance. Even those South Vietnamese who opposed Diem politically were offended, asserting that Sully's abuses were an affront to their national pride and that a people whose country was divided and at war ought to be entitled to at least some "special consideration."[44]

Ambassador Nolting entered the dispute at once, making all the arguments he had made before but adding that reporters rarely wrote picture captions. When Diem nevertheless ordered Sully to leave the country, the Saigon correspondents took the decree as a direct threat. When they met on the evening of 4 September to demonstrate their concern and to formulate a common response, they found themselves united in their bitterness toward Diem but unable to agree on a common course of action. Nonresident newsmen favored a moderate protest to Diem. The others wanted something sharper. Suspicion of Sully heightened the disagreement. At one point most of the newsmen present even turned to the reporter to ask if he had ever been a Communist or a French agent. Finally, only six Ameri-

[42] Francois Sully, "Vietnam: The Unpleasant Truth," *Newsweek*, 20 Aug 62, pp. 40–41.

[43] Picture, *Newsweek*, 20 Aug 62, p. 40.

[44] Airgram, Saigon Mission to State, 27 Aug 62, and CIA Study, sub: The Press in South Vietnam, in Msg, Saigon 572 to State, 13 Jun 61, both in FAIM/IR; Mecklin, *Mission in Torment*, pp. 132–35.

Francois Sully

can correspondents agreed to sign a strongly worded letter to Diem. The rest composed a more moderate protest the next day.[45]

Although reporters disagreed on the subject of Sully, they were united in the belief that the whole affair had been an attempt to intimidate them. The Diem regime failed to dispel the impression. When Secretary of State for the Presidency Nguyen Dinh Thuan invited the six protesting newsmen to his office to assure them that his government's actions against Sully had never been meant as a threat to the entire press corps, the newsmen interpreted the minister's words as yet another attempt at coercion. They even questioned the U.S. embassy's role in the affair. They knew that Nolting had argued on Sully's behalf, but they contended nevertheless that the ambassador's failure was intentional, one more instance of the United States placating Diem.[46]

The U.S. mission lost stature in Diem's eyes as well. Strengthened by his success in circumventing American wishes, secure in his knowledge that the United States would do nothing substantial to deter him, and angered by a 24 September article in *Newsweek* critical of his government, he began immediate preparations to ban the magazine from South Vietnam permanently.[47]

Despite a deep concern for secrecy and a desire to foster South Vietnamese sovereignty, the State Department once more refused to concur. Asserting that "U.S. policy is firm in supporting the principle of a free press and cannot overlook damage to it, no matter how irritating or unfounded press reports frequently are," it instructed Nolting to protest. The department framed a convincing appeal to South Vietnamese pride for him to use. Any attempt to banish an internationally prominent magazine such as *Newsweek* would give the world the idea that the South Vietnamese were too weak to stand criticism and that Diem wished to "cover up" his failures. The American public's confidence in the soundness of its government's policy of strong support for South Vietnam might in turn begin to waver.[48]

[45] Mecklin, *Mission in Torment*, p. 134; Msg, Saigon 255 to State, 6 Sep 62, FAIM/IR.

[46] "The Press in Vietnam," *Washington Post*, 6 Sep 62; Msg, Saigon 255 to State, 6 Sep 62; Mecklin, *Mission in Torment*, p. 134.

[47] Msg, Saigon 349 to State, 25 Sep 62, sub: Summary of GVN Information Directorate Communique, and Msg, State 363 to Saigon, 25 Sep 62, both in FAIM/IR.

[48] Msg, State 363 to Saigon, 25 Sep 62.

The argument worked. On 26 September the Diem regime dropped its plan to ban *Newsweek*, announcing instead that it would scrupulously review each issue of the various foreign publications entering the country in order to stop the sale of those that offended South Vietnamese tastes. Since selective censorship of that sort had always been a feature of the Diem regime's approach to the foreign press and since Nolting seemed to have persuaded the South Vietnamese that a relaxed attitude toward the press was better than continued animosity, the decision gave every appearance of being a victory for the United States.[49]

The skirmish had nevertheless only postponed the inevitable. Although Diem retreated, he still had no concrete proof that the United States would do anything but talk if he followed his inclination to leash American newsmen. On 25 October he therefore moved against another offending reporter, James Robinson of NBC News, who like Sully had managed to offend the Ngo family, but innocently and without malice. Robinson's first slip had come months before, when he had remarked to an official interpreter after a long private interview with Diem that the session had been "a waste of time." The comment had filtered back to the president, who had taken an immediate dislike to the reporter but had done nothing because of his preoccupation with Bigart and Sully. Robinson's second mistake came some months later, when he inadvertently reminded Diem of his presence by calling the Ngo family a "clique." Shortly thereafter, he found himself accused of a technical violation of South Vietnamese visa regulations.[50]

The U.S. mission once more began the familiar round of appeals and negotiations—this time making the point that since the Cuban missile crisis was in progress the United States and its allies ought to turn a unified face to the enemy. The argument failed, as did the pleas of Diem's own advisers. Within days Robinson found himself on the way to Hong Kong. The official communique that accompanied his departure read, "We in Vietnam have had enough of the calumnies and insults that the unscrupulous and unreliable heap upon our Chief of State, our Government, our Army, and our youth while we are fighting a ruthless war." The president's brother, Ngo Dinh Nhu, amplified the statement a short while later by telling an American visitor that the Diem regime intended to expel any correspondent who dared to belittle either the Ngo family or South Vietnam's ability to win the war.[51]

The moment called for an American response. U.S. officials in both Saigon and Washington began to see a pattern in Diem's actions and to fear that if they stood by passively in the face of continued provocation they would only prompt Diem and Nhu to carry out their threats. Messages passed between Washington and Saigon on the subject. Nolting began to draft a cable to the State Department requesting permission publicly to repudiate the Robinson expulsion by revealing that the United States had done everything to dissuade Diem but change the United States' policy of support for South Vietnam. The director of the State

[49] Msg, Saigon 354 to State, 26 Sep 62, FAIM/IR.
[50] Mecklin, *Mission in Torment*, pp. 137–38.
[51] Ibid.

Department's Vietnam Working Group, Chalmers Wood, even suggested that, as a last resort, Nolting might consider coercing Diem by threatening either a reduction or a cutoff of U.S. support for South Vietnamese cultural programs.[52]

In the end, nothing was done. The American news media failed to react to the expulsions of Robinson and Sully because, as Wood put it, neither man was among the "outstanding" members of his profession, and NBC was unsure whether Robinson deserved all-out support. In the absence of outside pressure Nolting's own inclinations rapidly asserted themselves. Reasoning that any American dispute with South Vietnam would be of more benefit to the Viet Cong than to NBC, he canceled his planned public statement and continued to avoid friction with Diem. For their part, the Saigon correspondents remained convinced that Robinson had been wronged. Seeing little firsthand evidence of the embassy's strenuous pleading in the case, they concluded once more that the U.S. mission had sided with the South Vietnamese against them.[53]

Confrontation: Ap Bac, January 1963

The policies governing the release of information on the war to newsmen in Saigon reinforced that conclusion. Since the United States had given the South Vietnamese government control over news of South Vietnamese military operations, the Diem regime would at times require U.S. military advisers to cooperate in its attempts to silence newsmen. An example of this practice occurred in mid-October 1962, about the time Diem moved against Robinson. The Diem regime had long objected to critical news stories based upon interviews with American advisers and South Vietnamese field officers, contending that most soldiers were too unsophisticated to deal properly with American reporters. On 13 October the chief of the South Vietnamese Joint General Staff, General Le Van Ty, tried to remedy the problem by ordering all American correspondents visiting field units to submit their questions in writing to field commanders, who were to respond with written answers cleared through South Vietnamese official channels. No informal questioning would be allowed. Shortly after Ty's order appeared, the South Vietnamese 7th Division interpreted it as a ban on all visits by newsmen to the field and began to insist that reporters obtain special permission before covering operations in the Mekong Delta.[54]

Recognizing that the directive would jeopardize relations with the Saigon correspondents and undermine American public and congressional support for the war, the State Department immediately instructed Nolting to seek suspension

[52] Ltr, Nolting to Wood, 1 Nov 62, draft cable enclosed; Ltr, Wood to Nolting, 16 Nov 62, FAIM/IR.

[53] Mecklin, *Mission in Torment*, pp. 138-51; Ltr, Wood to Nolting, 16 Nov 62.

[54] Msg, Saigon 536 to State, 21 Nov 62; Msg, State 513 to Saigon, 12 Nov 62; and Msg, Saigon 503 to State, 10 Nov 62. All in FAIM/IR. Ltr, Frederick G. Dutton, Assistant Secretary of State, to Senator Philip A. Hart, 27 Dec 62, copy in CMH files.

of the prohibition. General Ty complied, but only on 11 December, over a month too late. By then the Saigon correspondents were in full cry. On 9 November the Associated Press transmitted a dispatch outlining Ty's order and newsmen's reaction to it. David Halberstam followed on 21 November with an article entitled "Americans Under Order To Withhold War News." Ignoring the fact that American advisers could still contact reporters in Saigon if newsmen were barred from the field, the reporter all but blamed the American military command in Saigon for what had happened. He linked the order to the Sully and Robinson expulsions and quoted a U.S. military information officer in Saigon to the effect that both Ty's directive and the 7th Division's order had resulted from a "misunderstanding." Whatever the reason, he added, newsmen had yet to be allowed to reenter the Mekong Delta.[55]

Halberstam wrote a more vituperative memorandum to his editors at the *Times*. Restrictions on the press in South Vietnam, he said, had become "intolerable." Reporters had experienced increasing difficulty contacting U.S. officials, especially military officers, and the South Vietnamese government had continued to exercise "a general veto" over the press by using its prohibition on access to the delta to keep reporters from getting the news. The *Times* forwarded his arguments to the State Department, whose chief of public affairs told the newspaper that wartime conditions often imposed limitations upon what reporters could write. Although American officials considered the restrictions unduly harsh and were trying to have them rescinded, correspondents reporting from South Vietnam would have to understand that the United States was a guest of the South Vietnamese and had to be circumspect in commenting on their war.[56]

The argument may have satisfied the *Times*, but it did little good in the field. Although the press appears to have suffered little from Ty's restrictions and Halberstam himself continued his liberal use of military sources, correspondents once more concluded that the entire American establishment in South Vietnam was against them. Depending less upon official sources and more upon the word of resentful aircraft crews, angry local officials, and sincerely indignant U.S. Army advisers, they began to seek the evidence they needed to prove their contentions that Diem was inept and that the United States required a freer hand in running the war.[57]

The reporters found that proof shortly after General Ty lifted his restrictions on access to the delta. Early in January 1963 intelligence revealed a Viet Cong radio station operating near the village of Ap Bac in Dinh Tuong Province with an estimated reinforced guerrilla company guarding it. Expecting an easy victory, the South Vietnamese 7th Division immediately mounted an attack to destroy the station. The operation that ensued pitted two infantry battalions, an airborne battalion, a mechanized company, a ranger company, and fifty-one U.S. advisers

[55] Msg, State 513 to Saigon, 12 Nov 62; Msg, State 562 to Saigon, 30 Nov 62; Msg, Saigon 591 to State, 11 Dec 62, FAIM/IR; Halberstam, "Curbs in Vietnam Irk U.S. Officers."
[56] Msg, State 562 to Saigon, 30 Nov 62.
[57] Msg, State 532 to Saigon, 30 Nov 62, FAIM/IR.

against the well-trained and motivated but numerically inferior *514th Viet Cong* (regular) *Battalion*—four hundred men at most. Although surprise and preponderant strength favored friendly forces, South Vietnamese commanders allowed the enemy to escape. That failure presented the Saigon correspondents with just the *cause celebre* they were seeking.[58]

Nothing seemed to go right in the battle. Within a single stretch of five minutes, 5 U.S. helicopters were lost: 2 to enemy ground fire, 1 to mechanical malfunction, and 2 when their pilots flew, gallantly if unwisely, into the enemy's guns to rescue downed comrades who were in fact already safe behind South Vietnamese lines. As the afternoon progressed, U.S. advisers called for an airborne drop to the east of the village to plug the one escape route open to the Viet Cong, only to see the paratroopers drop to the west, where some were killed by friendly fire. When U.S. advisers requested a heavy artillery barrage against the enemy's positions, they could obtain no more than four rounds per hour. Finally, at dusk a South Vietnamese air strike accidentally hit a friendly unit, causing an undetermined number of casualties. Even though the enemy escaped during the night, mopping-up operations the next day went little better. South Vietnamese artillerymen accidentally shelled their own troops for ten minutes, killing three, wounding twelve, and forcing Brig. Gen. Robert York to find cover face down in the mud and dung of a rice paddy.[59]

Although much of what went wrong at Ap Bac was attributable either to bad luck or to South Vietnamese inexperience, the poor coordination and missed chances were symptomatic of what was wrong in all of South Vietnam. Concerned that a victorious army might produce an opposition leader capable of challenging the status quo, President Diem had long attempted to limit his army's initiative by severely reprimanding field commanders who took more than a few casualties in any given engagement. Both the officers with the troops at Ap Bac, none above the rank of captain, and their superiors farther to the rear knew that Diem frowned upon casualties and were too insecure to contest his will. Allowing air strikes and artillery to do most of their work, they ignored U.S. advice to attack and, in the opinion of U.S. advisers, used the confusion on the ground to mask their decision to let the enemy escape. So adept were they at delay that a company of armored personnel carriers took three and one-half hours to advance 1,500 yards against only small arms fire.[60]

The Saigon correspondents knew nothing about the battle when it started. Driving out to Ap Bac only after the first day's fighting had ended, they learned from angry American advisers that South Vietnamese commanders appeared to have thrown away a chance to win a major victory. Over the next few days they took copious notes on those officers' pungent remarks. One adviser told them

[58] Senior Adviser, 7th ARVN Division, After Action Report for the Battle of Ap Bac (hereafter cited as Ap Bac After Action Report), 9 Jan 63. See also Charles V. P. von Luttichau, The U.S. Army Role in the Conflict in Vietnam (hereafter cited as U.S. Army Role in the Vietnam Conflict), CMH MS, pp. 68–77; Interv with Col John Paul Vann, adviser to the 7th ARVN Division. All in CMH files.

[59] Ap Bac After Action Report; Interv with Col Vann.

[60] *Pentagon Papers*, 2: 134–35; Interv with Col Vann; Ap Bac After Action Report.

U.S. Advisers Inspect Battle Gear *after the battle of Ap Bac.*

the battle had been a "miserable damn performance." Others spoke of the lack of South Vietnamese aggressiveness, asserting that if the 7th Division had taken the initiative it would have suffered fewer casualties in the long run. A few mistakenly told of how one of three Americans killed, Capt. Kenneth N. Good, had died while begging reluctant infantrymen to advance. One of the newsmen counted the bullet holes in a downed helicopter and found the wallet and family pictures of a dead American pilot. Several others accidentally overheard a classified briefing convened for General Harkins in an unsecured area. In the end, the reporters had enough information to make their dispatches read like official after action reports. They used most of it.[61]

The first accounts of the battle that appeared in the United States were factual. In the rush to put the basic story on the wire before press runs began back home, newsmen had little time to analyze what had happened. East coast newspapers in the United States such as the *Washington Star*, the *Baltimore Sun*, and the *Washington Daily News*—all with early deadlines—thus went to press with the story of the battle but without negative commentary. Only the *New York Times* carried a critical story. Although terming Ap Bac "by far the worst day for American helicopters in Vietnam since the American build-up," the paper then kept mainly to the facts.[62]

Coverage became more interpretive late on 3 January and early on the fourth, when the full dimensions of what had happened began to emerge. On the third, benefiting from a deadline later than those of eastern papers, the *Chicago Daily News* blamed the defeat on "bad luck and disorganization." The paper quoted the adviser's comment that the battle had been "a miserable performance" and then observed that "the guerrillas held their ground and fought even after fighter-bombers . . . reached them, . . . while the Vietnamese forces . . . showed little

[61] Msg, Saigon 656 to State, DAIN 5969, 8 Jan 63; Msg, Saigon 726 to State, DAIN 14863, 5 Feb 63; quote from Ltr, Porter to Lewis, 3 Nov 71, in Lewis, "The Year of the Hare," pp. 47–48. For some of the advisers' statements, see Malcolm W. Browne, "U.S. Weighs New Tactics in Vietnam," *Philadelphia Inquirer*, 6 Jan 63; "Mistakes, Luck Trip Up Vietnam," *Chicago Daily News*, 3 Jan 63; "Reds Eluding Pursuit by Vietnamese," *Baltimore Sun*, 9 Jan 63.

[62] "Battle in Vietnam Leaves Hundreds Hurt," *Washington Star*, 3 Jan 63; "Three Americans Are Killed by Viet Cong," *Baltimore Sun*, 3 Jan 63; "Vietnam Casualties Soar in Furious Fight," *Washington Daily News*, 3 Jan 63; David Halberstam, "Viet Cong Downs Five U.S. Copters," *New York Times*, 3 Jan 63.

interest in chasing the communists.'' The same themes reechoed the next day elsewhere in the press, this time spiced by the sensational details the reporters had gleaned at the scene of battle. The *Washington Daily News* called Ap Bac a South Vietnamese ''humiliation.'' The *Baltimore Sun* observed that the guerrillas were ''slipping away . . . ahead of half-hearted Vietnamese pursuit.'' David Halberstam and the *New York Times* noted that ''what made the defeat particularly galling to Americans and the Vietnamese alike was that this was a battle initiated by government forces in a place of their own choice, with superior forces and with troops of the Seventh Vietnamese Division, which is generally considered an outstanding one in the country. Today the government troops got the sort of battle they wanted, and they lost.''[63]

The U.S. mission in Saigon handled Ap Bac strictly according to policy. When questioned by the press, American spokesmen limited their comments to events directly involving American personnel and helicopters and left the rest to the South Vietnamese. The only U.S. statement on the battle of any consequence came from General Harkins, who told newsmen at the scene that the 7th Division had sustained unusual losses but appeared to have the Viet Cong surrounded. A U.S. Information Service spokesman in Saigon also conferred with the press, noting, as official guidance stipulated, that ''This is war. Occasional setbacks are inevitable and normal. Yet the GVN continues to gain overall, and this series of strong Viet Cong reactions may indeed reflect the enemy's realization that he is in trouble.'' Only on 8 January, five days after the battle had ended, did Ambassador Nolting consent to a background session with newsmen. By then the time for correction and perspective had passed. The press, both in Vietnam and the United States, had the basic story of the battle and was using it to criticize the conduct of the whole war in Southeast Asia.[64]

Hanson W. Baldwin was one of the more perceptive of those who commented on the action. Assailing what he considered an overdependence on helicopters at Ap Bac and in South Vietnam in general, he observed that ''legs are a soldier's chief weapon'' and that neither helicopters nor soldiers ''who descend briefly from the sky'' could ever succeed in controlling South Vietnam. Government troops would have to learn to ''live and march and fight in the jungle'' if Saigon was to win. Arthur Krock of the *New York Times* said much the same thing, adding that Ap Bac had proved that ''no amount of U.S. military assistance can preserve independence for a people who are unwilling to die for it.'' The *Detroit Free Press* meanwhile asked rhetorically how a harsh dictatorship such as the one exercised by Diem could give peasants any motive at all for fighting.[65]

[63] ''Mistakes, Luck Trip Up Vietnam''; ''Vietnamese Humiliated,'' *Washington Daily News*, 4 Jan 63; ''Reds Eluding Pursuit by Vietnamese''; David Halberstam, ''Vietnamese Reds Win Major Clash,'' *New York Times*, 4 Jan 63.

[64] The U.S. Information Service is the field extension of the U.S. Information Agency. Quote from Msg, Saigon 656 to State, DAIN 5969, 8 Jan 63.

[65] Hanson W. Baldwin, ''Foot Soldier Holds Key to Victory in Vietnam,'' *Kansas City Star*, 7 Jan 63; Arthur Krock, ''Help to People Who Won't Fight Doomed to Failure,'' *Houston Chronicle*, 9 Jan 63; ''The Mess in Vietnam Calls for a Hard Look,'' *Detroit Free Press*, 5 Jan 63.

The *Washington Daily News* went farther than either Baldwin, Krock, or the *Free Press*. It called into question the practice of allowing the South Vietnamese to control the way the war was fought. According to U.S. policy, the paper said, South Vietnam was a sovereign nation with control over all commands issued on its battlefields. Although South Vietnamese officers were sensitive about their national pride and unwilling to surrender authority to foreigners, they would have to understand that Americans were sensitive, too, and unwilling to pay the price of someone else's irresponsibility, especially with American lives. The *Fort Worth Star-Telegram* put the matter more succinctly. Observing that changes would have to be made if South Vietnam was to survive, the paper suggested that "a fine new slogan for the South Vietnamese would be, 'Better led than red.'"[66]

Realizing that the uproar over Ap Bac would harm relations with the Diem regime, the United States moved to soften the effects of the controversy. Although cables between the U.S. mission in Saigon and the State Department affirmed that South Vietnamese forces had failed in the battle, spokesmen at the State Department attempted to put a good face on what had happened by announcing that, on the whole, South Vietnamese forces had fought with courage and determination.[67] General Harkins adopted much the same approach. "Anyone who criticizes the fighting of the Armed Forces of the Republic of Vietnam," he told newsmen in Saigon, "is doing a disservice to the thousands of gallant and courageous men who are fighting in the defense of their country."[68]

Later he called the battle a victory. So did the Commander in Chief, Pacific, Admiral Harry D. Felt. Arriving in Saigon at the height of the outcry, Felt added that the differences American advisers were experiencing with South Vietnamese commanders were comparable to a family quarrel. "There are times in your own family when you have disappointments with your wife," he said. "Generally we understand each other. It is only the exception when we become a little bit angry."[69]

The effort to reassure the South Vietnamese and to lower the volume of press reporting on Ap Bac had just the opposite effect. In reply to General Harkins the *Detroit Free Press* pointed out that although a commanding officer "had to go along with official policy or quit," it only added to the bleakness of the situation in South Vietnam when a general felt it necessary to "apply such thin and unconvincing whitewash." Other papers agreed, claiming that the American public had yet to be fully informed about the situation in South Vietnam. In fact, the *Milwaukee Journal* avowed, "We have an estimated 10,000 American military personnel in Vietnam. Our advisers are accompanying combat missions, flying

[66] "A Costly Adviser System," *Washington Daily News*, 11 Jan 63; "Reluctant Pupils," *Fort Worth Star-Telegram*, 9 Jan 63.

[67] Msg, Saigon 656 to State, DAIN 5969, 8 Jan 63. See also Peter Lisagor, "Military Advisers' Criticism of Troops Spurs State Department," *Chicago Daily News*, 8 Jan 63.

[68] "Harkins Lauds Vietnamese," *Washington Star*, 10 Jan 63.

[69] "Visiting South Vietnam," *Washington Post*, 10 Jan 63; "Vietnam Planes Back New Ground Operation," *Washington Post*, 11 Jan 63.

fighter planes, and ferrying Vietnamese troops into battle areas. We have had casualties from the start. Our men are in a war.''[70]

It was a European, Richard Hughes of the *London Sunday Times*, however, who put the contentions of the press about Ap Bac and South Vietnam most succinctly. The war was costing $400 million a year, Hughes said in an article reprinted by the *Washington Post*, and more than fifty American servicemen had already died. Despite that effort, the government of President Diem had failed to carry out the reforms it had pledged in return for increased American aid and had refused to allow U.S. advisers to improve the discipline and fighting spirit of its army. American officers were going to have to take command of the war, if South Vietnam was to survive. Even then, the conflict promised to become a ten-year struggle to uphold a ''reactionary, isolated, unpopular regime.'' Remarking that the situation in South Vietnam bore an alarming resemblance to the one that had confronted General George C. Marshall in China at the end of World War II, Hughes concluded that the United States might find the inclusion of the Communists in a coalition government in South Vietnam preferable to prolonging the war.[71]

Although they agreed with most of what Hughes had said, few American correspondents in Saigon at the time would have accepted the reporter's contention that a compromise with the Communists was possible. The American press believed that the war in South Vietnam was open to American manipulation and that the United States needed only to take control from Diem to succeed. Some U.S. newspapers had wondered whether the United States could truly fathom the Asian mentality and others had doubted the will of the South Vietnamese to fight, but none had ever challenged the basic assumptions that had brought the United States to Southeast Asia or questioned whether the war was beyond the American ability to win. Instead, at Ap Bac and elsewhere, they disagreed with tactics, arguing in favor of efficiency and American know-how.[72]

The vehemence of the news media's reaction to Ap Bac was, indeed, explainable only in the light of the whole climate of opposition the press had encountered in South Vietnam. For months American reporters had felt the wrath of the Diem regime, and for months American editors had been hearing about it and about all the things the Diem regime was doing wrong. The situation seemed so alien to all that the newspapermen considered proper that when the story on Ap Bac broke Diem had hardly a friend in any newsroom in the United States. As one official commentator in Saigon observed in a staccato message to the State Department, ''What happened looks from here like savagely emotional delayed reaction to ousters of Sully and Robinson, Mme. Nhu's charge that whole American press is 'communist,' and every other harassment over past six months. Ap Bac was reported as major GVN failure at cost of American lives, and it appears from here that American editorial writers, commentators, columnists licked their

[70] ''Duty's Demand on a General,'' *Detroit Free Press*, 12 Jan 63; ''Basic Problem in Vietnam,'' *Milwaukee Journal*, 12 Jan 63.

[71] Richard Hughes, ''U.S. Combat Command Over Vietnamese Urged,'' *Washington Post*, 13 Jan 63.

[72] Msg, Saigon 726 to State, DAIN 14863, 5 Feb 63.

chops with delight and reached for simplest adjectives they could muster.''[73]

American newspapermen indeed turned Ap Bac into a *cause celebre*, but they still exercised considerable restraint in what they wrote, refusing to report many of the truly extravagant remarks American advisers at the scene had made. They published factual errors—most notably the story, received from angry American advisers, that Captain Good had died while trying to persuade South Vietnamese troops to advance when he had in fact been reconnoitering forward positions—yet even the U.S. mission's Public Information Office had to admit that their reports appeared to be ''perhaps two-thirds accurate.'' Working from partial information on an emotional subject, they had done quite respectably. Their stories hurt but were little worse than could have been expected under the circumstances. Ap Bac was sensational in its own right.[74]

Much of the controversy over the battle must, indeed, be attributed to the way the U.S. mission handled the press. Having given Diem sovereign rights over information on South Vietnamese troops and operations, American military and civilian information officers failed to brief reporters on the battle until too late, a tactic that forced newsmen to rely almost completely upon emotional, firsthand sources. Then, in an attempt to reassure the South Vietnamese, General Harkins and Admiral Felt called Ap Bac a victory when everyone knew it had been a failure, in effect providing newsmen with more evidence that the U.S. mission was deluding itself and that U.S. policy in South Vietnam was bankrupt. Undisposed either to sympathy or to cooperation, reporters concluded yet again that everyone in authority was against them.

That was ironic, for in the months preceding Ap Bac, U.S. agencies in both Saigon and Washington had obviously begun swinging away from Diem and toward the press. During March, when Diem had first attempted to expel Sully, and again during September, when the South Vietnamese had finally removed the reporter, the State Department had recognized that assaults upon the Saigon correspondents jeopardized its policy of sustaining congressional and public support for the war and had begun stressing the need both to support the principle of a free press and to protect newsmen whatever the quality of their work. Ambassador Nolting was of the same mind. He told Chalmers Wood of the State Department during November that he held no brief for erroneous, discourteous reporting but that he would uphold the right of newsmen to report as they saw fit and would continue to assert that the best antidote for criticism was success rather than repression.[75]

Even if the full extent of the U.S. mission's attempts to influence the South Vietnamese had come to light, however, the United States would still have been at a disadvantage. For although Nolting did prevent Diem's first attempt to dislodge Sully and did stop the banning of *Newsweek*, his actions in each case only

[73] Ibid. See also Msg, Saigon 656 to State, DAIN 5969, 8 Jan 63.

[74] Msg, Saigon 726 to State, DAIN 14863, 5 Feb 63.

[75] Msg, State 1131 to Saigon, 23 Mar 62; Msg, State 363 to Saigon, 25 Sep 62; Ltr, Nolting to Chalmers Wood, 1 Nov 62.

General Harkins

preserved the status quo. Given Diem's antipathy toward the press, reporters were certain to decide in the long run that, whatever the United States said and did, it supported what was happening. In that sense, some sort of outcry, over Ap Bac or something else, was bound to occur.

In the event, Ap Bac and the controversy surrounding it marked a divide in the history of U.S. relations with the news media in South Vietnam. Before the battle newsmen criticized Diem, badgered American officials, and argued for more U.S. control of the war, but were still relatively agreeable. After it, correspondents became convinced that they were being lied to and withdrew, embittered, into their own community. Although Ambassador Nolting and General Harkins professed to be accessible to the press at any time, most senior American civilian and military officials in South Vietnam limited their contacts with newsmen to formal occasions such as news conferences and briefings, where they turned an ever more optimistic face toward their critics.

As time passed, the enmity between the two groups became emotional. At one point David Halberstam is reported to have driven past General Harkins' Saigon quarters, shaking his fist and vowing, "I'll get you, Paul Harkins." In the same way General Harkins' Assistant Chief of Staff for Plans, Maj. Gen. Milton B. Adams, U.S. Air Force, declared in an official debriefing that the policy which required official tolerance for newsmen was the only real frustration he had encountered during his tour of duty in South Vietnam.[76]

The U.S. mission's Chief of Public Affairs, John Mecklin, discerned the problems that were developing and attempted to correct them before they got out of hand. In reply to a State Department request for information on the adequacy of press coverage of the war, he told his superiors that reporting was about as good as could be expected, given the complex circumstances at work in South Vietnam and the fact that only United Press International, the Associated Press, and the *New York Times* considered the war important enough to station full-time correspondents in Saigon. Much of what was happening could be attributed to misunderstandings on all sides. While the Diem regime "pridefully" resented any form of hostile criticism, young reporters—"average age 27"—failed to note that the mark of a great nation was "tolerance and understanding of such tor-

[76] Jim Fain, "News in Vietnam Tough Chore," *Atlanta Journal*, 22 Mar 71; MACV Historical Office Interv with Maj Gen Milton B. Adams, USAF, Jul 65, p. 15, CMH files.

tured people as the Vietnamese,'' who often resorted to ''petty, pathetic maneuvers to save face.'' Over all, Mecklin said, ''routine'' official optimism and the practice of withholding bad news had not only soured relations with the press but had also lowered the status of truly good news.[77]

Although Mecklin briefed President John F. Kennedy on the subject during a trip to Washington, his appeal to reason was insufficient to dispel the antagonisms operating in Saigon. When he began to push for concessions to the press he alienated those of his colleagues within the mission who considered reporters enemies. They retaliated by attempting to undermine his standing with the press, leaking portions of his memorandum that were critical of newsmen without revealing his many favorable comments. As a result, the Saigon correspondents came to distrust Mecklin, whom they dubbed ''Meck the Knife.''[78]

Mecklin's attempt to gain backing for a policy of moderation was more successful in Washington. Recognizing the validity of the approach, the State Department in May issued a directive that stipulated the fullest possible cooperation with the press in order to justify ''our large human and material investment'' in Southeast Asia. Shortly thereafter, military information officers were assigned to each of South Vietnam's four corps tactical zones to serve as the eyes and ears of MACV's Office of Information. Communications links between units in the field and Saigon news briefers were also improved so that the terse operational summaries released to reporters could include the latest word on what was happening. Weekly press conferences began at the same time, featuring experts who briefed the press on areas of the war it might otherwise overlook.[79]

Those improvements nevertheless failed to remedy what was wrong. Most high-level officials of the U.S. mission in Saigon still refused to give credence to the objections of reporters, and reporters still remained ill disposed to any compromise with officialdom short of a change of staff at the top of the mission. Because no meeting ground existed between the two groups, the only possibility for remedying the problem fell to the South Vietnamese, who, as events would shortly prove, saw little need for any relations with the press at all.

[77] Msg, State 729 to Saigon, 24 Jan 63, FAIM/IR; Msg, Saigon 726 to State, DAIN 14863, 5 Feb 63; Mecklin, *Mission in Torment*, pp. 147–48.

[78] Mecklin, *Mission in Torment*, p. 148.

[79] Msg, State to CINCPAC, 21 May 63, FAIM/IR; MFR, JCS 2343/257, sub: Report of Krulak Visit to Vietnam, 25 June to 1 July 1963, p. 15, CMH files.

2

The Buddhist Crisis, 1963

The situation in South Vietnam worsened dramatically on 8 May 1963, when a large crowd of Buddhist priests and laymen assembled at the government radio station in Hue to protest an order by the Diem regime banning the display of religious flags on the birthday of the Buddha. The group accused the regime of anti-Buddhist discrimination because Roman Catholics had been permitted to parade with flags only a few days earlier to celebrate the birthday of Archbishop Ngo Dinh Thuc, Diem's brother. The crowd refused to disperse at the order of the city's Catholic commandant. Instructed to do his duty by his superiors who refused to admit they had made a mistake, that officer turned what had been a purely religious issue into a political crisis by ordering his troops to fire on the demonstrators with live ammunition. Nine people died.[1]

First Phase, May–August 1963

Buddhist leaders contacted the U.S. embassy in Saigon for support. When they realized that American officials were mainly interested in stabilizing the Diem regime and that coverage by the foreign press would give any action they took a worldwide audience, they decided instead to cultivate the Saigon correspondents. A relationship of trust sprang up between the two groups. The Buddhists provided reporters with easy access to the top leaders of their organization and with the dates and places of their next demonstrations. The reporters, in turn, kept their knowledge a secret, denying advance warning of the Buddhists' plans to both the Diem regime and the U.S. mission.[2]

[1] Dennis J. Duncanson, *Government and Revolution in Vietnam* (New York: Oxford University Press, 1968), pp. 327–38; Mecklin, *Mission in Torment*, pp. 153–54.
[2] Mecklin, *Mission in Torment*, p. 163.

Thich Quang Duc Immolates Himself

The news dispatches that followed electrified the world. Thich ("priest" or "reverend") Quang Duc burned himself to death on 11 June 1963, as the only newsman to heed Buddhist advance notices, Malcolm Browne of the Associated Press, photographed the scene. The pictures that resulted won front-page attention in newspapers everywhere. Over the next several months sensation followed sensation as the Buddhists marched and the South Vietnamese police reacted with violence.

Although Diem attempted to justify his government's extreme measures by protesting that Buddhist discontent was Communist inspired, available evidence contradicted that claim. Investigators found, in fact, that shortly after Buddhist leaders had repudiated an outright offer of aid from North Vietnam, an agent of the Diem regime had been apprehended attempting to plant incriminating enemy propaganda leaflets in a downtown Saigon pagoda where the secret police were certain to find them. While some individual Buddhists might indeed have been involved with the Communists, the conclusion seemed inescapable that the regime's attitude toward its antagonists was the basic cause of the problem.[3]

[3] Research Memo, RFE-75, U.S. Dept of State, I&R, 21 Aug 63, sub: Diem vs. the Buddhists, FAIM/IR; Robert Udick, "Diem Expects Victory in 2 to 3 Years," *Washington Post*, 29 Jul 63; CIA Information

Diem was adamant in his opposition to the Buddhists. Stating privately that he would never negotiate with those "pirates," he told his friends that he would shoot his enemies down with machine guns if they continued to demonstrate.[4] When an eleven-man delegation appointed by South Vietnam's National Assembly submitted an objective report on the crisis, Diem rejected it out of hand, insisting that it be "rewritten to conform to the facts." His brother, Ngo Dinh Nhu, supported him. The report, said Nhu, "represented a version of events accepted by foreigners that was largely fabrication."[5]

This action helped rather than hindered the Buddhists. In the face of Diem's intransigence, officials at all levels of the South Vietnamese bureaucracy ceased trying to accomplish anything positive and, in some cases, began to collaborate actively with the demonstrators. The Buddhists thus acquired valuable inside information about the government's intentions.[6]

Diem's attitude also destroyed any chance the South Vietnamese government might have had to work constructively with American newsmen. The regime considered the press an enemy and was unwilling to communicate its side of the story effectively. Worse, its attempts to intimidate the Saigon correspondents succeeded only in arousing their active opposition. While South Vietnamese policemen assaulted reporters and cameramen who attempted to cover Buddhist demonstrations, secret agents shadowed correspondents suspected of antigovernment tendencies. During August 1963 officials began censoring news dispatches, deleting not only sensitive information but also routine background material such as descriptions of Saigon's city life. In the end, the climate became so hostile to reporters that the U.S. mission warned David Halberstam to move out of his rented house to a more secure hotel so that he could avoid becoming too easy a target for the secret police.[7]

Recognizing the possible effect that harassment of the press would have on public opinion in the United States, American diplomats were once more torn between the policy of upholding Diem and the need for good public relations. Unhappy in the extreme to have American-armed and -trained South Vietnamese soldiers brutally repressing Buddhist demonstrators while U.S. television news teams watched, yet also aware that any strong official protest in public might seem a withdrawal of U.S. support for Diem, they chose again to compromise.

Rpt, 17 Jun 63, sub: Status of VC Efforts To Exploit Buddhist Situation, Gard Papers, CMH; Memo, U.S. Dept of State, I&R, 28 Jun 63, sub: International Repercussions of Vietnamese Buddhist Tensions, FAIM/IR; CIA Information Rpt, 21 Aug 63, sub: Government Attempt To Plant VC Leaflets, Gard Papers, CMH; Msg, Saigon 224 to State, 14 Aug 63, FAIM/IR.

[4] CIA Information Rpt, 6 Jun 63, sub: Report of a Discussion Between Members of the Central Vietnam Faction and Can Lao Party, Gard Papers, CMH.

[5] Ibid., sub: Report of Assembly Investigation of Hue Situation, Gard Papers, CMH.

[6] Ibid., 6 Jun 63, sub: Indications of Disaffection With the Diem Regime, and 15 Jun 63, sub: A Field Appraisal of the Buddhist Crisis, both in Gard Papers, CMH; SNIE 52-2-63, 10 Jul 63, sub: The Situation in South Vietnam, doc. 125, *Pentagon Papers*, 2: 729-33.

[7] Duncanson, *Government and Revolution in Vietnam*, pp. 334-35; Mecklin, *Mission in Torment*, pp. 153-65; Msg, Saigon 327 to State, 24 Aug 63, and Msg, Saigon 351 to State, 27 Aug 63, both in FAIM/IR; George W. Goodman, "Our Man in Saigon," *Esquire*, January 1964.

They made the fullest possible representations in private conferences with South Vietnamese functionaries but in public attempted to appear conciliatory and unthreatening.[8]

This approach did little to improve the situation. U.S. officials argued fervently and relentlessly behind closed doors for concessions to the Buddhists and the press, even threatening to disavow Diem if he failed to come to terms with his antagonists. In public, however—beyond allowing the Voice of America to broadcast an unadorned version of the news to South Vietnam and intervening on behalf of newsmen who ran afoul of the South Vietnamese police—they did little to give their warnings any weight. Instead, the U.S. military command stopped flying newsmen to Hue lest the South Vietnamese suspect that it was participating in pro-Buddhist agitation.[9]

Cautious optimism also continued in both Saigon and Washington. When he returned to South Vietnam in mid-July after an absence of seven weeks, Ambassador Nolting told newsmen gently that although unity of purpose with the South Vietnamese would become unattainable if internal dissension continued, the anti-Communist cause would prevail in the end. President Kennedy said much the same thing at a 17 July news conference. Questioned on whether the Buddhist crisis had been an impediment to American aid to South Vietnam, he told reporters, "Yes, I think it has. I think it is unfortunate that this dispute has arisen at the very time when the military struggle has been going better than it has been going for many months. I would hope that some solution could be reached for this dispute, which certainly began as a religious dispute, and because we have invested a tremendous amount of effort and it is going well."[10]

At first the U.S. policy of private anger and public forbearance seemed to work. On 16 June Diem signed an agreement with the Buddhists that appeared to yield to their demands and that caused a number of laudatory comments in the U.S. press. The move was nevertheless more a reaction to the pressures generated by Buddhists within the South Vietnamese bureaucracy after the suicide of Quang Duc than a response to American protestations. The regime's true orientation surfaced within days. While the state-controlled *Times of Vietnam* attacked the United States and taunted the Buddhists, Madame Nhu began a series of inflammatory public statements, calling the Buddhists "murderers" and asserting that her family would "ignore the bonzes, so that if they burn thirty women we shall go ahead and clap our hands."[11] When the government failed to silence either

[8] Msg, Saigon 252 to State, 27 Aug 63, FAIM/IR; Msg, State 1173 to Saigon, DAIN 53640, 3 Jun 63, Army Staff Communications Center files, Army War College; David Halberstam, "U.S. Aides Balked in Vietnam Crisis," *New York Times*, 10 Jun 63.

[9] SNIE 52-2-63, 10 Jul 63, sub: The Situation in South Vietnam, doc. 125, *Pentagon Papers*, 2: 729–33; Msg, Saigon 297 to State, 21 Aug 63, and Msg, Saigon A-127 to State, 9 Aug 63, sub: Security Office July 1963 Report, both in FAIM/IR; Halberstam, "U.S. Aides Balked in Vietnam Crisis"; Stanley Karnow, "The Newsman's War in Vietnam," *Nieman Reports*, December 1963, p. 7.

[10] "Nolting, Back in Saigon, Predicts Victory," *New York Times*, 11 Aug 63; quote from Kennedy News Conference, 17 Jul 63, *Public Papers of the Presidents: John F. Kennedy, 1963* (Washington, D.C.: Government Printing Office, 1964), p. 569.

[11] "Vietnam's First Lady," *New York Times*, 11 Aug 63.

Buddhist Demonstration in Saigon, 1963

the *Times* or Madame Nhu and began disingenuously to protest that individuals have a right to speak their minds in an open society, the Buddhist leadership concluded that Diem had no intention of living up to his promises and began publicly to voice its desire to bring him down.[12]

Although newsmen appreciated the embassy's backing and knew that the U.S. government was trying to persuade the Diem regime to adopt a policy of moderation, they remained more concerned with the U.S. government's continuing optimism than with making peace with Diem. That attitude led them to jump to easy conclusions without sufficient reflection. On one occasion, for example, they came across presumed evidence that the South Vietnamese Army had used "blister gas" to disperse a Buddhist demonstration. Aware that the United States had never given that kind of gas to the South Vietnamese, the U.S. mission pleaded for a delay in publication until it could investigate. Although most of the newsmen complied, one pushed ahead with his account. Inquiry shortly revealed that degenerated tear gas manufactured for use in World War II had been to blame,

[12] Memo of Conversation, U.S. Dept of State, 4 Jul 63, sub: Presidential Briefing, the Situation in South Vietnam, FAIM/IR; CIA Information Rpt, 8 Jul 63, sub: Staff Appraisal, Vietnam, Gard Papers, CMH.

but by then the wrong story was out and the damage done. Reporters also asserted again and again that 70 percent of South Vietnam's population was Buddhist, making the uprising seem a nationwide movement to repudiate Diem, when most South Vietnamese were in fact ancestor worshippers and the Buddhists' protests, while serious, were mainly an urban, politically oriented phenomenon.[13]

Although distorted news stories increased the difficulties of the U.S. mission in Saigon, they were only part of the problem. Far more serious was the fact that while the top levels of the mission were inordinately closemouthed around reporters, other officials, especially those who disagreed with the policy of supporting Diem, lacked such inhibitions. By leaking delicate American discussions with Diem to the press they embarrassed the president and helped to thwart the embassy's vigorous efforts to win an end to anti-Buddhist repressions.

The most flagrant case involved mission Charge d'Affaires William Trueheart's negotiations with the South Vietnamese government during Nolting's June and July absence. When he failed to move Diem toward reason, Trueheart had flailed the president with strong language and had warned that the United States might have to dissociate itself publicly from his anti-Buddhist activities. Of the utmost sensitivity because it amounted to almost a direct command from the United States, the statement was certain to humiliate Diem if it became public by tending to verify Communist assertions that he was little more than an American tool.[14]

Yet Trueheart had hardly finished speaking before an American official in Washington leaked what had happened to Max Frankel of the *New York Times*. The *Times* put the story on page 1 of its 14 June edition. Reading as though Frankel had seen the State Department's file of classified cables on the subject, the article outraged Diem and confirmed his suspicions that the United States government secretly agreed with what the Saigon correspondents were writing. As Halberstam later observed, it also destroyed the charge's ability to deal with Diem by convincing the Ngo family that Trueheart was "pro-Buddhist."[15]

High officials within the Department of Defense considered newspaper stories such as the one by Frankel unfair and resolved to stop them. While no one realistically expected to eliminate the type of high-level leak that had led to Frankel's article, the problem at lower levels seemed amenable to correction if the proper pressures were applied. For months U.S. Army advisers in the field had generalized in public about what was wrong with South Vietnam, its government, and its army, and for months newsmen had mined those statements for the sensations they contained. During May 1963 Secretary of Defense McNamara requested that the chairman of the Joint Chiefs of Staff take action to limit the practice.[16]

[13] Memo of Conversation, U.S. Dept of State, 4 Jul 63, sub: Presidential Briefing, the Situation in South Vietnam; "Vietnam's First Lady"; CIA Information Rpt, 8 Jul 63, sub: Staff Appraisal, Vietnam.

[14] Mecklin, *Mission in Torment*, p. 171.

[15] Max Frankel, "U.S. Warns Diem on Buddhist Issue," *New York Times*, 14 Jun 63; David Halberstam, "U.S. Dilemma in Saigon," *New York Times*, 5 Aug 63; Mecklin, *Mission in Torment*, p. 172.

[16] Ltr, CINCPAC 3010, ser. 157, to SECDEF et al., 8 May 63, sub: Record of the Secretary of Defense

The agency responsible for training officers destined for South Vietnam, the U.S. Continental Army Command (CONARC), shortly thereafter issued policy guidance to remedy the problem. Despite the drawbacks present in a policy of allowing newsmen open access to field units, the command noted, any Army attempt to reduce contacts between correspondents and American advisers would bring on public relations problems by seeming to prove that the United States had something to hide. A middle course seemed more advisable. Accordingly, U.S. Army personnel newly assigned to South Vietnam were to confine their conversations with newsmen to "areas of personal responsibility and knowledge" and to avoid the natural tendency to talk in generalizations. Soldiers in the field were meanwhile to leave broad estimates of progress and decline to high officials, who in theory had a better view of the total war.[17]

The approach was in many ways commendable. By restricting advisers to comments on areas of the war they knew and by warning against statements uttered in haste and anger, it sought to forestall situations such as the one at Ap Bac, where the advisers' intemperate rhetoric had made a bad affair worse. It might also have compelled newsmen to rely more upon sources who had a true overview—a necessity in an environment as complicated as the one in South Vietnam.

The Continental Army Command nevertheless failed to stop with general guidelines. To avoid an adversary relationship between reporters and the Army in South Vietnam, the command went on to spell out how the advisers should approach the press.

> You must remember that whether you wear one stripe or six, one bar or silver eagles you automatically become an "Army spokesman" when you are approached by the press. Within 24 hours the words of that Army spokesman can be flashed worldwide, particularly if they can be construed as criticism of the American or Vietnamese effort. Everything you say should have the ultimate aim of furthering that effort. Your approach to the questions of the press should emphasize the positive aspects of your activities and avoid gratuitous criticism. Emphasize the feeling of achievement, the hopes for the future, instances of outstanding individual or unit performance and optimism in general. But don't destroy your personal credibility by gilding the lily. As song writer Johnny Mercer put it, "You've got to accentuate the positive and eliminate the negative."

The statement concluded by underscoring official concern for public opinion. A soldier serving in South Vietnam was an "oracle," it cautioned. He was thus in a position to influence both the press and the worldwide reaction to U.S. policy. By confining his comments to his responsibilities and by emphasizing what was positive, he could "make a constructive contribution to public understanding."

In issuing the memorandum the Continental Army Command had sought to make soldiers recognize that they were part of the Army, to note that the Army's

Conference Held 6 May 1963 at HQ, CINCPAC, I-35588-63, ISA 337, 67A4660, box 7, Washington National Records Center (WNRC).

[17] This section is based on U.S. Continental Army Command, Orientation on Press Relations for Personnel Destined for Vietnam [June 1964], an inclusion in the packet of information on the news media given to General William C. Westmoreland before he left for South Vietnam in 1964—see Westmoreland History, bk. 2, tab F, in CMH.

view of the war differed from that of the press, and to encourage them to reflect the Army's view rather than their own in conversations with newsmen. Yet by quoting Johnny Mercer and by emphasizing the need for an accent on the positive, the memorandum's authors had used language that was easily misconstrued. For even though the memorandum was an internal document never intended for distribution to the press, someone was bound to bring it to public light sooner or later. When he did, given the rising tensions in South Vietnam, newsmen were certain to read the worst possible connotations into its more rhetorical passages and to conclude that the Army was out to salvage what it could of a bad situation by attempting to curtail free speech.

Problems arose shortly after the directive began to circulate, when David Halberstam acquired a copy from an inside source and made it the substance of a damning dispatch to the *Times*. Subtly avoiding the words *news management*, the reporter instead quoted a cover letter accompanying the document which said, "Indoctrination of military personnel in the importance of suppressing irresponsible and indiscreet statements is necessary." Since *indoctrination* connoted "brainwashing" in the minds of most Americans, and since the reporter was able to couple the passage with others from the body of the directive that appeared to reflect the same attitude, unsavory inferences were unavoidable.[18]

Halberstam's criticism was also damaging because it caught the Army in an indiscretion. To justify the need for a more restrained approach to the press, the memorandum had used one of Halberstam's dispatches as an example of "distorted" reporting. The reporter was able to reply that the authors of the directive had taken whole sections of his work out of context, making them appear broader and more critical than they actually were.

Despite the conflict between the press and the U.S. mission in Saigon and despite newsmen's occasional errors, most of the commentaries in the press on the Buddhist crisis were reasonably accurate. Although marred at times by rhetoric and mistaken facts, they often probed to the heart of the crisis.

Halberstam, indeed, was one of the more astute critics. Although he insisted incorrectly that the uprising was a matter of "the government on one side and most of the population, Catholic and Buddhist, on the other," he still grasped the larger context surrounding the event, the real significance of what was going on. He characterized Buddhist complaints as "a spearhead for all kinds of other lingering discontent. . . . The government's reaction to this protest is not an isolated episode but part of a pattern in which its strong qualities—true anti-Communism, stubbornness, resilience—are no longer enough. Observers feel that its limitations—suspicion of its major ally, suspicion of its people instead of sensing and reacting to them—are now greater than its positive abilities, and that it has virtually neutralized itself at a time when it desperately needs to harness all resources in this country."[19]

[18] David Halberstam, "G.I.'s Told Not To Criticize Vietnam," *New York Times*, 24 Jun 63.

[19] David Halberstam, "Religious Dispute Stirs South Vietnam," *New York Times*, 16 Jun 63; Duncanson, *Government and Revolution in Vietnam*, pp. 334–35; Mecklin, *Mission in Torment*, p. 172.

As for Diem's complaint that the Communists were involved in the Buddhists' struggle, the *Washington Post* agreed with Halberstam, commenting, "Of course the communists will exploit Buddhist grievances. And why not? It is Mr. Diem's regime itself that is gratuitously serving communist purposes by policies that are morally repugnant and politically suicidal." Worst of all, the harassment of the Buddhists caricatured U.S. contentions that South Vietnam offered a free alternative to totalitarian North Vietnam. No one could sensibly expect model democratic traditions to prevail under a state of siege, but the Diem regime's handling of the Buddhist crisis underscored the fact that South Vietnam's government was "dictatorial without being competent, arrogant, without being right."[20]

Ngo Dinh Nhu *(fifth from left), flanked by his brothers, Archibishop Ngo Dinh Thuc and Ngo Dinh Diem. Mme. Nhu is second from left.*

The *Chicago Tribune* was more rhetorical. In a scathing attack on Madame Nhu, who had just referred to a Buddhist suicide as "another monk barbecue-show," the paper asserted that if the decision to extend massive U.S. military support to South Vietnam was sound, "the simultaneous decision to support the dictatorship was unsound," with each day of continued support amplifying that bad judgment. There was "no diplomatic or humanitarian reason" for supporting Diem. Any number of South Vietnamese officials could replace him. The time had come for a change.[21]

U.S. News & World Report was more dispassionate than most of its competitors. It refused to accept that the Buddhist demonstrations stemmed from religious persecution and pointed out that denunciation of the Diem regime had become a custom among both American correspondents and South Vietnamese intellectuals. The magazine then described the dilemmas facing the United States. Although the U.S. Army had made "fair progress" in building a competent South Vietnamese military machine and in winning the peasantry to the side of the government, the Buddhist disturbances showed that the Diem regime had failed at what should have been one of its major undertakings—winning the support of the country's intelligentsia. Despite talk of religious discrimination, there had been so little persecution of Buddhists in South Vietnam's past that few competent observers believed persecution was the issue. Instead, a politically motivated

[20] "Quagmire in Vietnam," *Washington Post*, 20 Jun 63.
[21] "The Infamous Mme. Nhu," *Chicago Tribune*, 8 Aug 63.

attempt to topple Diem was in progress, with Diem handling his defense so poorly that his repressive tactics had begun to compound his problems. In the end the United States would confront three equally difficult choices: withdraw from South Vietnam and leave the country to the Communists, assume command of the war and commit U.S. troops to direct combat, or stay with existing policy and "face up to the prospect that it is going to be a long, tough haul." Because an American withdrawal from South Vietnam would "send shivers" through non-Communist Asia and since an injection of U.S. combat troops would prompt Communist cries of "imperialism," the magazine concluded that only the third choice seemed plausible. To succeed at it, however, the United States would have to adopt "a tougher, more direct line in dealing with Diem."[22]

The analysis, especially the statement that a stronger approach to Diem was necessary, agreed generally with what the Saigon correspondents had been saying all along; but *U.S. News & World Report*'s assertion that the South Vietnamese Army was improving and that South Vietnamese peasants were turning increasingly to the government found little support among the newsmen. Most reporters would concede that conditions seemed to have improved in the northern and central provinces, where the U.S. Army Special Forces had succeeded in breaking down Montagnard suspicion of the South Vietnamese government and were using the tribesmen to harass enemy supply routes. They nevertheless believed that the war would really be decided in the delta, where most of the country's agriculture and much of its peasant population were located and where the Viet Cong seemed to be solidly entrenched. The government of the United States, for its part, disagreed with much of what the correspondents were saying. While readily admitting that the Buddhist crisis would pose dangers if allowed to fester too long, the U.S. embassy, the Military Assistance Command, and the Department of State all believed that the correspondents had greatly overstated the situation.[23]

Overstated or not, the newspapers' comments were sometimes useful to American diplomats, who used them time and again to demonstrate to Diem that their advice was sound. During August 1963, for example, Madame Nhu's denunciations of the Buddhists became so shrill that the U.S. Department of State began to fear that Diem was on the verge of attacking the Buddhists' pagodas. The State Department instructed Nolting to warn Diem that any move of the sort would force the United States to denounce his government "promptly and publicly." Although Ngo Dinh Nhu responded shortly thereafter with avowals that he supported Diem's 16 June compromise "fully and with both hands," another of Madame Nhu's Buddhist-baiting outbursts appeared in the pages of the *New York Times* the very next morning.[24]

[22] "The Truth About a War Americans Aren't Winning," *U.S. News & World Report*, 5 Aug 63, pp. 47–49.

[23] David Halberstam, "Picture Is Cloudy in Vietnam's War," *New York Times*, 28 Jul 63; Msg, Saigon 228 to State, 14 Aug 63, FAIM/IR.

[24] Msg, State 173 to Saigon, 5 Aug 63; Msg, Saigon 189 to State, 7 Aug 63; Msg, State 178 to Saigon, 8 Aug 63. All in FAIM/IR files.

The contrast between Nhu's conciliatory statement and the vehemence of his wife angered and confused the Americans. Ambassador Nolting informed Diem that he would have either to repudiate his sister-in-law's remarks or become known before the world as a man tied to a woman's apron strings. For emphasis, Nolting showed Diem editorials in the *Washington Post* and the *New York Times* that condemned Madame Nhu while suggesting that the Saigon government was transforming itself into the Nhu government. Neither the slur on Diem's manhood nor the hostile newspaper articles had any effect.[25]

Official Optimism, Summer 1963

Ironically, while the United States used trenchant newspaper editorials to press Diem toward compromise, advising him to conciliate the press by exercising the utmost candor, American officials in South Vietnam were continuing to harden their own attitudes toward what the Saigon correspondents were saying. For months the U.S. mission had contended that the Buddhist crisis was a civilian movement with no real influence upon either the South Vietnamese Army's ability to fight or its will to win the war, and for months the American newsmen had followed the official line while maintaining a careful watch for signs of poor morale among South Vietnamese troops. As summer came to an end, the reporters began to see conditions that they thought verified their fears. In direct conflict with the official view that the war was progressing and that everything would end well, their warnings prompted the U.S. government to reaffirm its optimism at the very moment when events were about to alter the situation drastically.

An article by David Halberstam provided the occasion. For some time the reporter and his associates, Neil Sheehan of the Associated Press and Merton Perry, a stringer for *Time*, had been researching stories on conditions in the Mekong Delta in order to establish once and for all whether the war was going well. Fast-breaking news kept Sheehan's account from being published and *Time* refused to accept Perry's because it contradicted the magazine's pro-Diem policies, but Halberstam's, fat with statistics and concrete details, appeared on page 1 of the 15 August edition of the *New York Times*.[26]

Drawing upon interviews with South Vietnamese sources, American civilian officials, and junior members of the U.S. military advisory staff—no one in the U.S. mission's top echelons would submit to an informal interview—Halberstam challenged the U.S. government's cautious optimism by stating apodictically that the Viet Cong had become "almost cocky" in their confrontations with South Vietnamese troops in the delta. A year before the enemy had avoided battle with South Vietnamese regulars; now he was picking fights. A year before the Viet

[25] Msg, State 178 to Saigon, 8 Aug 63; Msg, State 180 to Saigon, 9 Aug 63; Msg, Saigon 204 to State, 10 Aug 63. All in FAIM/IR files.

[26] Halberstam, *The Making of a Quagmire*, p. 191.

David Halberstam, Malcolm Browne, and Neil Sheehan

Cong had only nineteen battalions of 200 men each; now they had twenty-one, each containing over 400 men armed with increasingly superior weapons.[27]

Statistics showed the results of that trend, the reporter continued. South Vietnamese casualties had increased by 33 percent, while those of the enemy had decreased by almost the same proportion. Government weapon losses, a major source of Communist armament, had risen by 20 percent, but those of the enemy had fallen by 25. South Vietnamese Army units had abandoned eighty crew-served guns during 1963; the Communists, only fifteen.

Halberstam went on to argue that the deterioration had spread to the government's civic action programs. South Vietnamese administrators tended more and more to believe that the strategic hamlets, by drawing people out of less settled areas into more crowded ones, enhanced the enemy's ability to move at will through remote sections of the delta. As a result the Viet Cong had been able to establish over thirty fortified villages in the center of Vinh Long Province, an area that the government had once hoped to secure with relative ease. Concluding that the Viet Cong had finally learned to counter U.S. helicopter tactics and

[27] This section is based on David Halberstam, ''Vietnamese Reds Gain in Key Areas,'' *New York Times*, 15 Aug 63.

seemed on the move, Halberstam quoted an anonymous American official who appeared baffled by the whole situation. "Frankly," he had said, "we civilians don't have the answer yet and the military doesn't either. I'm just not sure what it is."

Coming on top of the multitude of pessimistic editorials that had accompanied the Buddhist crisis and containing a galling, quasi-official declaration that neither the Department of State nor the U.S. Army had a grip on events, Halberstam's article drew an immediate response from supporters of American policies in South Vietnam. Secretary of State Dean Rusk was the main spokesman for the government. He told a Washington news conference that Halberstam was wrong, that all evidence favored the official interpretation. Communist sabotage and propaganda incidents were becoming less rather than more frequent, while large-scale enemy attacks had decreased in number. Meanwhile, Rusk said, the strategic hamlet program was drawing additional areas of South Vietnam under government control. Among news commentators, Marguerite Higgins of the *New York Herald-Tribune* objected most vehemently to Halberstam's charges. How, she asked her readers, could the Viet Cong ever conduct the "mobile warfare" Halberstam attributed to them when they had "no vehicles and no airplanes" of their own?[28]

The reaction to Halberstam's report continued behind closed doors long after the public response had ended. Shortly after the article appeared, the State Department asked the U.S. mission in Saigon for an evaluation of Halberstam's main points. General Harkins' staff responded with a classified memorandum designed both to provide material for public statements on the subject and to quiet any doubts that the article might have caused within official circles. Although the correspondent had indeed touched on problems, the staff noted, quantitative measures of progress proved that the South Vietnamese armed forces had made gradual, general advances in the Mekong Delta. Roads were open, rice deliveries were reaching Saigon, and the percentage of the population under enemy control continued to decline. While low-level reports indicated that the enemy was massing larger forces, those reports had never been confirmed. Statistics on enemy weapons lost and improvements in Communist equipment likewise made little difference. The enemy could neither find every rifle dropped on a battlefield nor ever surmount the enormous tactical advantage American helicopters gave government troops.[29]

The memorandum's authors saw similar error in Halberstam's analysis of the strategic hamlets. Although no one could deny that the effort in the delta had failed to succeed as quickly as the one farther north, whether the fact should be attributed to overextension or to lagging government effort remained a matter of conjecture. The desire of the South Vietnamese government to expand control as rapidly as possible to keep the Viet Cong from strengthening their grip on

[28] "Excerpts From Rusk News Conference," *New York Times*, 15 Aug 63; Marguerite Higgins, "Vietnam, Fact and Fiction," *New York Herald-Tribune*, 28 Aug 63.

[29] This section is based on Msg, Saigon 261 to State, DAIN 85011, 19 Aug 63.

the people had constantly to be weighed against the risk that setbacks would occur if the effort moved too fast. That the existence of the hamlets could afford the enemy freedom of movement, however, was contrary to the whole "principle and experience" of the program. When the hamlets had developed sufficiently to free militia units for active deployment in the field, they would interfere materially with enemy capabilities.

Although MACV's refutation corrected Halberstam's sometimes overeager acceptance of marginal statistics, it failed to contradict the reporter's main point—that the war in the delta was going against the South Vietnamese. For if the Viet Cong were neither as mobile nor as omnipotent as Halberstam believed, they were dangerous and growing more powerful by the day.

Some internal analyses supported Halberstam's contentions. Eight months before Halberstam published his report, the Senior Adviser in South Vietnam's IV Corps Tactical Zone, Col. Daniel B. Porter, had told General Harkins that the operational capability and efficiency of the South Vietnamese armed forces had shown vast improvement over the previous year but that all progress had to be measured against the poor condition of those forces at the outset. Although organization, equipment, and training were indeed much better than ever before, the professionalism of the army's leadership had failed to advance apace, affecting the entire effort against the Viet Cong. As long as commanders continued to refuse to demand obedience from their subordinates, South Vietnamese troops would lack the motivation and willingness to close with the enemy and destroy him.[30]

Three months later an Australian adviser on counterinsurgency attached to Harkins' staff, Col. F. P. Serong, wrote a report that could have been a model for Halberstam's analysis. While optimistic for the future if the United States acted to correct the problems he outlined, Serong noted improvements in the weight and quality of Viet Cong armaments. The Communists captured large numbers of the South Vietnamese Army's good weapons, while government forces captured mainly inferior ones. More and more 57-mm. recoilless rifles, for example, were appearing in enemy hands.

Serong went on to criticize the strategic hamlets. The United States was basing claims of the program's success on statistics compiled by local officials who had a stake in producing the best possible picture for their superiors. The whole program was becoming "superficially more imposing and actually more dangerous." The attempt to construct a large number of hamlets within a short time had led the Diem regime to concentrate its efforts in areas of the country that paralleled the main roads and to neglect totally areas that were less accessible. The regime thus owned the country's arteries but failed to dominate the intervening spaces, a condition that allowed the enemy to control large segments of the delta and to move at will through territories extending to Saigon's suburbs.[31]

Although willing to concede that the delta was the most difficult to master

[30] Memo, Col Daniel B. Porter for Gen Harkins, 13 Feb 63, sub: Final Report to General Harkins, 69A702, box 1, WNRC.

[31] Rpt, Col F. P. Serong to Gen Harkins, 14 Mar 63, 69A702, box 1, WNRC.

Helicopter Downed by the Viet Cong Is Retrieved

of South Vietnam's four corps tactical zones and that the effort would require programs tailored to the region's particular problems, General Harkins and his staff refused to believe that opinions such as those of Porter and Serong were "a valid yardstick," as Harkins told Admiral Felt, "against which to measure either our accomplishment or the tasks remaining." Contending that the South Vietnamese needed hope rather than more criticism and that government troops won against the Viet Cong whenever they faced the enemy "man to man," Harkins especially continued to believe that good training and equipment would stimulate self-confidence within the South Vietnamese Army. Above all, the United States' own best interests dictated that it work through local authorities rather than attempt to take on the burden of command itself.[32]

Harkins' approach required time to be effective, but time was no longer available. On 16 August Henry Cabot Lodge took Frederick Nolting's place as U.S. ambassador to South Vietnam. The State Department had told Lodge to expect an initial period of grace from the Saigon correspondents because all recognized the difficulties he faced. The ambassador was thus at best partially prepared

[32] Ltr, Harkins to Felt, 12 Aug 63, 69A702, box 1, WNRC. See also Msg, Saigon 261 to State, DAIN 85011, 19 Aug 63; Interv with Col Greene, 6 Jun 65, p. 17.

for the situation he found when he arrived in Saigon. For Diem and Nhu had perceived the interlude between Nolting's departure and Lodge's arrival as a moment of opportunity and had chosen that moment to settle accounts with the Buddhists.[33]

The Assault on the Pagodas

The attack came during the early morning hours of 21 August, after a full day of meetings between Diem and Nhu and South Vietnam's military chiefs. Fearing that a continued crisis might affect the morale and fighting ability of South Vietnamese forces, the chiefs requested permission to declare martial law and to return monks from outlying areas to their home pagodas. Diem and Nhu agreed to the plan, but Nhu, without the generals' knowledge, added a twist of his own. Because he knew that many of his countrymen believed that the army was sympathetic to the Buddhists and that rumors of military plots against the regime were circulating privately, Nhu decided to discredit the officer corps by making everyone believe the army had crushed the Buddhists on its own. After the generals had signed and publicized their decree of martial law, he sent elite, U.S.-trained police and special forces units disguised as regular army soldiers charging brutally into pagodas all across South Vietnam. They beat and arrested more than 1,400 monks.[34]

Nhu's tactic had a second, less subtle objective. Since American diplomats had warned that a crackdown on the pagodas would invite a strong statement of condemnation from the United States, Nhu hoped to hide the role of his personal shock troops in order to divert at least part of the American anger away from the regime and toward the army. As a first step, to make the United States dependent upon him for word of what had happened, he cut the telephone lines to the U.S. mission and the homes of American diplomats. Then, on the day after the attack, he instructed the minister of the interior, the official in charge of all police in South Vietnam, to tell the Americans that neither the police nor any members of the ministry had participated in the operation and that the military had planned and executed the whole affair. Nhu followed with a personal disavowal of his own. The question of martial law had never even crossed his mind, he told an American observer. He was unaware of any plan to attack the Buddhists in their pagodas.[35]

[33] Msg, Saigon 200 to State, for transmission to Lodge, 9 Aug 63, FAIM/IR.

[34] Research Memo RFE-75, U.S. Dept of State, I&R, 21 Aug 63, sub: Diem vs. the Buddhists; Msg, Saigon 292 to State, 21 Aug 63; CIA Information Rpt, 24 Aug 63, sub: Maj. Gen. Tran Van Don Details the Present Situation in South Vietnam and the Plan To Establish Martial Law; Msg, Saigon 320 to State, 24 Aug 63. All in FAIM/IR files.

[35] Msg, State 173 to Saigon, 5 Aug 63; Msg, Saigon 293 to State, 21 Aug 63; CIA Information Rpt, 23 Aug 63, sub: Nhu's Statements on the Government's Actions Against the Buddhists. All in FAIM/IR files.

The ploy came close to working. With little reason at first to disbelieve Nhu, the U.S. mission allowed the Voice of America to broadcast news that the attack had been strictly an army operation. Since the Voice was a much trusted source of unvarnished news in South Vietnam, many South Vietnamese immediately began to blame the army for what had happened.[36]

The credibility American officials accorded Nhu also handicapped the U.S. government's attempt to distance itself from the raids once it began to suspect that the Diem regime might have been behind them. Censuring the attack conditionally, "on the basis of incomplete information" and "if this information is correct," the State Department drafted a declaration that termed the raids an unpardonable violation of assurances that the South Vietnamese government was following a policy of reconciliation.[37] The mission's Charge d'Affaires William Trueheart immediately saw the flaws in the document. In a wire to the State Department he protested that information already in the department's possession made it "altogether clear" that the Diem regime had instituted serious repressions. Since there was little reason to qualify anything, he added, "I should personally have thought that stronger language than 'cannot be condoned' was in order."[38] State then issued a slightly less ambiguous declaration. "On the basis of information from Saigon, it appears that the Government of the Republic of Vietnam has instituted serious repressive measures against Vietnamese Buddhist leaders. The action represents a direct violation by the Vietnamese government of assurances that it was pursuing a policy of reconciliation with the Buddhists. The United States deplores repressive actions of this nature."[39]

That statement may have satisfied the need of American diplomats to chastise Diem without breaking with him, but it made little impression upon educated South Vietnamese who opposed the regime. When they saw government troops using U.S. vehicles, radios, and weapons against the Buddhists, they came to the practical conclusion that if the Americans were providing Diem with equipment, advice, and money, they must also agree with his tactics. People who normally avoided discussing political topics began to implicate the United States in the events of the previous week and to declare that only the Americans could stop what was happening.[40]

American newsmen in South Vietnam had long before come to the same conclusion. As their ranks began to swell from an original nucleus of six reporters to a contingent of over sixty shortly after the raids, they constituted a massive threat to Nhu's plans. In an attempt to deal with the problem, Nhu closed Saigon's Post, Telephone, and Telegraph office after the declaration of martial law

[36] CIA Information Rpt, 24 Aug 63, sub: Maj. Gen. Tran Van Don Details the Present Situation in South Vietnam.

[37] Msg, State 225 to Saigon, 21 Aug 63.

[38] Msg, Saigon 286 to State, 21 Aug 63, FAIM/IR.

[39] Msg, State 225 to Saigon, 21 Aug 63; Msg, Saigon 286 to State, 21 Aug 63; Msg, State 226 to Saigon, 21 Aug 63, FAIM/IR.

[40] Msg, U.S. Army Attache, Saigon, to DA, DAIN 88075, 23 Aug 63; Msg, Saigon 355 to State, 26 Aug 63, FAIM/IR.

and inaugurated rigid censorship of all news dispatches. Seeking to intimidate reporters, he also condoned the arrest and interrogation of newsmen and news photographers. Harassment became so intense that on one occasion the police seized an automobile hired by Bernard Kalb containing $8,000 worth of television camera equipment while it was parked in broad daylight directly in front of MACV headquarters. In the end, only Joseph Fried of the *New York Daily News* was able to arrange an accommodation with the Nhus. After gaining the right to interview Madame Nhu by allowing her to censor his story personally, he saw his dispatch waved through censorship almost unscathed, with the addition only of the adjectives *despicable* and *miserable* in front of all his references to monks and Buddhists.[41]

Nhu and his planners had nevertheless accounted for neither the U.S. mission's preoccupation with American public opinion nor the ingenuity of the Saigon correspondents. As soon as the raids began and the extent of South Vietnamese censorship became apparent, on the theory that a news blackout in Saigon would lead to damaging speculation in the United States and around the world, the U.S. embassy opened its official lines of communication to the press. Later, the State Department decided to end the practice rather than jeopardize the right of American diplomats to use reserved channels of communication in countries more anti-American than South Vietnam. Newsmen then began employing "pigeons"—travelers, members of U.S. military aircraft crews, or anyone else leaving the country—to smuggle their reports to cable offices outside of South Vietnam. Although the MACV Information Officer, Lt. Col. B. Lee Baker, U.S. Air Force, announced that members of the U.S. command aiding newsmen in that manner were committing courts-martial offenses, the restriction had little effect. The reporters merely switched to employees of civilian airlines.[42]

The version of events they wrote was sometimes more informed than the ones dispensed by official American spokesmen. Days before the raids had begun, a disaffected South Vietnamese information officer working closely with Diem had informed newsmen that an attack on the Buddhists was imminent. On the evening of the twentieth an anonymous telephone call to Halberstam, relaying information supplied to the monks by the sympathetic wives of Nhu's combat policemen, had confirmed the rumor and had warned that the operation would begin that night. With ample time to prepare themselves, the reporters had thus witnessed firsthand much of what had occurred.[43]

Both the reporters and the U.S. mission in Saigon agreed on the general outline of what had happened. The main point of divergence centered upon the

[41] William P. Bundy, Notes on MACV Briefing, 25 Sep 63, Chron files, CMH; Msg, CINCPAC to Secy of State, 21 Aug 63; Msg, Saigon 327 to State, 24 Aug 63; and Msg, Saigon 288 to State, 21 Aug 63, all three in FAIM/IR files; [AP], "Three Newsmen Detained at Saigon Student Rally," *New York Times*, 25 Aug 63.

[42] Msg, Saigon 252 to State, 27 Aug 63; Msg, CINCPAC to Secy of State, 21 Aug 63; Interv, author with Col Rodger Bankson, 6 Sep 73, CMH files; Halberstam, *The Making of a Quagmire*, pp. 228–29.

[43] Halberstam, *The Making of a Quagmire*, pp. 228–29.

role of Nhu. Although the correspondents suspected from the beginning that Nhu had planned the operation, they held their accusations until they had proof. By checking among their many sources, they rapidly determined the truth and included it in their reports. The U.S. mission, on the other hand, based its conclusions on Nhu's denials and the fact that the attackers had been dressed in regulation military uniforms. Its deduction that regular South Vietnamese forces were to blame became the basis for official news releases.[44]

Confusion resulted. State Department spokesmen in Washington, citing "highly trustworthy sources," told newsmen that Diem had approved the raids only reluctantly after a lengthy meeting during which his generals had argued persuasively for strong action to save the nation. The Saigon correspondents disagreed. From their own "highly reliable sources" they drew information that the attack had been planned and executed by Nhu without the knowledge of the army. The reporters claimed that the top levels of the U.S. mission were guilty of negligence because they had failed to listen to lower-echelon warnings of a possible raid. The contrast between the two versions so impressed the *New York Times* that its editors put them side by side on page 1 of the paper's 23 August edition with the comment that they exemplified "the confusing situation in South Vietnam."[45]

Then the *Times* and other newspapers compounded the confusion. Relying upon leaks from within the federal bureaucracy and slighting official protestations that the Kennedy administration had no wish to break with Diem, they began to speculate on the possibility that the United States would see the need for a coup and hail the ouster of South Vietnam's leaders. The result was a flock of news stories that placed official affirmations of the continuing nature of U.S. policy next to leaked avowals that much thought was being given to the outlook for internal changes in South Vietnam.[46]

The situation remained fluid until the U.S. government at last constructed a valid picture of what had happened. On the morning after the raids the chief of staff of the South Vietnamese Joint General Staff, General Tran Thien Khiem, had denied emphatically that the attack had involved regular army units. Nhu's word had carried more weight at the time, but three days later, as other high-ranking South Vietnamese generals began speaking out, Khiem's statement took on more meaning. On the twenty-fourth the public relations deputy to General Tran Van Don, who commanded the South Vietnamese Army for the duration of martial law, told Rufus Phillips of the U.S. Operations Mission that Nhu had tricked the army into declaring martial law and that the attack on the pagodas

[44] A file of the reporters' early dispatches transmitted by the State Department may be found in the Gard Papers, CMH.

[45] Tad Szulc, "Kennedy Weighs Policy," *New York Times*, 23 Aug 63; David Halberstam, "Plan Said To Be Nhu's," *New York Times*, 23 Aug 63; "U.S. Problem in Saigon, Attack Called Surprise to Top Officials," *New York Times*, 24 Aug 63.

[46] "U.S. Would Hail Ouster of Diem," *New York World-Telegram-Sun*, 22 Aug 63; "U.S. Sees Need for Coup," *Washington Daily News*, 22 Aug 63; Tad Szulc, "U.S. Reviewing Its Policy on South Vietnam," *New York Times*, 24 Aug 63.

had been carried out by troops under Nhu's control. Later that day General Don himself added ominous overtones to Khiem's report. Noting that the Voice of America's announcements blaming the army for the raids had hurt the army's standing with the people, he told an American observer cryptically that "things could not revert back to what they were before" and that "some of the ministers had to be changed."[47]

With the truth finally apparent, the United States acted both to clarify the public record and to reassess its approach to Diem. Since the attitude of the South Vietnamese public was all-important if the war was to be won, the Voice of America moved immediately to repudiate its earlier errors. In a statement that characterized the army's imposition of martial law as an attempt to solve a difficult problem amicably, it divorced that move entirely from the Buddhist raids and their brutal execution. At the same time the State Department prepared new instructions for Lodge, who had arrived in South Vietnam on the day after the raids. "The U.S. government cannot tolerate a situation in which power lies in Nhu's hands," it cabled the ambassador. "Diem must be given a chance to rid himself of Nhu. . . . If he remains obdurate and refuses . . . we must face the possibility that Diem himself cannot be preserved."[48]

Although U.S. officials sought to find alternatives to Diem, even contacting the generals to probe the possibility of a coup, the new policy failed. The generals as a group lacked the cohesion necessary for a coup, and the Kennedy administration remained undecided about whether Diem really had to go. Seeking more information, President Kennedy dispatched Maj. Gen. Victor H. Krulak and a senior Foreign Service officer, Joseph A. Mendenhall, on a four-day tour of South Vietnam. The two were to assess the effect of recent events on both the conduct of the war and the attitudes of the South Vietnamese people.[49]

Accompanied by John Mecklin and Rufus Phillips, Krulak and Mendenhall reported to the National Security Council shortly after their return to Washington. Krulak was optimistic about the future. Although there was much fighting yet to come in the Mekong Delta, he asserted that the effects of the Buddhist crisis upon the morale of the South Vietnamese armed forces had been negligible. "The shooting war" was going ahead "at an impressive pace," and General Harkins had most problems under control.[50] Mendenhall, Phillips, and Mecklin disputed that view, arguing that Diem was losing the allegiance of his people and that both the strategic hamlet program and the war in the delta were a shambles. The United States, the three concluded, would never win in South Vietnam with Diem at the helm.[51]

Events in South Vietnam had meanwhile become even more complicated. On

[47] Msg, Saigon 292 to State, 21 Aug 63; Msg, Saigon 320 to State, 24 Aug 63; CIA Information Rpt, 24 Aug 63, sub: Maj. Gen. Tran Van Don Details the Present Situation in South Vietnam.

[48] Msg, State 243 to Saigon, 24 Aug 63, doc. 126, *Pentagon Papers*, 2: 734. See also Msg, State 244 to Saigon, 24 Aug 63, FAIM/IR.

[49] *Pentagon Papers*, 2: 236f.

[50] Trip Rpt, Maj Gen Victor H. Krulak, sub: Visit to Vietnam: 7–10 September 1963, CMH files.

[51] *Pentagon Papers*, 2: 243.

23 August students in the Schools of Medicine and Pharmacy at the University of Saigon had begun to demonstrate to gain the release of arrested Buddhists and to demand what they called "the reinstatement of religious freedom." When the government responded by closing the university and arresting the demonstrators, high school students, a group assiduously cultivated by the Viet Cong, began to riot. Mass arrests followed, deepening middle-class resentment against the regime because many of those detained were the children of civil servants and military officers. As the first week of September passed, disorders continued; students refused to work in their classrooms and harassed government security forces in the streets.[52]

The deepening crisis accentuated divisions within the U.S. mission. General Harkins and other senior officials closely identified with early pro-Diem policies remained steadfast in their assertions of progress, finding little evidence that the regime's loss of popular support threatened the military effort. "As everyone else seems to be talking, writing and confusing the issue here in Vietnam, it behooves me to also get into the act," Harkins cabled General Maxwell D. Taylor at the White House. "From most of the reports and articles I read," he said, "one would say Vietnam and our programs here are falling apart at the seams. Well, I just thoroughly disagree."[53]

The Coup Against Diem, September–November 1963

Ambassador Lodge, on the other hand, had arrived in South Vietnam at a moment of extreme recalcitrance on the part of the Diem regime. An astute politician who had immediately improved the embassy's relations with the Saigon correspondents, he listened carefully to anyone critical of the government. Concluding that only the fall of Diem could remedy the situation but that no opportunity then existed to take action, he listed his misgivings in a long cable to President Kennedy that recounted a "very private" conversation he had held with Maj. Gen. Duong Van Minh. Minh believed the enemy was gaining in strength because Diem continued to alienate more and more of the South Vietnamese people, especially the students. Corruption within the country's bureaucracy meanwhile remained endemic, extending even to the theft of American aid. As far as the effort to defeat the Viet Cong on the battlefield was concerned, Minh said the heart of the army was not in the war. "All this by the Vietnamese No. 1 General," Lodge told Kennedy, "is now echoed by Secretary of Defense [Nguyen Dinh] Thuan . . . , who wants to leave the country."[54]

[52] CIA Information Rpts, DAIN 89531, 26 Aug 63, sub: Law Students' Anti-Government Demonstration on 24 August 63, and DAIN 95460, 4 Sep 63, sub: Position and Planning of Vietnamese Students; Msg, Saigon 438 to State, DAIN 98201, 7 Sep 63.

[53] Msg, Harkins to Taylor, quoted in *Pentagon Papers*, 2: 246.

[54] Msg, Saigon 544 to State, for President from Lodge, 19 Sep 63, *Pentagon Papers*, 2: 747. See also *Pentagon Papers*, 2: 254.

Ambassador Lodge Confers With Diem, 1963

With time the disagreement between the U.S. embassy and the Military Assistance Command in Saigon began to take on public dimensions. Reiterating all the old arguments about the war in the delta and noting that high American officials were at last challenging erroneous military estimates, Halberstam and his colleagues endorsed Lodge's push for drastic action to save the situation. Widespread editorial criticism ensued in the United States as newspaper commentators began to assert that the United States had been "outmaneuvered" by Diem and that the Kennedy administration was flailing blindly in search of a new policy. The theme reechoed in Congress, where Senator Frank Church of Idaho introduced a resolution calling for an end to U.S. aid if the South Vietnamese government continued its inept policies. Senator Mike J. Mansfield of Montana warned publicly that an unresolved policy rift within the Kennedy administration would reduce official effectiveness and risk disaster on the battlefield.[55]

Responding to the pressure, President Kennedy dispatched Secretary McNamara and General Taylor to Saigon on yet another fact-finding mission,

[55] Halberstam, "U.S. Civilian Aides in Vietnam Press for a Decision on Diem," *New York Times*, 15 Sep 63; Digest of Opinion in Recent US Press, in Fact Book—Vietnam, Sep 63, CMH files; "Senator Mansfield Criticizes Policy Rift," *New York Times*, 21 Sep 63; "The Church Resolution," *New York Times*, 13 Sep 63; Hilsman, *To Move a Nation*, p. 505.

a decision that prompted vigorous preparations on the part of the agencies most involved with formulating U.S. policy for South Vietnam. The Defense Department took the lead, compiling for McNamara and Taylor a 135-page fact book on the country, its people, the state of the counterinsurgency campaign, and the charges against the Diem regime. It devoted fully one-fourth of the document to refutations of the Saigon correspondents. A long quotation from a 20 September 1963 article in *Time* introduced the section on the press. Observing that the Saigon correspondents were so confident of their own convictions they dismissed any other version of events as "the fancy of a bemused observer," the magazine noted that many of the newsmen seemed "reluctant to give splash treatment to anything that smacks of military victory in the war against the communists." The authors of the fact book then launched a thorough, 24-page, item-by-item attack on Halberstam's delta report, labeling it a product of "preconceived opinions and judgments . . . replete with inaccuracies, many of which must be attributed to the reporter himself."[56]

When they arrived in South Vietnam, McNamara and Taylor received the usual optimistic picture of the war from the embassy and the Military Assistance Command. They nevertheless began to encounter contradictory testimony almost immediately. A professor with many contacts in South Vietnam, P. J. Honey, began the litany. He observed confidentially that Diem had "aged terribly" during his years in power, was "slow mentally," and would not last "twenty-four hours" without the aid of Nhu. On the other hand, while Nhu handled bribes and manipulated the regime's power base, he still needed "the cloak of Diem's prestige" to maintain his grip on the country. Since neither man was capable of changing and since change was essential for victory, Honey could only conclude that the United States could never win with such a combination.[57]

Other private visitors added weight to Honey's assessment. The papal delegate, Archbishop Salvatore Asta, noted that Diem's machinery for dominating the country was as perfect as that of a Communist police state. As a result, the people tended to prefer the devil they did not know to the present evil. Vice President Nguyen Ngoc Tho, the man most often mentioned in U.S. circles as a possible replacement for Diem, also spoke out. He questioned the success of the strategic hamlet program and asserted that increased Viet Cong strength had to be attributed to widespread disaffection among the peasantry. Tho charged that the United States had never used its strength and influence intelligently in South Vietnam and so had failed to prevent the current political deterioration.[58]

Most of those statements represented educated opinion, but McNamara also got a firsthand taste of what his informants were talking about. Shortly after his arrival in Saigon, Madame Nhu, who was touring the world on behalf of the

[56] Msg, White House CAP 63516 to Lodge, 17 Sep 63, doc. 136, *Pentagon Papers*, 2: 745; Msg, State 431 to Saigon, 18 Sep 63, doc. 137, *Pentagon Papers*, 2: 746; quotes from Fact Book—Vietnam, Sep 63, CMH files.

[57] Record of McNamara Conversations, 27 Sep 63, CMH files.

[58] Record of McNamara Conversation With Monsignor Asta, 30 Sep 63, CMH files. Tho's remarks are in *Pentagon Papers*, 2: 249.

regime, told newsmen in Rome that the younger officers attached to the U.S. mission were "little soldiers of fortune" whose irresponsible behavior was forcing senior officers toward a confused policy. Responding with the U.S. government's first public reprimand of a member of the Diem regime, Ambassador Lodge termed the statement a cruel and incomprehensible assertion that revealed no sympathy for the fact that U.S. advisers were dying side by side with South Vietnamese soldiers. In a meeting with Diem himself, McNamara encountered the same obdurate attitude displayed by Madame Nhu. Diem dismissed the regime's repressions as mainly the product of inexperience and attributed his problems to "vicious" attacks in the U.S. press against himself, his family, and his government. Madame Nhu, he said, was only defending herself against the abusive reporting of American newspapermen.[59]

Although convinced that the political crisis could have a dire effect upon the military effort in South Vietnam, McNamara steered his mission's final report between that view and general support for Harkins' assertions. While he claimed that the war was progressing well enough to allow for the withdrawal of most U.S. personnel by 1965 if Diem took immediate hold of his problems, he warned that continued political tension in Saigon could erode favorable trends. He then advised a shift of South Vietnamese military strength to the delta to counter increasing enemy pressures in the region, but also recommended a suspension of long-term development aid to South Vietnam to prod Diem toward reform.[60]

The Kennedy administration followed McNamara's advice, withholding funds destined for the South Vietnamese Commodity Import Program and terminating support for the special forces units that had carried out the pagoda raids, but Diem and Nhu failed to react. South Vietnamese civil servants remained under instructions to avoid all contact with Americans; students continued to be arrested and detained for the most trivial offenses; and the repression of bonzes went on unabated. On 5 October a Buddhist monk burned himself to death in protest with three forewarned American newsmen in attendance. Diem's police smashed the reporters' cameras, causing another round of unfriendly comment in American newspapers.

The divisions within the U.S. government worsened as both sides of the embassy-MACV disagreement hardened their positions. During meetings with Harkins and Lodge, McNamara and Taylor had stressed the need for cooperation between American agencies in Saigon, but Ambassador Lodge continued to keep his own counsel, cutting Harkins' advice progressively out of his thinking. In the past, "Fritz [Nolting] would always clear messages concerning the military with me or my staff," Harkins complained in messages to Taylor. "This is not so today."[61] Reports with major military implications thus went to Washing-

[59] Halberstam, "Lodge Deplores Mrs. Nhu's Views of U.S. Officers," *New York Times*, 27 Sep 63; Memo of Conversation, Diem, Lodge, McNamara, Harkins, 29 Sep 63, 69A702, box 1, WNRC.

[60] Memo for the President, sub: Report of the McNamara-Taylor Mission to South Vietnam, 2 Oct 63, doc. 142, *Pentagon Papers*, 2: 751; *Pentagon Papers*, 2: 250f; Maxwell D. Taylor, *Swords and Plowshares* (New York: W. W. Norton, 1972), pp. 248–300.

[61] Msg, Harkins MAC 2028 to Taylor, 30 Oct 63, *Pentagon Papers*, 2: 784–87.

ton without the general's knowledge, and important incoming messages often failed to reach his desk. In the same way, when the State Department's Bureau of Intelligence and Research produced an analysis of war statistics which argued that downhill trends had started more than six months previously, the Defense Department refused to concur. In a stiffly worded reply placing great reliance upon an increase in the number of strategic hamlets and a rise in the pace of government-initiated attacks, it contended that the unfavorable indicators State had cited were actually signs of progress. Compressed into progressively smaller areas of the country, the Viet Cong had less territory to defend and could therefore temporarily concentrate their manpower. In the process they created a semblance of winning when they were actually losing.[62]

The disagreement continued into late October, when a group of South Vietnamese generals, interpreting the Kennedy administration's stronger line toward Diem as a signal that a coup might be welcome, contacted U.S. officials in Saigon to see what the American position would be if they indeed took action. Lodge and Harkins immediately clashed over the prospect. Believing that only a coup could remove Diem, Lodge favored change and pushed for it, but Harkins remained convinced that an unknown, untested group of generals could never replace a leader of Diem's strength of character. Concerned lest an unsuccessful revolt be laid to American influence despite careful attempts to avoid direct engagement in the plot, the Kennedy administration nevertheless sided with Lodge, opting for a policy of benign noninvolvement toward a coup. Lodge told the generals that the United States would support any regime that would attract the allegiance of the South Vietnamese people while fighting Communists effectively. He thus made it plain to the generals that the U.S. mission in Saigon would neither participate in their plotting nor thwart their plans.[63]

With that affirmation in hand, the generals staged a *coup d' etat* in which both Diem and Nhu died, effectively ending the debate over the war's progress. Within days, as the populations of Hue and Saigon took to the streets in wild celebration of their government's fall, dozens of jails across South Vietnam emptied and hundreds of former political prisoners, many with tales of torture and mutilation to tell, began to talk freely of their experiences. Viet Cong units in the countryside meanwhile moved to cut off and swallow as many of Nhu's weak strategic hamlets and poorly positioned military bases as they could, on the theory that the new regime would soon retrench, depriving their forces of supplies and weapons that had always seemed available for the taking.

By December the truth was obvious. Shortly after the assassination of President Kennedy, Secretary of Defense McNamara informed newly installed Presi-

[62] Research Memo, RFE-90, U.S. Dept of State, I&R, 22 Oct 63, sub: Statistics on the War Effort in SVN Show Unfavorable Trends, doc. 147, *Pentagon Papers*, 2:770; Memo, McNamara for the President, 21 Dec 63, doc. 156, sub: Vietnam Situation, *Pentagon Papers*, 3: 494.

[63] Msg, Harkins MAC 2028 to Taylor, 30 Oct 63, and Msg, Harkins MAC 2033 to Taylor, 30 Oct 63, both in *Pentagon Papers*, 2: 784–87. Msg, White House CAP 63560 to Lodge, 5 Oct 63, doc. 143, *Pentagon Papers*, 2: 766; Msg, CIA CAP 74228 to Lodge, 6 Oct 63, doc. 145, *Pentagon Papers*, 2: 769.

dent Lyndon B. Johnson that the picture in South Vietnam was gloomy both in the provinces around Saigon and in the delta. Government programs were "seriously over-extended." The enemy had meanwhile destroyed many strategic hamlets, and a high percentage of the country's population remained under Viet Cong control. The situation had, indeed, been deteriorating since July, "to a far greater extent than we realized because of our undue dependence upon distorted Vietnamese reporting."[64]

Although McNamara's assessment might have been construed as a victory for the Saigon correspondents, it was more a comment on the U.S. mission's failure to assess the situation accurately. For months, drawing upon dissenting arguments available to the mission itself, reporters had asserted that the war was going poorly. Caught between the newsmen's harsh judgment and a sincere desire to inspire South Vietnamese self-confidence, Ambassador Nolting and General Harkins had rejected the reporters' contentions out of hand. Trusting in misleading statistics and citing the publication of harmful leaks as evidence of the newsmen's ill will, the two focused upon the reporters' errors of detail, while disdaining virtually every criticism of the war that appeared in a newspaper. In doing so, they and their sympathizers demonstrated their conviction that war should be left to experts and that reporters on the battlefield were often at best a nuisance and at worst a menace.

One of the most persuasive critics of the press at the time, *New York Herald-Tribune* reporter Marguerite Higgins, would have disagreed with that judgment, but she argued all the same for Harkins' and Nolting's point of view. In a series of articles on the war during 1963 and 1964 and in a later book, she accepted with little question the U.S. mission's assertion that the war was making at least fitful progress under Diem and decried the negative point of view of the Saigon correspondents. The mistaken, sensationalized news stories that resulted, she charged, so overwhelmed the more optimistic assessments of the U.S. mission that they came to infect the deliberations of policy makers in Washington. In that sense, the Saigon correspondents contributed greatly to the decision to bring down Diem and shared responsibility for the war that followed.[65]

The role of the press in the events leading up to the fall of Diem was nevertheless far more complex than the analysis Higgins proposed. As an incident related by Associated Press correspondent Peter Arnett implies, the U.S. government and the American news media in South Vietnam were caught up in a dilemma of major proportions. One hot noonday Arnett stood outside the Saigon market watching a Buddhist monk squat on the pavement, squirt himself with gasoline, and flick a cigarette lighter. "I could have prevented that immolation by rushing at him and kicking the gasoline away," the reporter said later. "As a human being I wanted to. As a reporter I couldn't. . . . If I had stopped him, the secret police

[64] Memo, McNamara for the President, 21 Dec 63, sub: Vietnam Situation, doc. 156, *Pentagon Papers*, 3: 494.

[65] Marguerite Higgins, *Our Vietnam Nightmare* (New York: Harper & Row, 1965), pp. 186–87.

who were watching from a distance would have immediately arrested him and carried him off to God knows where. If I had attempted to prevent them doing this, I would have propelled myself directly into Vietnamese politics. My role as a reporter would have been destroyed along with my credibility.'' Instead Arnett photographed the man burning in the street, beat off half-a-dozen policemen as he dashed back to his office to file his pictures, and ended by doing the very thing he had sought to avoid. For by releasing what he had seen to the world, he intervened in South Vietnamese politics just as surely as if he had lit the bonze's lighter.[66]

The U.S. government argued that newsmen should have shown more restraint in their reporting and that adverse press coverage only strengthened Diem's unyielding attitude; but if reporters had held back, they would have collaborated in Diem's repression of the Buddhist movement, another form of intervention into South Vietnamese politics. Just as the Kennedy administration's refusal to participate one way or the other in the coup against Diem created conditions favorable to the act by giving the generals a free hand, so the Saigon correspondents, by their very presence in South Vietnam, altered the context of the war and created conditions that, in concert with Diem's refusal to reform, helped ensure the downfall of American policy.

President Kennedy might have avoided the problem by permitting censorship of the press or by allowing Diem to evict the Saigon correspondents from South Vietnam, but he and his administration were captive to the belief that the news media had a hold on American public opinion. To have suppressed the correspondents without the support of the American public and Congress, so the reasoning went, might spark accusations in Congress and the press that the president was spending American lives and treasure on a ''clandestine'' war in Southeast Asia—almost certain political suicide. This concern, which reappeared throughout the war, whatever the administration in power, altered the way the war was fought and contributed again and again to the frustration of American aims.

[66] Peter Arnett, ''Reflections on Vietnam,'' *Nieman Reports* 26 (March 1972): 8.

3

Maximum Candor

The overthrow of Diem and the rise of a triumvirate led by Maj. Gen. Duong Van ("Big") Minh inaugurated a brief period of optimism among American officials. Minh seemed genuinely popular among the South Vietnamese, and his regime appeared to promise the sort of people-oriented administration the U.S. government had long considered essential for victory. Although a mood of caution necessarily prevailed, David Halberstam soon concluded that "a week ago this war looked tough and demanding, and it still looks that way. But there is one major change. The change is a hope . . . that the repressive political climate that weighed heavily on the population and on the army has been lifted for good. . . . The pessimism that reached into many high American places in recent months is gone."[1]

This optimistic mood changed within days. Although the Diem regime had not been a model of efficiency, its officials had at least maintained an appearance of normalcy. With the advent of Minh, even that facade fell away. As power struggles ensued within the new administration, politically acceptable men supplanted province chiefs identified with the old regime and many major military commands changed hands. For a time administrative chaos reigned and collapse seemed imminent.[2]

On 30 January 1964, Maj. Gen. Nguyen Khanh overthrew Minh, compounding the confusion. Although the U.S. government threw its whole weight behind the new regime in hopes of discouraging further disorders, another round of ousters and realignments followed. By March thirty-five of forty-one province chiefs had been replaced; the enemy controlled half the land area in twenty-two of forty-three provinces; large segments of the population were once more showing signs of apathy; and security in the IV Corps Tactical Zone had become so

[1] David Halberstam, "Americans in Saigon Draw New Hope From Coup," *New York Times*, 6 Nov 63. See also Lyndon B. Johnson, *The Vantage Point* (New York: Holt, Rinehart & Winston, 1971), p. 43.
[2] *Pentagon Papers*, 2: 304.

bad that the Viet Cong directed all aspects of peasant life outside government-controlled administrative centers. Khanh seemed an able man within his experience, Secretary of Defense McNamara told President Johnson. He took well to American advice, showed energy and comprehension, and had enough of a chance of taking hold to merit continued American support. But he possessed no wide political following. His standing with the army seemed uncertain, and he had thus far failed to counter the problems of morale and organization that afflicted his administration. While no effective opposition to him existed, another coup could occur at any moment.[3]

General Minh

Solidifying Public Opinion: First Attempts, March–June 1964

Reflecting all of the pessimism but little of the optimism of McNamara's assessment, American public opinion paralleled the downward course of events in South Vietnam. In November 1963, after the fall of Diem, 57 percent of the people sampled by a Harris poll approved of their government's handling of the war. By March 1964, four months after Kennedy's death, the number of those approving had fallen to 43 percent—a figure all the more telling because the new president, Lyndon B. Johnson, received very high scores in every other category: 80 percent for "working for peace," 71 percent for "keeping the economy healthy," 65 percent for "moving the country ahead."[4]

Obviously dissatisfied with the way the war was going, the American public was nevertheless undecided about what to do. Thirty-five percent of those queried by Harris favored establishment of a neutral government in South Vietnam, while only 28 percent opposed the step. Yet by a score of 56 to 18 percent, they favored continuing the policy of supporting an anti-Communist regime in Saigon. Harris concluded that "in terms of public opinion, . . . administration policies in Viet-

[3] Memo, McNamara for the President, 21 Dec 63, sub: Vietnam Situation, *Pentagon Papers*, 3: 494. See also Memo, McNamara for the President, sub: Vietnam and Southeast Asia, *Pentagon Papers*, 3: 496.

[4] Louis Harris, "U.S. Handling of Vietnam Issue Has Public Confused, Cautious," *Washington Post*, 30 Mar 64; Harris, *The Anguish of Change* (New York: W. W. Norton, 1973), pp. 53–55.

nam are treading a cautious tight-rope fraught with much doubt but also with no clear alternative."[5]

Although the polls were discouraging they were also more than a little misleading. For if the American public disliked the war, it was, as a body, hardly overconcerned. The year 1964 was one of unprecedented prosperity in the United States. The gross national product had risen $112 billion over the previous three years. President Johnson's "Great Society," the debate over the Civil Rights Act of 1964, and the Beatles all shared space with South Vietnam in the pages of the press, more often than not overshadowing it. The *Washington Post* caught the flavor of the times when it reported in May that 63 percent of the people interviewed in a recent public opinion survey had said that they gave little or no attention to events in Southeast Asia. If the public seemed critical of the war on one level, it was obviously operating on another, perhaps in the hope that the whole issue would go away.[6]

The Johnson administration was well aware of the American public's ambivalence. Noting that the war "has been going badly since last summer and will probably get worse during the next six to nine months," a working group within the Department of Defense had already addressed the problem in a January 1964 memorandum to the Deputy Assistant Secretary of Defense for International Security Affairs, Henry S. Rowen. "Perhaps the most critical factor of all is U.S. determination," the group had asserted. "That is, whether the United States is prepared—especially during an election year—to resist pressures for neutralization, to accept sizeable increases in U.S. casualties, and to live with a situation which, at best, will be discouraging for many months." During France's Algerian War shifting attitudes in Paris had undone practical progress on the village level. "It is important, therefore, for the administration to consider now the pros and cons of a major public statement by President Johnson on the continued U.S. commitment to victory over the insurgents." Such a declaration would clarify the status of American policy; put an end to rumors at home and abroad that the United States was considering some face-saving settlement on South Vietnam; and help to prepare the legal, political, and propaganda grounds that would be needed if the United States had to bring strong pressures to bear against North Vietnam.[7]

Although President Johnson intended to prepare the strongest political and military case for possible later action against North Vietnam and said that he would use every public opportunity to discourage talk of neutralization, he remained unconvinced that a major presidential address was necessary. Instead, on 26 March Robert McNamara addressed the National Security Industrial Association.

[5] Harris, "U.S. Handling of Vietnam Issue Has Public Confused, Cautious," *Washington Post*, 30 Mar 64; Harris, *Anguish of Change*, p. 55.

[6] "Year of Unprecedented Prosperity," *Life*, 16 Oct 64, p. 36; "The Gallup Poll: Less Than 40% of People Follow Vietnam Events," *Washington Post*, 27 May 64; See also Richard Harwood, "Lessons From the Pentagon Papers," reprinted in Laura Babb, ed., *Of the Press, By the Press, For the Press (and Others, Too)* (Washington, D.C.: The Washington Post Company, 1974), pp. 84f.

[7] Memo, OASD (ISA) for Rowen, 25 Jan 64, sub: Interim Report on Vietnam, CMH files.

The United States, he said, would never allow the neutralization of South Vietnam since that would be at best "an interim device to permit communist consolidation and eventual takeover." American withdrawal was equally unthinkable. The United States had no intention of abandoning an ally and was studying the implications of carrying the war to North Vietnam itself. The road ahead would be "long, difficult, and frustrating," McNamara concluded, but "when the day comes that we can safely withdraw, we expect to leave an independent and stable South Vietnam."[8]

At the time of the talk, government planners believed that a few well-placed articles and speeches by Lodge, McNamara, and other high administration officers, along with news of a few victories or of favorable political developments in South Vietnam, would be enough to prepare the way for a generally hard and realistic public viewpoint on the war. The reverse actually happened. As victories failed to materialize and political stagnation fed the chaos in South Vietnam, the Saigon correspondents plied their American readers with news of defeats and disunion. The climate of debate that resulted sparked an increasing number of congressional statements both favoring and criticizing the president's Southeast Asia policies. The administration was forced to recognize that Washington-based speeches could never substitute for an effective public relations program in the field. "It is easy in Washington to underestimate the cumulative effect of Halberston's [*sic*] *New York Times* reporting, as well as other recent . . . stories," President Johnson's adviser on national security affairs, Walt W. Rostow, told Secretary Rusk on 6 May; ". . . it may be wise to consider whether a low key campaign of public information [in both Washington and Saigon] may, even now, be in order."[9]

The realization came none too soon. As early as November 1963, trouble had broken out between the Saigon government and American correspondents, with Ambassador Lodge siding with the South Vietnamese. "The U.S. press should be induced to leave the new government alone," Lodge had said at the time. "They have exerted great influence on events in Vietnam in the past, and can be expected to do so again. Extensive press criticism, at this juncture, could be critical."[10]

The State Department soon found occasion to back its ambassador. On 3 January 1964, the Associated Press published a story based on a comment by a U.S. Army senior adviser which called the hard-core Viet Cong soldier "probably the best fighting man in South Vietnam. He forces us to maneuver in plain sight while he is perfectly hidden. . . . To defeat one of his strong points you must either be very lucky or accept losses of perhaps five government troops to one of his."

[8] OASD PA, News Release no. 249-64, 26 Mar 64, sub: Address by Secretary of Defense McNamara Before the National Security Industrial Association, CMH files. William Bundy wrote most of the presentation. See also Draft Memo, Bundy for the President, 18 May 64, Chron files, CMH.

[9] Draft Memo, Bundy for the President, 18 May 64; Memo, Walt W. Rostow for Dean Rusk, 6 May 64, sub: The Public View of Vietnam, I-7095, ISA 092, 68A306, box 41, WNRC.

[10] *Pentagon Papers* 2: 306.

Viet Cong Destroy a South Vietnamese Outpost

On 6 January another Army adviser told newsmen that South Vietnam's first military operation of 1964 had been a total failure. Several others repeated the point, describing the final night of the action, when a guerrilla unit had lobbed four mortar shells into a ranger training camp as an act of defiance. "The Viet Cong would not have given their position away so obviously . . . a few months ago," the advisers said. "Now they do it and laugh." Finally, on 7 January the *New York Times* described the same operation, observing that numerically superior government troops backed by heavy air support had allowed a surrounded Viet Cong battalion to escape. Applying gall to the wound, the paper then pointed out that the enemy unit in question had been the *514th Viet Cong Battalion*—the one that had escaped during the battle of Ap Bac almost exactly a year before. In the belief that the publication of such remarks damaged U.S. relations with South Vietnam, the State Department immediately reaffirmed its policy of total support for the South Vietnamese government and instructed Lodge to put an end to unrestrained comment by U.S. advisers.[11]

[11] "Hard-core Viet Cong Is Declared Probably Best Vietnam Fighter," *Washington Post*, 3 Jan 64; "Anti-Red Move Fails in Vietnam," *Baltimore Sun*, 6 Jan 64; "Reds Escape Trap in Vietnam Clash," *New York Times*, 7 Jan 64; National Security Action Memorandum (NSAM) 273 for Secy of State et al., 26 Nov 63, CMH; *Pentagon Papers*, 2: 306f.

General Harkins, who was responsible for the advisers, protested. Although he agreed that "all press briefings should be along objective lines and convey as much optimism as the situation warrants," he told the State Department that "in the heat of battle American advisers are going to express their true feelings without stopping to consider whether they are within earshot of a media representative." The spontaneous outburst of those feelings during the course of an operation was "uncontrollable without resorting to measures which in the long run probably would be more damaging to the effort than the news stories which result." Despite those misgivings, Harkins agreed to emphasize the need for restraint at the next MAAG Senior Advisers' Conference and during visits to the field. He cautioned, however, that it was not advisable to issue any written directives or memorandums.[12]

Failing to follow his own advice, Harkins then issued a memorandum that quickly found its way into the hands of the Saigon correspondents. Joseph Fried of the *New York Daily News* summarized the reporters' reaction. "American military personnel here have been told to muzzle criticism of South Vietnam's new government," he said. "In the past . . . American advisers' criticism of . . . Vietnamese military policies served newsmen as a balance to sometimes misleading government accounts of battles against the communists. Harkins' memo was seen here as limiting such expressions in the future."[13]

The *Chicago Sun-Times* took a more rhetorical stance. Recalling the CONARC memorandum of the year before, the paper observed that Harkins' directive marked the second time in ten months that a muzzle had been applied to military personnel serving in South Vietnam. "What Gen. Paul D. Harkins . . . has to fear that would cause him to circulate yet another military gag order is not known," the paper commented. "The basic truth about South Vietnam is pretty well established. The United States is in an unholy mess in that area and it is not the fault of the military . . . [but of] the politicians and foreign policy experts who put the military into an impossible position."[14]

Although critical in tone, Fried's and the *Sun-Times'* statements were less strident than the press commentaries of the year before. So was most of the reporting of the time. *Newsweek* and the *New Republic* censured Secretary McNamara for saying "the situation continues grave" on one day and then, after newspapers broadcast his pessimism, that "there has been a very noticeable improvement" on the next. *Time* commented that the situation in South Vietnam was so hazardous that the chief of Long An Province had received the country's highest military decoration for sleeping in the outlying hamlets of his province rather than in his well-protected provincial capital. Hanson W. Baldwin observed pungently that government troops were "eagerly on the defensive" in the delta. Yet the bitter rhetoric and personal recriminations of the year before were largely gone. Diem was dead and Khanh lacked the stature to take his place as the target of

[12] Msg, Harkins MAC J74 0236 to State, 10 Jan 64, Westmoreland History, bk. 2, Jan–Feb 64, CMH.
[13] Joseph Fried, "Harkins Curbs Yank Beefs on S. Viet Regime," *New York Daily News*, 3 Mar 64.
[14] "South Vietnam Gag," *Chicago Sun-Times*, 8 Mar 64.

the press; Madame Nhu no longer provided grist for spectacular news stories; Buddhist suicides had stopped; and official optimism seemed to have faded somewhat before a recognition that the war was slowly going bad.[15]

Ambassador Lodge's careful handling of the Saigon correspondents was also a factor. Although officialdom's concern for putting the best possible face on the war grew as optimism faded in Washington, Lodge knew that repeated attempts to restrict the press could only cause more friction. He therefore took personal responsibility for the U.S. mission's dealings with the press, making the leak a prerogative of the ambassador and providing newsmen with the stories he wanted to see in print. He likewise refused to confide his plans to even the closest of his associates and thus rarely found his secrets appearing in the newspapers. When the former deputy public affairs officer of the U.S. embassy in New Delhi, Barry Zorthian, replaced Mecklin as mission public affairs officer in January 1964, Lodge even went so far as to stipulate that the new man might continue Mecklin's duties as supervisor of psychological warfare against the enemy but was to have nothing to do with the press because he did that work himself.[16]

An example of the way Lodge applied his principles occurred in late January, when the *Washington Post* published an article by Neil Sheehan detailing the grave deterioration that had occurred in the Mekong Delta over the previous two years. Sheehan asserted that the government had, "after a faint and half-hearted struggle, handed its rural population over to the enemy" and that the war against the Communists had gone "a long way toward being lost." Only in the last two paragraphs of the piece did the reporter mention the new government's plans for restoring security to the region—and then he cast those plans as a major test of the Saigon regime's ability.[17]

A short time before the article appeared, Lodge had toured the South Vietnamese countryside, observing several situations where U.S. Army advisers believed the South Vietnamese were making progress. He reported his observations to the State Department, where the subsequent arrival of Sheehan's report prompted a flurry of ideas about how the ambassador could communicate his findings to the press. Newsmen should accompany U.S. and South Vietnamese officials during tours of the countryside, the agency's public affairs specialists informed Lodge, and should receive continual background briefings on favorable developments. "We are sure you will agree," they added, that "it would be helpful with respect to both public opinion in U.S. and morale in Vietnam if press reports showed beginnings of progress and slight note of optimism which has come through in your reports."[18]

[15] "Foreign Policy: L.B.J.'s Test," *Newsweek*, 10 Feb 64, p. 19; "Darkness on the Mekong," *New Republic*, 8 Feb 64, p. 3; "The Other Government," *Time*, 31 Jan 64, p. 31; Hanson W. Baldwin, "The War in Vietnam—Is Victory for the West Possible?" *New York Times*, 16 Feb 64.

[16] Memo, Carl Rowan for the President, 26 May 64, CMH files; Mecklin, *Mission in Torment*, p. 223; Interv, author with Barry Zorthian, 19 Dec 75, CMH files. See also Hilsman, *To Move a Nation*, p. 514.

[17] Neil Sheehan, "Neglect Erodes Vietnam's Strategic Hamlet System," *Washington Post*, 27 Jan 64.

[18] Msg, Saigon 1307 to State, DAIN 185026, 15 Jan 64; Msg, Saigon 1374 to State, DAIN 192337, 23 Jan 64; Msg, State 1132 to Saigon, DAIN 194621, 27 Jan 64.

The suggestion might have worked in other circumstances, but Lodge knew that the Saigon correspondents would suspect any official overture. Although he acknowledged that the U.S. embassy in Saigon had failed to include newsmen on official itineraries and that the wire services and the *New York Times* ought to be invited in the future, he refused to make any personal approach to the correspondents. "I have, of course, had it in mind to background the U.S. press," he told the State Department, "but from long experience I have learned that it is much better to wait for them to come to you instead of you sending for them. If I were to send for them to tell them how well the war was going they would not believe it, and I would suffer the same fate as so many others. Now, Sheehan of UPI and [Hedrick] Smith of the *New York Times* have both asked to see me, and I hope to be able to get some ideas across."[19]

Lodge pursued many of the same ends as Nolting, but his personal touch helped to heal the animosities that had characterized the U.S. mission's relations with the press during the Nolting years. Precisely because it was personal, however, his approach did little to improve the ability of either the Military Assistance Command or the embassy to deal effectively with newsmen. Since the ambassador had control over all contacts with correspondents, Zorthian and the U.S. Information Agency had no responsibility for relations with the press and could do little to assist reporters. Misinformation published in news dispatches often went uncorrected as a result. As Lodge became too busy to give more than partial attention to public relations, overall embassy direction of the MACV Public Affairs Office also began to drift. Small to begin with and largely staffed by inexperienced, overworked officers, that organization found itself at times lacking even a clear idea of who was responsible for what. In addition, more than forty correspondents were by that time present in South Vietnam. Although some were new to the country, the group contained a hard core of veteran reporters—Nicholas Turner of Reuters; Malcolm Browne and Peter Arnett of the Associated Press; Frank McCullough of *Time*; Robert Shaplen of the *New Yorker*; Merton Perry and Francois Sully of *Newsweek*; Neil Sheehan, now of the *New York Times*; Jack Foisie of the *Los Angeles Times*; and Beverly Deepe of the *Christian Science Monitor*; to name a few. All could be counted upon to ask the difficult questions.[20]

Examining the Information Program, June 1964

During March, while civilian officials in Washington still assumed a few well-placed articles and speeches would suffice as a public affairs strategy, pres-

[19] Msg, Saigon 1423 to State, DAIN 196127, 29 Jan 64.

[20] Memo, Rowan for the President, 26 May 64; Memo, Rowan for Secretary McNamara, 4 Jun 64, sub: Improvement of Informational-Psychological Program in South Vietnam, CMH files; Msg, MACV 2854 to JCS, 8 Jun 64, Reorganization of Information Program at MACV, for Gen Taylor and Secretary Sylvester from Westmoreland, Westmoreland History, bk. 2, May–Jun 64, CMH. There were

sures began to build within the Army and the Defense Department for a review of the information program in South Vietnam. At that time Army Chief of Staff General Earle G. Wheeler ordered an officer recently returned from South Vietnam, Brig. Gen. John M. Finn, to create a special working group to write a report on "all aspects of operations and administration that affect U.S.-GVN operations." Finn took Wheeler at his word, composing a study that covered every phase of the war from what he called its "lack of a common concept" to the necessity for an informed press.[21]

General Wheeler

In the report's detailed section on relations with the news media, Finn advised an expansion of the MACV Public Affairs Office to provide newsmen with "up-to-date, factual information on current operations and policies." He pointed out that the Saigon correspondents frequently went along on military operations and were thus "thoroughly knowledgeable" about the war. With that fact in mind, the U.S. Army ought to assign highly experienced information officers to positions in the field and to appoint a civilian to head them. More likely to gain the newsmen's confidence than a strictly military team, the group would begin its work by determining which reporter wielded the most influence over his fellows and enlist his assistance in correcting any problems that arose with the press. The U.S. commander in South Vietnam would meanwhile direct the press toward areas where favorable publicity was desired by conducting informal discussions with newsmen and by periodically soliciting their opinions.[22]

Finn's recommendations took on added emphasis shortly after they reached Wheeler's desk, when the chief of the MACV Public Affairs Office, Lt. Col. B. Lee Baker, U.S. Air Force, petitioned the Defense Department for a review of its restrictions on the release of information to the press. Baker said that the rules obscuring the U.S. Air Force's role in the war, the employment of Army and Marine Corps helicopters, the use of napalm, and the presence of jet aircraft in Southeast Asia were naive. Reporters knew that U.S. pilots flew many of the air strikes supposedly flown by the South Vietnamese and that helicopters were

at the time no full-time TV news reporters assigned to cover the war, see "TV's First War," *Newsweek*, 30 Aug 65, p. 32.

[21] Rpt, DCSOPS to the Chief of Staff, Army, 21 Mar 64, sub: Actions To Improve U.S.-GVN Operations in South Vietnam, CMH files.

[22] Ibid.

taking offensive action against the enemy despite official attempts to soften that fact. They had seen napalm in use and had only to visit the observation deck at Tan Son Nhut Airport to count the jet aircraft continually parked near commercial runways. Continued adherence to unrealistic restrictions, Baker warned, would only harm military credibility.[23]

The Assistant Secretary of Defense for Public Affairs, Arthur Sylvester, deferred action on Baker's request until June, and the Army's Office of the Deputy Chief of Staff for Operations decided to act only on those of Finn's recommendations that the Army could handle unilaterally. Events occurring late in March and continuing through April and May nevertheless proved the accuracy of Baker's prediction while adding urgency to the call for a revision of public affairs policy.[24]

On 28 March the *Indianapolis News* published the letters home of Capt. Edwin Gerald "Jerry" Shank, U.S. Air Force, who had recently died in battle. Shank had written his family regularly during the months before he died, sometimes twice a day. Two months after his death, his relatives released his letters to the *Indianapolis News*. *U.S. News & World Report* picked up the story, giving it four pages prefaced with the title "A Captain's Last Letters From Vietnam, 'We Are Losing. Morale Is Bad . . . If They'd Just Give Us Good Planes. . . .' " Shank's letters revealed the details of U.S. Air Force combat activities in South Vietnam. Although they dealt mainly with Shank's own vivid experiences in the war and contained conventional complaints about the quality of equipment, one detailed the pilot's responsibilities as a "trainer" of South Vietnamese airmen, the frustrations involved in fighting a war without recognition, and the anger that sometimes resulted.

> What gets me the most is that they won't tell you people what we do over here. I bet you that anyone you talk to does not know that American pilots fight this war. We—me and my buddies—do everything. The Vietnamese "students" we have on board are airmen basics. The only reason they are on board is in case we crash there is one American "adviser" and one Vietnamese "student." They're stupid, ignorant, sacrificial lambs, and I have no use for them. In fact, I have been tempted to whip them within an inch of their life a few times. They're a menace to have on board.[25]

Although the story failed to take hold at once, by mid-April it was a major concern of almost every important newspaper in the United States. Congress, too, took up the issue. Much of what the news media had to say centered upon a concern that American soldiers were fighting and dying without proper equipment, but everything took on added meaning because U.S. officials had dissembled about the character of American operations in South Vietnam. At a news

[23] Memo, B. L. Baker for Arthur Sylvester, 12 Mar 64, sub: Restrictions on Release of Information in RVN, DDI News from Vietnam file.

[24] Memo, DCSOPS for ACSFOR, Staff Plan [1964], sub: Report to the Chief of Staff on Action To Improve U.S. Efforts in South Vietnam; MFR, Arthur Sylvester, 1 Oct 64, sub: News Restrictions in Vietnam. Both in DDI News from Vietnam file.

[25] Memo, Col C. R. Carlson, USAF, Chief of Public Information Division, Office of Air Force Information, for the Director of Information, 10 Apr 64, sub: Capt. Shank's Letters Home, Air Force Clipping Service files.

Captain Shank Poses in Front of His Aircraft

conference on 22 April, House Republican Minority Leader Charles A. Halleck of Indiana cited the Shank letters as proof that Americans had been misinformed about the war. "Let's have the whole brutal business out on the table," he said. "Although the American public is repeatedly assured that our service men are only . . . instructors, there is mounting evidence that many of them are engaged in actual offensive operations." When *U.S. News & World Report* published the letters in full on 4 May, Senator Margaret Chase Smith of Maine inserted the article into the *Congressional Record* with the comment that "there is a genuine need, a desperate need, for the American people to be told the truth on the Vietnamese war. They are not getting the facts from their government." Further congressional comment followed on 8 May, when *Life* reprinted the letters under the heading "We Fight and Die, But No One Cares."[26]

The Defense Department responded to the charges, but to little effect. Although the Air Force defended the record of its aircraft in South Vietnam, noting that each had been rebuilt before consignment to Southeast Asia, and although Arthur

[26] Tom Lambert, "GOP Charges U.S. Deceives People on GI Role in War," *New York Herald-Tribune*, 22 Apr 64; "A Captain's Last Letters From Vietnam," *U.S. News & World Report*, 4 May 64, p. 46; U.S. Congress, Senate, *Congressional Record*, 88th Cong., 2d sess., 27 April 1964, p. 8889; "We Fight and Die, But No One Cares," *Life*, 8 May 64, p. 34B.

Sylvester demonstrated that *Life* had edited Shank's letters to make them appear more critical than they actually were, the controversy broadened. Originally applicable only to the Air Force, it became the concern of all the military services when a group of relatives of American soldiers and airmen killed in South Vietnam bought a full-page advertisement in the *Washington Star* to list the names of the 127 Americans who had died in South Vietnam since January 1961. "We believe this list is not complete," they charged, "and that many more Americans have been killed by communist bullets in Vietnam than has been reported by the Department of Defense."[27]

The director of the U.S. Information Agency, Carl Rowan, returned from a fact-finding trip to South Vietnam while the controversy was at its height. He told President Johnson that Lodge's one-man rule over the U.S. mission's public affairs program had harmed coordination of the overall public affairs effort and that Barry Zorthian should take control of the entire program. Although Zorthian would be unable to stop critical articles written by newsmen "who go out into the field, gain the confidence of our soldiers, and then pick up information . . . not at all helpful to our over-all mission," he could at least take the action necessary to end the confusion plaguing the public relations effort and inaugurate measures to balance critical war coverage with "the stories we want told."[28]

While Rowan's recommendation circulated between the White House, the State Department, and the Department of Defense, Arthur Sylvester took the first step toward a reinvigorated information program in South Vietnam. Predicting that problems with the press would worsen as the war went bad, he cut official red tape to bring to Washington one of the Army's most experienced public affairs officers, Col. Rodger Bankson. A veteran of the censorship program during the Korean War, Bankson was then serving as the chief of information for the U.S. Strike Command in Florida. Sylvester instructed Bankson to set up a Southeast Asia Division within the Office of the Assistant Secretary of Defense for Public Affairs. He wanted the organization to know everything it could about the war so that it could maintain liaison with the press corps in Washington while developing intelligent policy guidance for use in the field.[29]

Shortly after arriving in Washington, Bankson traveled to South Vietnam to conduct a six-week survey of MACV's public affairs operations. While he was away, on 2 June, a high-level conference chaired by Rusk and McNamara convened in Honolulu to consider the situation in South Vietnam. As part of that conference a subcommittee composed of Sylvester, Rowan, Zorthian, and a num-

[27] Ltr, Eugene M. Zuchert, Secy of the Air Force, to Honorable Richard Russell, Chairman of the Senate Armed Services Committee, 13 May 64, CMH files; Laurence Barrett, "Building in Viet Warplanes," *New York Herald-Tribune*, 14 May 64; Jack Raymond, "Air Force Backs Record of Its Planes in Vietnam," *New York Times*, 14 May 64; "Pentagon Hits 'Editing' of Dead Pilot's Letters," *Washington Star*, 23 May 64; Ted Lewis, "Capital Stuff: Kin of Dead GI's Pose a Question," *New York Times*, 13 May 64; Tom Lambert, "House Quiz for McNamara on Obsolete Planes Used in Vietnam," *New York Herald-Tribune*, 13 May 64.

[28] Memo, Rowan for the President, 26 May 64.

[29] Intervs, author with Col Rodger Bankson, 6 Sep 73 and 16 Jun 75, both in CMH files.

ber of other experienced information officers met to evaluate the information program. Finding conditions gloomy and unsatisfactory on many counts, the group reported that the Saigon correspondents were aware of everything that was happening in South Vietnam and had begun to boast that they had revealed the facts when U.S. officials were still "pretending" things were going well. Reporters would continue to write in a negative vein as long as South Vietnamese fortunes declined. The information program had yet to be devised that could make defeats look like victory or South Vietnamese lassitude appear as fiery enthusiasm.[30]

Barry Zorthian observed that the absence of victories was only part of the problem. The Saigon correspondents were "as skeptical and cynical a group of newsmen as he had ever seen," mainly because official spokesmen had misled them in the past. A program of creative press relations was of paramount importance in such a context, yet the handling of the news media in South Vietnam was so diffuse and the rules under which military information officers labored so unrealistic that little chance for originality in dealing with the press remained.[31]

The assembled information officers set about devising a set of suggestions to correct what was wrong. Officials at all levels in Washington and in South Vietnam, they said, had to understand that the information effort was an integral part of every program drawn up to meet the crisis in Southeast Asia. With that principle established, the job of improving official credibility could proceed in the proper context, and Washington agencies could begin to issue new guidance designed "to wipe out the several directives now on the books which some military information officers interpret as requiring them to lie." Since Colonel Baker had himself been discredited by those requirements, he too would have to go.[32]

Turning to the lack of cohesion within the overall information effort, the conferees repeated Rowan's earlier suggestion that one man take across-the-board authority for public relations in South Vietnam. That individual would sit in on all meetings and briefings and know as much as possible about the war. He would advise members of the U.S. mission on which newsmen to see and what points to make. Although he would report to the ambassador, he would possess "Czar" powers enabling him to marshal whatever resources he needed to the task of moving the positive side of the story to the news media of the world.

With that foundation in place, the conference directed its attention to the MACV information apparatus. Military members of the group argued that newsmen serving in South Vietnam required access to immediately available transportation. Colonel Baker was "bumming rides every day," they said, and could never be certain of his ability to get the press to a news development where report-

[30] Quote from MFR, William P. Bundy, 2 Jun 64, sub: Tuesday Afternoon Session at Honolulu, Chron files, CMH. See also "The War in Asia," *Newsweek*, 8 Jun 64, p. 25; Memo, Carl Rowan for Secretary Rusk, 4 Jun 64, sub: Improvement of Informational-Psychological Program in South Vietnam, CMH files.

[31] MFR, CINCPAC, 1 Jun 64, sub: Special Meeting on Southeast Asia, Plenary Session, I-36213/64 092SEA, 68A4023, box 5, WNRC.

[32] This section is based on Memo, Rowan for Rusk, 4 Jun 64, sub: Improvement of Informational-Psychological Program in South Vietnam.

ing could be in the national interest. The military services also had to give high priority to improving the quality of the military information officers they sent to South Vietnam. Truly qualified men seemed to consider service in Saigon a stigma on their careers. Several had even resigned rather than accept an assignment there. As a result, most of the public affairs officers in South Vietnam lacked either experience or the general ability to do the job.

The information officers' final recommendation addressed an old but basic issue. Claiming that most of the damaging articles appearing in the press were the result of military gripe sessions, the group called upon the military services to inaugurate a vigorous internal education program designed to reduce the numbers of incidents where soldiers sounded off to the press.

Although the participants in the conference believed that effective management and realistic information policies could do much to improve reporting of the war, they had few illusions about the immediate future. In the briefing for Secretary Rusk that followed the meeting, they predicted that their recommendations would have little if any effect within the next three to six months. Assuming that Khanh avoided assassination, the situation would either continue along much as it had or, more probably, deteriorate.[33]

Westmoreland and Zorthian Take Charge

President Johnson acted upon the information officers' recommendations shortly after the conference ended, appointing Barry Zorthian on 6 June to be the U.S. mission's chief public affairs officer. In addition to continuing as director of psychological warfare, Zorthian was to be the U.S. mission's overall counselor on relations with the news media. Subject only to the ambassador, he was to set policy; maintain liaison between the embassy, the Military Assistance Command, and the press; publicize information to refute erroneous and misleading press reports; and help newsmen cover the positive side of the war.[34]

Zorthian assumed his duties at once, coordinating his ideas both with Bankson and with General William C. Westmoreland, who had been Harkins' deputy since January and who was slated to become U.S. commander in South Vietnam on 20 June 1964. In the days that followed the three devised a plan to improve the U.S. mission's information program. Calling upon the U.S. Army to recognize that South Vietnam was no place to send fledgling information officers, they started by changing the name of the MACV Public Affairs Office to MACV Office of Information (MACOI), a semantic alteration that they hoped would lend the operation greater stature. The sole release point in South Vietnam for news of military operations, that agency was to have three administrative divisions. Troop

[33] Fact Sheet, 5 Jun 64, sub: South Vietnam Action Program, attached to Memo, William H. Sullivan for the Secretary of State, 5 Jun 64, sub: Measures To Strengthen Situation in SVN, CMH files.
[34] Msg, State 2192 to Saigon, 6 Jun 64, FAIM/IR.

Information was responsible for the command newspaper, the Armed Forces Radio Service broadcasting station, and all activities involved in the orientation and indoctrination of military personnel. The Press Relations Division was to handle news queries, press releases, weekly briefings for correspondents, and the monitoring of the South Vietnamese government's public relations where U.S. military interests were involved. Special Projects would develop and place material designed to offset erroneous stories filed by the news media. Of the three divisions, Westmoreland, Zorthian, and Bankson expected the Special Projects Division to be the catalyst in the development of a revitalized information program. They located the unit next to the U.S. Information Service and the office of the embassy's press attache in order to ensure supervision by Zorthian, and they gave it responsibility for counteracting negative, distorted reports in the press by finding objective stories and taking newsmen to them. In coordination with Zorthian and the U.S. mission, Special Projects was also to handle field trips for the news media, to supervise the activities of the MACV information liaison officers stationed in each of South Vietnam's corps tactical zone headquarters, and to collaborate with South Vietnamese government agencies on matters that might either attract the press or involve the Military Assistance Command in public relations problems.[35]

Convinced that more positive news reporting would result if the Military Assistance Command sped correspondents to the scene of a military action while fighting was still in progress, Westmoreland planned to assign a passenger-carrying CV-2 Caribou to the Special Projects Division for that purpose. When he learned that the Defense Department intended to send motion picture teams to South Vietnam to film Viet Cong atrocities for later release to television stations in the United States, he also assigned a helicopter for the use of official photographers.[36]

In the two weeks following the drafting of the plan, the Military Assistance Command and the U.S. Information Service set up a division of labor in Saigon. While Zorthian maintained overall contact with the press, passing on new policies and overseeing the South Vietnamese government's relations with the news media, the Special Projects Division assumed the more mundane tasks of establishing a press center and coordinating correspondents' trips into the field. The division assembled a file of developments in progress that could be used to tip reporters to often overlooked stories and began paying special attention to important correspondents. It likewise began seeking ways to increase the number of news media interviews with the ambassador and other high-ranking American and South Vietnamese officials and started providing reporters with specially pre-

[35] Msg, MACV 2854 to JCS, 8 Jun 64, sub: Reorganization of Information Program at MACV, for General Taylor and Secretary Sylvester from Westmoreland, Westmoreland History, bk. 2, May–Jun 64.

[36] Ibid.; Msg, State 2107 to Saigon, 28 May 64, FAIM/IR; Memo, John McNaughton for Arthur Sylvester, 30 Apr 64, sub: Securing Publishable Photos of VC Atrocities, I-6703/64, ISA 062VN, 68A306, box 40, WNRC.

pared news stories, U.S. mission-originated radio tapes, and film clips shot by official photographers.[37]

While those programs were taking shape, General Westmoreland began to include correspondents on his trips into the field. He made appearances at the weekly MACV press briefings and undertook special trips to locations and projects where his presence might attract newsmen to favorable stories. In concert with Ambassador Lodge, Zorthian meanwhile inaugurated a series of weekly, off-the-record background sessions at which knowledgeable U.S. experts attempted to educate newsmen in the subtleties of the American involvement in Southeast Asia. He also began consulting with the South Vietnamese government on ways to improve its facilities for transmitting news overseas and started negotiations with the government's minister of foreign affairs on ways to give a more professional character to South Vietnamese relations with the Saigon correspondents.

Since the products created by the new information program would have value only if the American people accepted them, Westmoreland and Zorthian urged better coordination between Washington agencies and the U.S. mission in Saigon to ensure that the information released in each place actually helped prepare a climate in the United States receptive to the official point of view. The mission and the Military Assistance Command were already attempting to identify positive, articulate soldiers for appearances before the American news media; they suggested that Washington agencies do something similar by bringing editors, businessmen, and other top opinion leaders to Saigon, where a special effort could be made to highlight the importance of what the United States was accomplishing. Name entertainers could also be invited. Their performances would help to improve military morale at the fighting level while providing material for later use at home, where the celebrities could make special appearances to talk about the significance of what they had seen.

Westmoreland and Zorthian's program corresponded closely to what the State and Defense Departments were already thinking and received ready approval. On 7 July the State Department added a final touch by issuing new public affairs guidance for South Vietnam that superseded all previous messages on the subject except the one appointing Zorthian "Information Czar." Because broad public and congressional support was a requisite for the success of United States policy in South Vietnam and support for the war would never survive in an atmosphere of distrust, State charged Zorthian with ensuring that the mission's public affairs activities promoted "maximum candor and disclosure consistent with the requirements of security."[38]

Since adequate press coverage required timely information as well as full disclosure, State gave the U.S. mission in Saigon charge of all decisions on the normal release of information to the press. Then, observing that credibility was

[37] This section is based on Msg, Saigon 2622 to State, 27 Jun 64, sub: Steps That Have Been and Will Be Taken To Improve and Expand Press Relations Effort, FAIM/IR.

[38] Msg, State 59 to Saigon, DAIN 339361, 7 Jul 64, Army Staff Communications Center files, Army War College.

Colonel Bankson

the key to the successful discharge of the mission's responsibility for press relations, the agency ordered members of the American team in South Vietnam to refrain from any activity that would tend to mislead the press or damage relations with the news media. Only one restriction was to govern, and it seemed little burden. The U.S. mission was to keep Washington agencies fully informed about what it was telling the press so that they could coordinate the information they released with what was being said in Saigon.

Although the mission might avoid every action suggestive of untruth and attempt to provide full and accurate information to the news media, both Zorthian and Westmoreland knew that the press had a long memory. In the hope of erasing some of the unfortunate script of past months and in the belief that new blood untouched by old indiscretions might have an advantage over men identified with earlier controversies, they began to push for a change of personnel at the MACV Office of Information. Baker was their principal target. Westmoreland believed that the officer had done "an excellent job within his capabilities" and was more than willing to have him finish out his tour of duty as chief of information; but Baker had become a symbol of the animus between the press and officialdom during the Harkins era and was, in that sense, a liability to future MACV dealings with the Saigon correspondents.[39] An Air Force officer, Baker also represented a point of view different from the one Westmoreland wanted to prevail in South Vietnam. The general believed that "a first class Army public information officer with training and experience in ground warfare" was better suited for telling the story of the war. During his orientation in Washington prior to becoming MACV deputy commander in January 1964 he had made the point to Arthur Sylvester and had apparently received assurances that a change would be made at an appropriate time.[40]

A possible replacement for Baker appeared on 17 July, when Bankson finished his six-week appraisal of the war and stopped by Westmoreland's office to brief the general before returning to Washington. Impressed with the officer's "grasp of the situation, . . . his alertness, and his obvious competence," Westmoreland

[39] Msg, Westmoreland MAC 3877 to Sylvester, 27 Jul 64, Westmoreland History, bk. 6, tab 2, CMH.
[40] Msg, Westmoreland MAC 3632 to Wheeler, 16 Jul 64, William C. Westmoreland Papers, CMH; Memo, Gen Winant Sidle for the author, 7 Nov 84, sub: Public Relations, CMH files.

sought Zorthian's concurrence and then asked Sylvester to approve Bankson as the next MACV chief of information.[41]

The Joint Chiefs of Staff had switched the MACV chief of information slot from Air Force to Army on 30 June in deference to what they had already known was Westmoreland's wish. When the request for Bankson arrived, however, Sylvester backed away from his earlier agreement and refused to release the colonel. He questioned instead whether the presence of an Army spokesman was all that necessary since, he said, the caliber of the man in charge was more important than his branch of service. Sylvester told General Wheeler that the change in service designation at the MACV Office of Information had slipped across his desk unnoticed and that it would upset the balance of responsibilities between the Army and the Air Force in South Vietnam if allowed to stand.[42]

Wheeler translated that remark for Westmoreland. Noting that the problem was larger than either the Office of Information or Sylvester, he observed that the Air Force and the Army were already feuding over which service should have control over helicopters in South Vietnam and that another confrontation might prove embarrassing. Feeling ran high, especially in Marine Corps and Navy circles and particularly in view of the decision to appoint an Army deputy commander for the Military Assistance Command, that the Army was attempting to cut the other services out of the war. Thus, while Westmoreland indeed had a point, Wheeler could only conclude with Sylvester that the ability of MACV's chief of information was more important than his service.[43]

Westmoreland conceded the issue for the time being rather than create, as he put it, a "cause celebre in inter-service wrangling."[44] He nevertheless continued to believe that an Army public affairs officer would be better suited to the war in South Vietnam. When he raised the matter again in August, Admiral Felt backed him. Felt advised the Joint Chiefs that since Westmoreland had charge of the information program in South Vietnam he should have his way. Finally, on 28 December, the Air Force Chief of Information, Maj. Gen. E. B. LeBailly, proposed a compromise. He offered to give in to Westmoreland's request, provided Air Force officers were appointed as deputy chief of MACV's Office of Information and director of the agency's Press Relations Division. Since the Press Relations Division dealt directly with the news media, he noted, it was the logical place to provide perspective on the air war and an excellent spot for an Air Force officer. In the same way, the presence of an Air Force deputy at the Office of Information would reflect the fact that U.S. operations in South Vietnam were a cooperative venture involving all of the military services. The Depart-

[41] Msg, Westmoreland MAC 3653 to Wheeler, 17 Jul 64, Westmoreland Papers, CMH.

[42] Msg, JCS to COMUSMACV, 30 Jun 64, CMH files; Memo, Sylvester for Wheeler, 9 Jul 64, DDI Policy file.

[43] Msg, Wheeler JCS 3497 to Westmoreland, 16 Jul 64, Westmoreland Papers, CMH; Ltr, Wheeler to Westmoreland, 17 Sep 64, Westmoreland History, bk. 6.

[44] Msg, Westmoreland MAC 3653 to Wheeler, 17 Jul 64.

ment of Defense accepted LeBailly's proposal, giving the MACV Office of Information the Army orientation Westmoreland had sought.[45]

Sylvester and Westmoreland shortly thereafter agreed on a plan to ensure that all future MACV chiefs of information had the fullest possible preparation. Leaving Bankson in Washington for the coming year to continue as special assistant for Southeast Asia, Sylvester appointed the Chief of Information, U.S. Army, Europe, Col. Benjamin W. Legare, as MACV Chief of Information. At the end of one year, Bankson was to succeed Legare, and Col. Winant Sidle, a former deputy chief of U.S. Army Information then serving as military assistant to the chairman of the Joint Chiefs of Staff, was to become special assistant in Bankson's place. Upon Bankson's departure from Saigon in 1967, Sidle would become the chief. Although Sylvester made no selections beyond Sidle, from then on each succeeding chief was to serve one year as special assistant for Southeast Asia before leaving for Saigon. The system held until 1970, when the Office of the Special Assistant was transferred to the Directorate for Defense Information. Until then, each incoming MACV chief of information was fully aware of all public relations policies and problems because he had spent at least a year working on them in the Pentagon.[46]

Few officials within the Department of Defense expected that the agreement with LeBailly would end interservice rivalry in South Vietnam, but there was at least the hope that compromise might help to control the tendency. No similar prospect existed for the policy of maximum candor. Conceived almost solely as a means of mollifying the press in order to obtain favorable news coverage, the program grew out of an assumption that the South Vietnamese would somehow demonstrate the viability of U.S. policy by overcoming their failure to achieve sustained victories. In that, it ran counter both to the reality of the war and to the first law of propaganda which states that even the best promotional buildup will ultimately fail to sell a questionable product.

[45] Msg, Westmoreland MAC 3632 to Wheeler, 16 Jul 64. Msg, COMUSMACV MAC J-1 8578 to CINCPAC, 23 Aug 64; Msg, CINCPAC to JCS, 2 Sep 64; Memo, E. B. LeBailly for Sylvester, 28 Dec 64; and Memo, Wheeler for the SECDEF, 17 Mar 65, all in DDI Policy file.

[46] Memo, Sidle for the author, 7 Nov 84, sub: Public Relations.

4

More Than Goodwill

When General Maxwell D. Taylor replaced Lodge as U.S. ambassador to South Vietnam in June 1964, he immediately moved to reaffirm the Johnson administration's public affairs policy. Asserting that "our relations with the press should be based on a principle of maximum candor and disclosure consistent with the requirements of security and responsible conduct," he directed General Westmoreland and other heads of American departments in Saigon to cooperate with Barry Zorthian in every way possible. Zorthian, in turn, was to ensure that the U.S. mission's dealings with newsmen were both "effective and responsible."[1]

Taylor's intentions notwithstanding, U.S. credibility in South Vietnam depended upon more than the U.S. mission's goodwill. International diplomacy, the political needs of the Johnson administration, and South Vietnamese instability were all considerations, and each in its own way conflicted with the concept of good public relations that "maximum candor" intended to promote.

An Impossible Position: Laos, June 1964

Diplomatic problems began even before Taylor arrived in South Vietnam, while the policy of maximum candor was still in formulation. They centered on Laos. The United States considered the Laotian conflict and the war in South Vietnam two aspects of the same problem and viewed the neutralization of Laos stipulated by the 1962 Geneva Agreements as an essential ingredient in any long-term settlement of the Indochina question. During April and May of 1964, that neutrality came under threat. In April a group of right-wing generals deposed Neutralist Premier Souvanna Phouma in a bloodless coup. Then in May the Com-

[1] Memo, Taylor for Deputy Ambassador U. Alexis Johnson et al., 4 Aug 64, sub: Mission Press Relations, Westmoreland History, bk. 7, tab 6; Msg, State 2192 to Saigon, 6 Jun 64, FAIM/IR.

Prince Souvanna Phouma Visits President Kennedy

munist Pathet Lao launched an offensive on the Laotian Plain of Jars that threatened to extend their hegemony far beyond boundaries acceptable to the United States.[2]

Souvanna's government was almost powerless, but, by virtue of the Geneva Agreements, it possessed a certain legitimacy. The United States on that account publicly condemned the coup and sent Assistant Secretary of State William P. Bundy to Vientiane. Bundy won the restoration of Souvanna's government, but the Communist offensive required stronger measures. President Johnson authorized American low-level jet reconnaissance of enemy positions in the battle area and temporarily permitted U.S. civilian pilots to fly Laotian Air Force fighter planes against enemy targets. On 21 May the U.S. Department of State announced officially that Souvanna's government had granted permission for reconnaissance flights, but since both Souvanna and U.S. policy makers wanted to keep American actions in Laos in low profile, it said nothing about American participation in combat operations.[3]

[2] This section is based on Vincent Demma, Review of U.S. Military Efforts in Laos, 1962–1965, CMH MS [1968], in CMH files. See also *Pentagon Papers*, 3: 158f.

[3] Draft Memo, William P. Bundy, 20 May 64, sub: Possible Actions in SEA, in a note for the SECDEF, 20 May 64, Chron files, CMH; Memo, MACV ACofS J-2 for Westmoreland [May 64], sub: Laos Low

The official position remained the same until 6 June 1964, when Communist gunners shot down an unarmed reconnaissance aircraft. To defuse any propaganda statements the Communists might make, U.S. spokesmen immediately confirmed the loss, adding that the United States was consulting with Laos on "measures for the protection of these flights." Within hours and without any announcement, armed escorts began to accompany U.S. reconnaissance missions over Pathet Lao territory.[4]

President Kennedy Discusses Laos *at a news conference.*

The next day Communist gunners claimed a second American aircraft, an armed escort actively engaging enemy antiaircraft batteries. In the belief that Communist propagandists would "charge us with firing and that we would be in an impossible position," the State Department quickly convened a confidential background briefing (called a backgrounder) for the press.[5] The department confirmed that the downed aircraft had indeed been armed, but said nothing about whether the plane had actually fired its guns. The public communique that followed took the same approach. Official briefers reaffirmed that the United States had been "undertaking such flights since 21 May to disclose information about Pathet Lao and Viet Minh activity which is in direct violation with the Geneva accords," but omitted any reference to possible U.S. infractions of the same agreement.[6] While giving the impression that American operations were designed to restore Laotian neutrality, the United States left the use of its armed escorts in protecting reconnaissance flights purposely ambiguous. That equivocation prepared the ground for a public relations crisis that rapidly began to develop.[7]

Shortly after the announcement, on 9 July, President Johnson ordered a U.S. retaliatory strike against enemy antiaircraft positions in Laos to demonstrate that overflights would continue by force if necessary. The attack sparked a Communist Chinese news service denunciation of U.S. offensive combat in Laos and pitted

Level Operations, in Book of Miscellaneous Facts, 67A4604, box 1, WNRC; Joint Chiefs of Staff Memorandum (hereafter cited as JCSM) 746-64, 26 Aug 64, sub: U.S. Armed Reconnaissance, CMH files; Department of State *Bulletin*, 29 June 1964, p. 994.

[4] Department of State *Bulletin*, 29 June 1964, p. 994.

[5] Msg, State 1158 to Vientiane, 8 Jun 64, FAIM/IR.

[6] Msg, State to Vientiane, ref. Embtel 1586, 10 Jun 64, FAIM/IR. State Department cables lacking message numbers are draft office copies.

[7] Msg, State to New Delhi, 8 Jun 64, FAIM/IR.

the Johnson administration's desire to promote a positive image for U.S. efforts against its need both to placate Souvanna and to continue desirable military operations.[8]

The *London Daily Mirror* took up the Chinese story as soon as it appeared. Other papers followed. U.S. reporters armed with leaked inside information meanwhile began to clamor for an explanation. Officials in Washington wanted to put the matter in its true light to avoid having sensational news stories quoted as fact, but Souvanna felt that any acknowledgment of firing would play into Communist hands. The State Department tried to reason with him, warning that the credibility of all American actions in Southeast Asia was at stake. Charges had already appeared in the press and Congress that the president had failed to inform the public and Congress clearly on what was happening in South Vietnam. "While we have not had the same problem with Laos we must do everything to avoid it if we are to maintain the desired degree of firmness without being accused of concealed and irresponsible action which could vitiate all our efforts."[9] Despite those reservations, when Souvanna remained adamant, the department decided it had little choice but to bow to his wishes, for the sake of maintaining his full cooperation.[10]

The public statement that followed was thus little more informative than earlier communiques. "There has been no change in the matter of photo reconnaissance flights. These flights have taken place and will take place at the intervals necessary for the purpose of obtaining information. . . . We have a clear understanding on this matter with the Laos government, and we are in agreement with that government also that it is not in the interest of the government of Laos or of those who undertake these hazardous missions that any operational part of their work should be discussed."[11]

American newspapers recognized the diplomatic issues underlying the official stance but saw little value in the tight-lipped news policies that accompanied it. When White House Press Secretary George Reedy referred reporters to the Departments of State and Defense on all questions involving operations in Laos and those departments sent them back to the White House empty-handed, sharp confrontations between officials and newsmen ensued, with the newsmen taking the position that patterns of secrecy had begun to emerge in the administration's handling of the war.[12]

The *Chicago Tribune* attributed the news blackout to election year politics. Douglas Kiker of the *New York Herald-Tribune* observed that the situation was mainly the result of an attempt by President Johnson to distance himself from events that could blow up in his face during the presidential campaign. *Aviation*

[8] Fact Sheet, DCSOPS for Chief of Staff, Army, 15 Jul 64, sub: Strike Against Anti-Aircraft Installation in Xieng Khouangville, CMH files.

[9] Msg, State to Vientiane, ref. Embtel 1586, 10 Jun 64.

[10] Ibid.

[11] Department of State *Bulletin*, 29 June 1964, p. 995.

[12] "U.S. To Continue Flights When Needed Over Laos," *New York Times*, 12 Jun 64; Carroll Kilpatrick, "Reedy Refused To Talk About Asia or Cyprus," *Philadelphia Inquirer*, 12 Jun 64.

Week magazine linked the crisis to the question of official honesty. It charged that Secretary McNamara's optimistic reports on the Vietnam War were "regularly contradicted by events" and that Arthur Sylvester's word had "sunk so low" most Pentagon reporters refused to believe a story "until it had been officially denied." The *Washington Post* said that the United States had come to a sad pass when it had to rely upon China's news agency for reports on covert military operations. "Does the government really have the naivete to believe that its hand in these operations can be concealed? If it is to conduct or sponsor such raids, then let the matter be decided openly in terms of whether American interests require it. But let there be no repetition of the humiliating sequence whereby communist China makes a fool, if not a liar, out of the United States." Most of all, the paper concluded, "Let there be an end to the week-long news vacuum about Southeast Asia . . . created by the official black out in Washington and filled by eager propaganda from Peking."[13]

Although the issue was neither naivete nor lies but the deliberate withholding of information in deference to the wishes of the politically threatened Souvanna, the *Post* was correct in asserting that the affair had done damage to U.S. credibility. By the end of June reporters attending news conferences at the Pentagon were regularly asking Arthur Sylvester whether there would be an announcement about Laos or "anything else of importance." When Sylvester responded with the official line, the military affairs correspondent of the *Washington Star*, Richard Fryklund, commented that "the public learns a lot less than it should about the basis for McNamara's decisions. . . . You can run a tight Pentagon that way, but you can't run an effective democracy."[14]

Conflicting Priorities

The crisis over Laos was still at its height when Barry Zorthian and General Westmoreland returned in June from Honolulu to begin their campaign to improve the Saigon mission's relations with the press. Just as the State Department seemed unable to placate Souvanna while observing maximum candor on Laos, so Zorthian and Westmoreland found themselves caught between the wish to create a positive image for U.S. efforts in South Vietnam and the fact that the war was becoming increasingly complex and difficult to manage.[15] (*Map* 2)

Ambassador Taylor's memorandum on relations with the press embodied the dilemma. It attempted to satisfy all the demands imposed by the war while yielding to the Johnson administration's desire to make the U.S. government speak

[13] "Still Managing the News," *Chicago Tribune*, 15 Jun 64; Douglas Kiker, "White House Blackout on Asian News," *New York Herald-Tribune*, 12 Jun 64; "The Credibility Gap," *Aviation Week Magazine*, 15 Jun 64; "The Price of Secrecy," *Washington Post*, 17 Jun 64.

[14] Fryklund is quoted in "New Pique," *Newsweek*, 29 Jun 64.

[15] Msg, MACV J-1 4719 to Dept of the Army, 9 Jun 64, CMH files.

NORTH VIETNAM
17th Parallel
Quang Tri
Hue
I CORPS
Thua Thien
Da Nang
Quang Nam
THAILAND
LAOS
Ho Chi Minh Trail
Mekong R
Chu Lai
Quang Tin
Quang Ngai
Kontum
Pleiku
Binh Dinh
Qui Nhon
Phu Bon
Phu Yen
CAMBODIA
TONLE SAP
Darlac
Ban Me Thuot
Khanh Hoa
Nha Trang
Cam Ranh
Quang Duc
Tuyen Duc
Ninh Thuan
Mekong
R
PHNOM PENH
Binh Long
Phuoc Long
Lam Dong
Binh Thuan
Tay Ninh
Long Khanh
Binh Tuy
II CORPS
SOUTH CHINA SEA
Xuan Loc
Phan Thiet
SAIGON
My Tho
Vung Tau
III CORPS
Kien Giang
An Giang
Vinh Long
Vinh Binh
Phong Dinh
CAPITAL SPECIAL ZONE
GULF OF THAILAND
Chuong Thien
Ba Xuyen
Bac Lieu
Quan Long
An Xuyen
IV CORPS
Con Son
SOUTH VIETNAM
1965
0
100 MILES
0
100 KILOMETERS

Map 2

with a single voice on the issue. Under the system it prescribed, official spokesmen were to carry the burden of press relations. Everyone else would confine his conversations with newsmen to areas of his own expertise and would report to his unit's information officer any discussions that touched on anything important. Since the United States was in South Vietnam at the sufferance of the South Vietnamese, cordial relations with those colleagues were likewise to take priority. No member of the U.S. mission was ever to offend his hosts by commenting publicly on their internal affairs.[16]

Although desirable from Taylor's standpoint, those rules were impractical in South Vietnam because the Saigon correspondents had well-developed sources of their own and were used to going their own way. Drawing upon official briefings for background and personal contacts for details, the more responsible among them would continue to work as they had. Less meticulous reporters also conformed to habit. Writing poorly researched stories either to beat their competition or to attract their editors' attention, they would reject as tainted any information from official channels.[17]

As for American official comments on South Vietnamese affairs, no one in Saigon believed that the practice could be avoided, if only because the South Vietnamese themselves refused to carry out the public affairs function. The chief of the South Vietnamese government's information apparatus objected to regular, candid briefings for American newsmen because he valued his own political anonymity. If his briefers said anything at all in public, he told Zorthian privately, he would become a party to innumerable political controversies and could never escape appearing to take one side or the other. If the government in power then fell to its opposition, his future would indeed be bleak.

The vacuum that resulted posed no disadvantages for large news-gathering agencies such as United Press International and the Associated Press. They had contacts. Yet the rest of the Saigon correspondents also had to file reports. If the U.S. mission failed to supply the information they needed, they were bound to resort to any expedient.

The mission confronted the problem forthrightly but immediately ran afoul of the Johnson administration's need to cushion the impact of the war upon Congress and the American public. During June, July, and August 1964 the MACV Office of Information canceled all the security restrictions that had caused trouble in the past. It also began to coordinate Department of Defense-sponsored visits for U.S. newsmen and inaugurated a series of wide-ranging backgrounders for the press by key members of the U.S. mission. Beginning in September, Zorthian authorized daily MACV briefings for the Saigon correspondents covering both U.S. and South Vietnamese topics. On his own, he also started freewheeling weekly background sessions with selected newsmen to discuss any subject the reporters wanted to raise. Yet if those efforts fulfilled the design laid down

[16] Memo, Taylor for Deputy Ambassador Johnson et al., 4 Aug 64, sub: Mission Press Relations.
[17] Zorthian made this point in retrospect in Msg, Saigon 4014 to State, 2 Jun 65, FAIM/IR. This message is the source of this section.

by the Honolulu Conference and promised to end angry official confrontations with newsmen, the Johnson administration from the very beginning questioned every news story that threatened the low profile it sought.[18]

During mid-June 1964, for example, the U.S. Army's senior adviser in the II and III Corps Tactical Zones, Col. Wilbur Wilson, gave an exceptionally frank background briefing to the Saigon correspondents. Although he asserted that the effectiveness of the South Vietnamese armed forces had increased by almost 100 percent since the fall of Diem, Wilson nevertheless contended that the rate of improvement was hardly enough to win the war. Many South Vietnamese generals had obtained their rank through political intrigue rather than military competence, he said. As a result they still had difficulty following American advice, and they were still no match for the enemy. "So many of these damned countries," Wilson concluded, "this country, for instance, along with Cambodia and Laos are right out of the Middle Ages—700, 800, 900, or 1000 years behind the times. The Communists have introduced a highly disciplined, 19th century technique for the purpose of seizing control. . . . They work at their jobs seven days a week and twenty-four hours a day, and they're convinced their cause is right and they will win. If we can arouse the same dedication among the Vietnamese, we can win the war."[19]

The Associated Press carried an account of Wilson's remarks the day after they appeared. Attributing them to "a ranking United States military adviser" who would allow himself to be identified only as an Army officer of the grade of major or colonel, the article noted that his opinions were similar to those of military advisers in the field but in sharp contrast to the comments of official U.S. spokesmen in Saigon. Since the Defense Department believed any comparison of that sort, coming at the height of the Laotian imbroglio, threatened to shake the U.S. posture in Southeast Asia, it cabled Saigon immediately.[20]

Secretary of the Army Stephen Ailes and General Wheeler were scheduled to appear before Congress on 20 June, the Department told Westmoreland. In anticipation of possible questions on the subject, the two men wanted the officer who had made the statement identified "(not for attribution or retribution)." They also asked for a transcript of the briefing with special emphasis on "comments made concerning the caliber, determination to win, and effectiveness of the Vietnamese soldier," together with Westmoreland's assessment of whether the Associated Press account of the session was accurate.[21]

Westmoreland played down the interview in his response. Wilson had spoken without a script, he said, but the Associated Press had still given an accurate representation of what had transpired. When Wilson had begun to speak about South Vietnamese corruption, the MACV information officer in attendance had suggested that any comments on the subject ought to be considered off the rec-

[18] Memo, Arthur Sylvester for the SECDEF, 1 Oct 64, DDI News from Vietnam file.
[19] Msg, Westmoreland MAC 3099 to Harris, 19 Jun 64, Westmoreland Papers, CMH.
[20] "Viet Cong Threat Worse, U.S. Aide in Saigon Says," *New York Times*, 17 Jun 64.
[21] Msg, Harris WDC 4106 to Westmoreland, 18 Jun 64, Westmoreland Papers, CMH.

ord; but Wilson himself had interjected, "I'm not telling anyone here anything they don't already know." Westmoreland made no comment on Wilson's accuracy, implying that he shared the colonel's feelings.[22]

Concern about the Wilson briefing subsided when nothing came of it, but officials in all of the Washington agencies involved with the war remained unsure of what was happening in South Vietnam and continued to become agitated every time unfavorable news appeared. During July the Saigon correspondents revealed a series of costly enemy ambushes that had seriously disrupted the movement of South Vietnamese Army convoys. "These reports, particularly those relating to sizeable friendly losses, are causing something of a stir here," General Wheeler immediately cabled Westmoreland. Secretary Rusk and the Director of the Central Intelligence Agency, John A. McCone, "among others of lesser governmental stature, have asked me what is wrong with the tactical security arrangements of the Vietnamese forces. . . . I realize that these unfortunate happenings are a great worry to you. Nevertheless, you should know that these stories are putting into people's minds a very poor impression of the alertness and the military capabilities of our Vietnamese allies."[23]

The anxiety Wheeler described centered on the American image as much as on that of the South Vietnamese. The U.S. government's inability to explain or counter such fast-breaking news stories gave weight to allegations in the press that the Johnson administration was withholding unfavorable information. Officials at the State Department became so concerned about the problem during August that they even began to consider setting up their own private news service to give advance warning of South Vietnamese failures and possibly critical news reporting. "A fast, unclassified reportorial cable on all military actions of sufficient dimensions to attract wire service coverage . . . is a priority policy need as well as a public affairs requirement," they told Ambassador Taylor. The cable would include an assessment of the reliability of casualty figures and other statistics and a preliminary analysis of the significance of whatever event it described.[24]

The proposal sought much more than the U.S. mission could deliver. Communications circuits within South Vietnam were already overloaded, and the Military Assistance Command lacked the personnel to sustain an effort of the size envisioned. Even if resources had been available, officials could never have gathered and transmitted correct information quickly enough to refute news stories composed from partial impressions by reporters who worked against deadlines. In the end, the State Department had to settle for a much less comprehensive system. When President Johnson realized during September that Communist successes and South Vietnamese losses were continuing to receive greater press coverage in the United States than South Vietnamese successes and enemy losses, the best he could do was order Ambassador Taylor to submit a weekly report on the

[22] Msg, Westmoreland MAC 3099 to Harris, 19 Jun 64.

[23] Msg, Wheeler JCS 3635 to Westmoreland, 24 Jul 64, Westmoreland Papers, CMH.

[24] Msg, State 478 to Saigon, 19 Aug 64, FAIM/IR. See also Draft Msg, Joint State/Defense/USIA to Saigon, 29 Dec 64, DDI News from Vietnam file.

military and political situation in addition to already required monthly reports.[25]

The Johnson administration's attempts to use the news media to communicate with Hanoi also figured into official concern about negative news reporting. The White House knew that the North Vietnamese read American newspapers and considered news stories an excellent means for signaling American intentions to the enemy. Yet the news media partially offset the effect by questioning every aspect of the American approach to the war that appeared open to doubt. The result was sometimes unfortunate. On one occasion North Vietnam's prime minister, Pham Van Dong, quoted American news reports, in particular articles by Walter Lippmann, to demonstrate that South Vietnam was falling apart. He even turned a favorite cliche of the period to his own advantage by concluding that there was "no light at the end of the tunnel" for the United States.[26]

The administration attempted to remedy the problem by carefully orchestrating each signal to North Vietnam for maximum effect. During June and July 1964, for example, it began an involved series of signals to Hanoi designed to forestall the enemy's presumed aggressive intentions toward all of Southeast Asia. The program included officially sanctioned leaks to the press that affirmed the American will to uphold treaty commitments, a public acknowledgment that the United States was maintaining military contingency stockpiles in Thailand, and an official announcement that the U.S. Air Force was operating out of a new base at Da Nang. To correct possible misunderstandings about U.S. policy stemming from conflicting news stories, the State Department then joined with Defense and the U.S. Information Agency in circulating a directive to American public affairs officers around the world to clarify the U.S. position. The notice indicated that President Johnson had delivered a carefully drafted expression of U.S. policy at his 23 June news conference and that the statement should become the "principal source of guidance" in briefing newsmen and others on the subject. The United States intended "no rashness" and sought "no wider war," but it was "determined to use its strength to help those . . . defending themselves against terror and aggression."[27]

"March North": June–August 1964

The attempt to create an aura of resolute nonbelligerence around the United States' Southeast Asian involvement seemed at first to work but fell into confusion when Premier Nguyen Khanh intervened. Khanh knew that he lacked the support of the South Vietnamese people and that the senior civilian and military

[25] Memo, Bankson for Sylvester, 23 Dec 64, and Memo, Bankson to Brig Gen G. C. Fogle, JCS, 20 Dec 64, both in DDI News from Vietnam file. Msg, Wheeler JCS 4593 to Westmoreland, 20 Sep 64, Westmoreland Papers, CMH.

[26] Msg, State 74 to Saigon, 11 Jul 64, FAIM/IR.

[27] Cir 89, Joint State/Defense/USIA, 6 Jul 64, CMH files. See also *Pentagon Papers*, 3: 145, 182.

members of his government had yet to develop unity of purpose in their fight against the Viet Cong. He decided to rally his people to his side with a public relations campaign that ran counter to the one the United States was waging.[28]

General Khanh

Khanh made his move on 14 July. Summoning *New York Herald-Tribune* correspondent Beverly Deepe to Da Lat, he told her that the Chinese had moved a regiment of their troops into North Vietnam and that the North Vietnamese had transferred three battalions—1,800 men—to South Vietnam. That, he said, constituted an "overt invasion." To reinforce the point, he had the *Saigon Post* publish an editorial asserting that both the South Vietnamese commander of the I Corps Tactical Zone, Lt. Gen. Nguyen Chanh Thi, and the MACV senior adviser in the region, Col. John H. Wohner, admitted that the enemy was moving organized military units into South Vietnam.[29]

With North Vietnamese escalation established, Khanh carried his campaign into its second phase. The United States had long been concerned about the lack of cohesion within his government and had been pressing him to sponsor a public rally where all the major figures of his regime could line up in the manner of the Russian politburo on May Day to demonstrate their solidarity. He now accepted the suggestion, adding a fillip of his own. At a 20 July rally in Saigon marking the anniversary of the signing of the Geneva Agreements, Khanh refused to deliver the low-key speech the U.S. mission had sought and instead shocked the Americans by criticizing the slowness of U.S. tactics in defeating the enemy. Avowing that his people demanded offensive operations against the Communist heartland, he then led the crowd in shouting, "To the North! To the North!"[30]

Two days later the commander of the South Vietnamese Air Force, Air Vice Marshal Nguyen Cao Ky, renewed the chant. At ceremonies opening the air base at Bien Hoa to the press, Ky kept his remarks guarded until a reporter asked a

[28] Msg, Saigon 414 to State, 14 Aug 64, sub: U.S. Mission Monthly Report for July 1964, and Msg, Saigon 506 to State, 21 Aug 64, both in Monthly Summaries file, CMH.

[29] Msg, Saigon 109 to State, 15 Jul 64, CMH files; Beverly Deepe, "N. Viet Troops Cross Border, U.S. Aides Say," *New York Herald-Tribune*, 14 Jul 64.

[30] For background of the Khanh speech, see Interv, author with Barry Zorthian, 20 Feb 76, CMH files. See also Peter Grose, "Khanh Leads Cry for War on North at Saigon Rally," *New York Times*, 20 Jul 64; "Two Generals," *New York Times*, 26 Jul 64.

leading question; then he launched into a series of startling revelations. Declaring that the only way to counter Communist aggression was to retaliate, he said that South Vietnam had for the previous three years sent sabotage missions into North Vietnam and that his pilots were at that moment training for possible large-scale attacks. The local press picked up his remarks, embellishing them with a drumfire of editorial comment that favored "March North" and criticized the United States for its soft, indecisive policies.[31]

Concerned that both the North Vietnamese and the American public might interpret Khanh's remarks as an indication that the United States was escalating the war, the MACV Office of Information acted to preserve President Johnson's policy of resolute nonbelligerency shortly after the premier made his first allegations. When reporters turned to the Military Assistance Command for an explanation of Khanh's remarks, the Office of Information convened a background briefing that focused on the character of enemy infiltration. MACV Chief of Intelligence Brig. Gen. Richard G. Stilwell acknowledged that individual North Vietnamese soldiers had indeed been entering South Vietnam for months. Although he voiced alarm at their presence in the country, Stilwell told newsmen that no evidence existed to imply that those soldiers were operating as organized units.[32]

The prompt rebuttal had the desired effect. When the State Department complained to Ambassador Taylor that the Military Assistance Command had failed to coordinate with Washington agencies before holding the backgrounder, Taylor retorted that from what he had seen the briefing had succeeded. The Associated Press' version of the story seemed reasonable, and Reuters' copy also appeared balanced. From then on, indeed, U.S. newsmen paid little attention to Khanh's assertion that North Vietnam was invading the South with organized military units.[33]

The U.S. embassy's attempts to bring Khanh to heel were less successful. Shortly after Ky made his speech, Ambassador Taylor; his deputy, U. Alexis Johnson; and Zorthian met with Khanh to request both a clarification of South Vietnamese intentions and a repudiation of Ky's remarks. Khanh responded vaguely that there were no basic differences between the American and South Vietnamese positions on the subject. He then baldly avowed that while "March North" might not be U.S. policy, it was South Vietnamese policy nonetheless. The session had hardly ended before a nameless South Vietnamese insider, probably at Khanh's behest, leaked what had transpired in the meeting to the Saigon correspondents.[34]

Taylor and Johnson did extract a promise from Khanh that the Ministry of

[31] Interv, author with Zorthian, 20 Feb 76; Peter Grose, "Sabotage Raids Confirmed by Saigon Aide," *New York Times*, 23 Jul 64. See also *Facts on File* (New York: Facts on File, Inc., 1965), 24: 250); Msg, Saigon 193 to State, 23 Jul 64; Msg, Saigon 414 to State, 14 Aug 64, sub: Mission Monthly Report for July 1964. Latter two in Monthly Summaries file, CMH.

[32] Msg, Saigon 109 to State, 15 Jul 64.

[33] Ibid. See also Msg, State 234 to Saigon, 24 Jul 64, FAIM/IR.

[34] Msg, Saigon 193 to State, 23 Jul 64; Msg, Saigon 232 to State, 27 Jul 64, CMH files; Peter Grose, "Khanh-Taylor Clash," *New York Times*, 24 Jul 64; Interv, author with Barry Zorthian, 10 Dec 75, CMH files.

Defense would repudiate Ky's remarks, but what resulted was also far from satisfactory. Two hours after the meeting ended, a South Vietnamese Army major appeared before Zorthian with a draft statement reading, "A general may have declared to a few newsmen that South Vietnam has the capability for retaliatory attacks against the military forces of North Vietnam, and may have expressed the opinion that in order to end the aggressive war started by the Communists it is necessary to envisage military actions right on enemy territory. The Defense Ministry affirms that the above pronouncements are only the personal opinion of the military leader and do not reflect the Ministry's thinking on the matter."[35]

The statement fell short of U.S. wishes by failing to soften Ky's semiofficial confirmation of covert attacks against the North and by leaving questions of possible Chinese involvement unanswered. Zorthian then proposed an addition to the communique to assert that the general's remarks did not "refer to operations of the armed forces of the Vietnamese government" and to affirm that "no confirmed action by Chinese communist aircraft against Vietnamese aircraft" had occurred during the raids.

The major told Zorthian that the minister of defense would probably approve the suggestion but returned later with word that the statement would have to stand as originally drafted. Zorthian immediately contacted the South Vietnamese deputy minister of information to tell him that the additions were "important to meet questions that will be raised in U.S. opinion," only to learn that Khanh himself had issued direct orders barring any change. When he requested a postponement to give Taylor time to take up the issue with Khanh himself, he met with the cold response that the government of South Vietnam intended to handle the situation "in its own way." The deputy minister later told Zorthian in confidence that Khanh and Ky were merely answering their public's call for action. Having stated that they were moving to counter North Vietnamese aggression, they were in no position to deny it. They were also unwilling to appear to be puppets who reversed their statements at the whim of the United States.

Although Khanh refused to repudiate Ky's statement, he did reluctantly nod to U.S. pressure. At the end of the month he issued a clarification suggesting that he had never envisioned a massive military assault against North Vietnam. That said, he nevertheless added enigmatically that his army would continue its energetic efforts to remove the Communist scourge, leaving U.S. officials wondering whether the idea of an invasion had actually lost all appeal to him.[36]

American policy makers could do little more to curb Khanh, but they did manage to turn to good purpose the uncertainty his statement caused. On 2 August and 4 August North Vietnamese gunboats launched attacks against the U.S. destroyers *Maddox* and *Turner Joy* in the Gulf of Tonkin.[37] When President Johnson

[35] Unless otherwise indicated this section is based on Msg, Saigon 193 to State, 23 Jul 64.

[36] Msg, Saigon 414 to State, 14 Aug 64, sub: U.S. Mission Monthly Report for July 1964.

[37] Neither the press nor the Johnson administration doubted the validity of the attacks at the time. For a description of the incidents and the controversies that later developed, see Edward J. Marolda and Oscar P. Fitzgerald, *The United States Navy and the Vietnam Conflict: From Assistance to Combat,*

Secretary McNamara Briefs the Press

retaliated by authorizing a counterattack against enemy naval bases from which the attacks had originated, the Communists granted an exclusive interview to Australian journalist Wilfred Burchett. In it they avowed that they would tolerate no more attacks on their territory. One hundred thousand veterans of the war with France stood poised above the 17th Parallel, they told Burchett. If the United States abrogated that boundary with further aggressive bombing, any excuse for restraining those troops would be gone. An area-wide conflict involving both Laos and South Vietnam would result. The State Department considered the interview a Communist ploy to deter further bombing raids against the North and doubted that Hanoi intended to invade South Vietnam directly. Yet on the theory that the North Vietnamese would seize any opportunity to make the United States appear the aggressor in Southeast Asia, the U.S. Information Agency had already instructed its worldwide information apparatus to turn "March North" into an American propaganda theme. When foreign newsmen asked U.S. public affairs officers whether the United States would support South Vietnamese attacks against North Vietnam, the agency responded that since North Vietnam had for years provoked the people of South Vietnam, it was hardly surprising that South Vietnam's leaders should talk of military operations against the North. Such action on their part would not be "aggression" but "understandable retaliation for years of cruel and vicious attacks . . . indisputably directed and supported by the communist authorities in North Vietnam."[38]

The United States thus succeeded in making Khanh's rhetoric serve some purpose, but it was never able to generate the sort of favorable news coverage it had envisioned when it adopted its new public affairs policies. The contradictions endemic to South Vietnam were the reason. The Honolulu Conference had warned that victories over the enemy and stability in Saigon were the only remedy for problems with the press, yet as 1964 progressed the South Vietnamese became so mired in political intrigues that neither victory nor stability seemed attainable. News reports from the field merely reflected that fact.

1959–1965 (Washington, D.C.: Naval Historical Center, Government Printing Office, 1986), pp. 437–62.

[38] Msg, Tokyo 547 to State, DAIN 379034, 13 Aug 64, Army Staff Communications Center files, Army War College; Msg, State 462 to Saigon, 17 Aug 64, FAIM/IR. Quote from USIA Talking Paper No. 21 to All Principal USIS Posts, CA-339, 4 Aug 64, CMH files.

At first there seemed some hope that the Johnson administration might gain the free hand in Southeast Asia that it had long sought. The Gulf of Tonkin incident gave the president the opportunity he needed to win congressional approval for a resolution supporting his policies in South Vietnam, and the subsequent reprisals against enemy naval bases galvanized the American public. Only a few weeks before the attack 58 percent of the voters polled by Louis Harris had said they disapproved of Johnson's handling of the war; suddenly 85 percent said they favored it.[39]

Events nevertheless shattered the mood. On 7 August Premier Khanh took advantage of the confusion surrounding the Gulf of Tonkin crisis to declare a state of emergency. The South Vietnamese people reacted favorably at first, perhaps hoping that U.S. raids would be the prelude to a genuine march north, but within a week they realized that the United States intended only limited action. Discontent grew, exacerbated by Khanh, who chose that moment to promulgate a new constitution and to have himself declared president, all without preparing public opinion. By 17 August Buddhist and student opposition to the arrangement had spread, causing civil disturbances in most of South Vietnam's major population centers. Urban discontent overflowed into rural areas, where the continuing influence of the Catholic minority over South Vietnam's political life once more became an issue. Violent rioting ensued, bringing so much pressure to bear upon Khanh that he finally decided to annul his constitution and resign the presidency.[40]

On 26 August Khanh retired to the resort town of Da Lat north of Saigon, but no new leader emerged to take his place. Chaos deepened. Fighting broke out between Buddhists and Catholics in Saigon as one caretaker government succeeded another. By 3 September Taylor, Westmoreland, and several South Vietnamese military officers had prevailed upon Khanh to return to Saigon to resume the premiership, but that step brought only a semblance of order. Ten days later a group of dissident generals staged a coup that aborted at the last moment, when a number of powerful young officers rallied to Khanh.[41]

The picture of South Vietnamese demoralization that emerged during those days raised serious doubts among American officials about the willingness of the South Vietnamese people to continue the fight against the Communists. Shortly after arriving in Saigon, Ambassador Taylor had recommended that the United States wait for Khanh to impose a measure of stability on the country before inaugurating any concerted attempt to pressure the North with direct American attacks. With instability growing and the enemy becoming bolder, he now began

[39] Louis Harris, "Public Solidly Behind Johnson on Vietnam," *Los Angeles Times*, 10 Aug 64; Harris, *Anguish of Change*, p. 56.

[40] This section is based on Msg, Saigon 872 to State, 16 Sep 64, sub: U.S. Mission Monthly Report for August 1964, Monthly Summaries file, CMH.

[41] *Pentagon Papers*, 2: 334; Msg, Saigon 1125 to State, 13 Oct 64, sub: U.S. Mission Monthly Report for September 1964, Monthly Summaries file, CMH.

to advocate pressures against the North as a way to stiffen the South Vietnamese people's morale and gain time for Khanh.[42]

The Saigon correspondents and the rest of the U.S. press were far more dubious about the efficacy of further American action. Shortly after Khanh made his first moves, *Newsweek* commented that during six months of rule the premier had managed to maintain his own power but almost nothing else. By 20 August the crisis was deepening and the *New York Times* was asking rhetorically whether there could be any hope for a stable government in South Vietnam. Shortly thereafter, Stanley Karnow of the *Saturday Evening Post* published a long article scoring the spreading chaos, the bureaucratic mentality of the country's military and civilian leaders, and the fact that inept officers were ruining the work of "tough" South Vietnamese troops. Karnow quoted an angry U.S. Army adviser to make the point that "We've thrown in helicopters, aircraft, artillery, and with each new machine the ante goes up. Nobody wants to fight because some new gadget is supposed to be coming along to win the war painlessly." Finally, in early September, Peter Kalischer of CBS News began comparing South Vietnam to Humpty Dumpty, and Peter Grose of the *New York Times* commented that Khanh had failed because Asian countries in disarray were obviously unsuited to "neat, American-style" solutions.[43]

As the crisis continued, it began to affect the Saigon correspondents' opinion of "maximum candor." The newsmen had sided openly with the Buddhists during the disturbances preceding the fall of Diem, but by mid-1964 they had become convinced that all sides were trying to manipulate them. The pronouncements of the South Vietnamese government had always been suspect, *Newsweek* asserted in an article summarizing their objections. Yet the Buddhists and students were trying just as hard to use the American news media. Meanwhile, the State and the Defense Departments continued to insist that reporters were far too defeatist in their commentaries on the war. Truth under circumstances of that sort, the magazine concluded, could only be illusory.[44]

Newsweek went on to catalog the Saigon correspondents' reaction to the Military Assistance Command's expansion of facilities for the press. Few reporters doubted that the command's liberalized information policies and sponsorship of trips to South Vietnam by stateside journalists were anything less than attempts to erase the feeling of suspicion that had hampered official relationships with newsmen. Yet the effort, "Operation Candor," appeared to be producing the opposite effect. Jack Raymond of the *New York Times* believed a ten-day visit too short for an outsider to acquire "a feel" for the war, and Malcolm Browne asserted that "free junkets" of the sort being offered the press actually created a psychological obli-

[42] Msg, Saigon 872 to State, 16 Sep 64, sub: U.S. Mission Monthly Report for August 1964; *Pentagon Papers*, 2: 335.

[43] "South Vietnam: First-half Report," *Newsweek*, 10 Aug 64; Peter Grose, "Pace of Fighting Holds in Vietnam," *New York Times*, 20 Aug 64; Stanley Karnow, "This Is Our Enemy," *Saturday Evening Post*, 22 Aug 64; Kalischer's and Grose's remarks are quoted in "The Viet Beat," *Newsweek*, 7 Sep 64.

[44] "The Viet Beat."

gation on the recipient to follow the official line, no matter how hard he tried to be objective. In the same way, *Newsweek* continued, reporters were convinced that the dedication of but a single helicopter by MACV's Office of Information for use by the press had actually served to channel news reporting and to limit the ability of newsmen to move freely about the country. In the past, a space-available policy had given newsmen access to rides on any official aircraft that had room for them. Under the new policy only a few seats on a single helicopter were available. Resident reporters might agree that the difficulties were mainly the result of poor judgment and inexperience on the part of information officers, but *Newsweek* could only conclude that "Press facilities have not been expanded. They have been formalized."[45]

Although the press complained, it still benefited greatly from the relaxation of tensions and the increased flow of information that the new policies provided. When the South Vietnamese government closed commercial cable facilities during the September coup, the Military Assistance Command and the U.S. embassy once more arranged for newsmen to file their copy through official channels. General Westmoreland even authorized a special B-57 flight to the Philippines to move a huge sack of news dispatches and film around bureaucratic obstacles imposed in Saigon.[46]

Despite their protestations to the contrary, the reporters had very little trouble moving about the country. Time and again they obtained telling stories despite the deteriorating political conditions. By October they were so attuned to what was happening that their dispatches began to resemble the classified assessments the U.S. mission was sending back to Washington. On 3 October, for example, Zorthian complained to Ambassador Taylor that items from U.S. newspapers available to the mission gave the overall impression of rapid decay in South Vietnam, indicating that U.S. efforts there would soon reach an impasse. A comment by the *Washington Star* was typical. "Each day, the situation assumes a new dimension of chaos. Each day, the chance of restoring a minimum of effective government becomes dimmer." Yet shortly after Zorthian made his comment, the U.S. mission transmitted to Washington a Monthly Assessment of Military Activity that said almost the same thing. "The month of September was characterized by political turbulence, uncertainty as to the future of the Government of South Vietnam, and confusion resulting from a lack of decisive and firm central control. Viet Cong incidents increased. Government military operations decreased in all categories excepting small unit operations. More aircraft were lost or damaged by enemy action. Additionally, the manpower picture continued to be unsatisfactory and there was an increase in combat units rated ineffective. In summary, there is little or no evidence of overall progress during the month."[47]

[45] Ibid.

[46] Memo, Rodger Bankson for Arthur Sylvester, 19 Jan 65, DDI News from Vietnam file.

[47] Memo, Barry Zorthian for Ambassador Taylor, 3 Oct 64, sub: Evaluation of Media Coverage for the Week of September 27 to October 3, 1964, Westmoreland History; MACV, Monthly Assessment of Military Activity, September 1964, 8 Oct 64. Both in CMH files.

The reporters had taken their story not from documents and cables leaked by disgruntled officials but from their own experience of the war. By November, election time in the United States, many members of the American public had reached the same conclusions. Although the electorate swept Johnson into office with an unprecedented 61 percent of its vote, no more than 42 percent of the people polled by Louis Harris gave the president high marks for his handling of the war. By December that support had dwindled to 38 percent. Maximum candor may have succeeded in improving the quality of war reporting, but it had obviously failed to achieve its primary object: the creation of a climate of opinion favorable to the Johnson administration's ends in South Vietnam.[48]

[48] Harris, *Anguish of Change*, p. 57.

5

Keeping the Options Open

The final months of 1964 brought no end to the chaos in South Vietnam. Buddhist and student unrest continued. An abortive coup occurred on 13 September, and Montagnard tribesmen revolted in the western highlands six days later. The *New Yorker*'s Southeast Asia correspondent, Robert Shaplen, concluded that dissension and self-destruction seemed permanent features of the country's landscape. "Individual is pitted against individual and group against group," he wrote, "the motivating impulse in nearly every case being greed for money or power, or the desire for self-preservation rather than preservation of the country in time of war."[1]

Dean Rusk shared Shaplen's assessment. Although the secretary of state held out hope that some remedy existed for South Vietnam's problems, he observed to Ambassador Taylor that "The picture of petty bickering among Vietnamese leaders has created an appalling impression abroad. . . . We have tried to exercise the greatest patience . . . but patience and understanding are being drained away. . . . The American people are already beginning to ask what are we supporting and why."[2]

Taylor and Westmoreland attempted to communicate Rusk's concern to South Vietnamese officials, but their remonstrances had little effect. Athough a few halting signs of progress appeared during October, when General Khanh ratified a provisional charter of government and reestablished civilian rule under a new prime minister, Tran Van Huong, the decline at best slowed. Exploiting South Vietnam's difficulties politically and militarily, the enemy continued to expand and strengthen his areas of control throughout the country.[3]

[1] Robert Shaplen, "Letter From Saigon," *New Yorker*, 19 Sep 64, p. 183.

[2] Msg, State 654 to Saigon, for Ambassador from the Secretary, 14 Sep 64, FAIM/IR.

[3] The attempts at communication are in Memo, Westmoreland for Taylor, 1 Oct 64, sub: Your Memorandum of 16 September re: Talking Paper on U.S. Attitude, Westmoreland History, bk. 8, tab 52, CMH. For the status of the situation in October, see Msg, Saigon 1495 to State, DAIN 461510, 13 Nov 64, sub: U.S. Mission Monthly Report for October 1964, Army Staff Communications Center files, Army War College.

As the full extent of the country's demoralization became apparent, the Departments of State and Defense began seriously to consider the possibility that the Saigon regime might lose all ability to rule. Weighing a range of options from total withdrawal to outright American intervention, the analysts concluded that a program of direct, gradually increasing air attacks against North Vietnam held the best chance of improving the situation. Besides boosting South Vietnamese morale and limiting North Vietnam's support for the Viet Cong, a campaign of that sort seemed likely to strengthen the American negotiating position while demonstrating to the world that the United States stood by its commitments.[4]

President Johnson was less certain. Besides doubting that air strikes would be effective without a solid government in South Vietnam, he had an upcoming presidential election in the United States to consider. Since his platform stressed restraint while Republican candidate Barry Goldwater was arguing for U.S. escalation of the war, he had little wish to appear to be conceding the issue to his opponent. There was also American public opinion and his domestic agenda—his so-called Great Society—to consider. Surveys indicated that one out of four Americans was oblivious to the war in South Vietnam and that many more were unprepared for the difficult choices in the offing. In so uncertain an environment, if the president moved against North Vietnam the public might either turn on him, destroying his political future and the prospects for his domestic programs, or push so vehemently for all-out war that few options remained. Choosing a middle course, Johnson ratified the concept of the strikes in principle but shied away from potentially irrevocable action. Although prepared to take large risks if necessary, he intended to keep his options open.[5]

Mutual Cooperation

The task of preparing the American public fell in great part to Barry Zorthian and other information officers in Saigon. Zorthian believed that openness was a practical necessity to protect official credibility. In the absence of fixed guidelines that would almost certainly have followed a firm presidential commitment to action, he plied the press with far more information than might otherwise have been possible. At times, he told his associates, correspondents would land in areas information officers considered undesirable. Unless military security was at issue, they were to respond to the reporter's self-determined needs. Only in that way

[4] Msg, DIA to CINCPAC, DIAAP-2F 70205 [Sep 64], Courses of Action 1964 file, and Msg, CINCPAC to JCS, CMIN 95041, for Wheeler from Sharp, 26 Sep 64, General Estimates of the Situation file, both in CMH; *Pentagon Papers*, 2: 328–30; Luttichau, U.S. Army Role in the Vietnam Conflict, ch. 4.

[5] *Pentagon Papers*, 3: 193–95; Luttichau, U.S. Army Role in the Vietnam Conflict, ch. 4; NSAM 314, 10 Sep 64, doc. 195, *Pentagon Papers*, 3: 565; Council on Foreign Relations, Public Opinion Study, Nov 64, quoted by Richard Harwood, "Lessons From the Pentagon Papers," reprinted in Laura Babb, ed., *Of the Press, By the Press, For the Press (and Others, Too)* (Washington, D.C.: The Washington Post Company, 1974), p. 84.

would they establish the mutual cooperation that would make reporters willing to cover subjects officials believed important.[6]

Newsmen's suspicion of every new MACV policy that seemed to curtail the freedom of the press and the conviction of many military officers that reporters already had far too much freedom nevertheless complicated the information officers' task. Although tension between the two groups eased as maximum candor became the basic policy of the command, minor irritations continued to threaten the type of mutual cooperation Zorthian had in mind.

During 1962 and 1963, when the press corps in Saigon had rarely numbered more than twelve, information officers had usually briefed reporters on upcoming South Vietnamese military operations. At the end of 1964, however, with more than forty newsmen serving in Saigon, the practice no longer seemed advisable. Yet when the Military Assistance Command, out of concern for military security, announced that it would end the briefings, it immediately kindled resentment among reporters who had come to depend upon the notification. Norman Sklarewitz of *U.S. News & World Report*, for one, charged that while the U.S. command would usually provide transportation if a correspondent asked to go to a specific area, it would rarely assist newsmen as it had in the past to accompany preplanned operations. The result, the reporter said, was far fewer eyewitness opportunities.[7]

The senior MACV adviser in South Vietnam's III Corps Tactical Zone, Col. Jasper J. Wilson, disagreed. Reporters, he said, continued to concentrate on failures, mistakes, rumors, and gossip and should on that account have been cut off long ago. Yet they still had the run of the region around Saigon. Their arrival was generally known in advance elsewhere in South Vietnam, but III Corps was so close to the capital that the press could roam at will, hitchhiking on any available U.S. Army helicopter. Reporters often learned of what was happening from U.S. pilots and South Vietnamese soldiers, "(presumably for favors or other more direct benefits)," long before official dispatches could reach higher headquarters.[8]

Information officers were sympathetic to complaints of that sort but tended to side with the reporters. Sklarewitz's contentions to the contrary, no one (Zorthian, in particular) was interested in restricting eyewitness opportunities. Wilson and other advisers might experience some inconvenience and news stories unflattering to the South Vietnamese might appear, but shortcomings were bound to surface. Since the truth was seldom as bad as hearsay, getting it into the open where it could awaken the American public seemed the correct thing to do.[9]

[6] Msg, Saigon 4014 to State, 2 Jun 65, FAIM/IR; Interv, author with Barry Zorthian, 13 Apr 76, CMH files; Msg, Saigon 1776 to State, 10 Dec 64, Westmoreland History, bk. 11, tab 12, CMH.

[7] U.S. Congress, Senate, Committee on Foreign Relations, *Hearings, News Policies in Vietnam*, 88th Cong., 2d sess., 31 Aug 66, p. 68; Interv, author with Rodger Bankson, 28 Aug 75, CMH files; Norman Sklarewitz, "Official Obstacles to Vietnam War Coverage Growing," *Overseas Press Bulletin* 18 (12 November 1964).

[8] Col Jasper J. Wilson, Report to the Chief of Staff of the Army on the Vietnam War [1965], Special Forces file for 1964, CMH.

[9] This section is based on Intervs, author with Zorthian, 13 Apr 76, and with Bankson, 28 Aug 75.

Ambassador Taylor *(center) and other top U.S. officials meet with correspondents from* Life.

Although information officers wanted the press to seek out news on its own, they remained convinced that reporters would neglect broad perspectives to concentrate on problems. Rather than leave newsmen entirely to themselves, therefore, they used all the means at their disposal, from off-the-record intelligence briefings to interviews for direct quotation, to acquaint the press and the American public with the official point of view.

Zorthian's reaction to a request in early November by *Life* magazine for an on-the-record interview by four of its correspondents with top mission officials revealed the way the information officers worked. *Life*'s editors wanted to devote the better part of an issue to the war and needed a detailed, high-level summary of how things were going. Zorthian supported the idea, arguing that the interview would allow mission spokesmen to state their point of view far more coherently than in the past. The occasion also provided a rare opportunity to publish extended official quotations without editorial misreadings and imprecise paraphrasing.[10]

[10] Interv, author with Zorthian, 13 Apr 76. See also "A *Life* Panel: The Lowdown From the Top U.S. Command in Saigon," *Life*, 27 Nov 64, p. 46.

The panel met on 14 November 1964 with *Life* correspondents Lee Hall, Marshall Smith, Robert Morse, and John Flynn interviewing Ambassador Taylor, Deputy Ambassador U. Alexis Johnson, General Westmoreland, AID Director James S. Killen, and Zorthian. Taylor began the session by defining the American purpose in South Vietnam. Calling attention to South Vietnamese responsibility for achieving victory, he sketched the social and political dimensions of the problems confronting the United States and pointed out that progress remained difficult because North Vietnam continued to reinforce the Viet Cong. Asked whether the South Vietnamese were losing the war, he set the theme for the rest of the session by responding gravely that the issue was very much in doubt and that the victor would be the one with the ultimate will to win.[11]

The other members of the panel elaborated on Taylor's leads, stressing that American involvement in South Vietnam had to be seen as a whole. In remarks later edited out of the published version of the interview, Zorthian applied the principle to the press. Newsmen had to explain the war properly to the people of the world. Too often, he said, reporters attempted to judge events in South Vietnam by the standards of the past: daily victories and defeats, ground won and lost, and statistical measures of progress. Those indicators might be valid in a conventional war, but they failed to give more than a partial picture of what was happening in South Vietnam.[12]

All of the panel's official participants mentioned areas of progress, but only General Westmoreland was forthrightly optimistic. Citing the types of statistics Zorthian wanted the press to avoid, he asserted that South Vietnamese regular and paramilitary forces were fighting well and that the regulars were maintaining "morale, esprit, and pretty good discipline." Political instability had indeed led to a rise in desertion rates, but those figures were beginning to drop. Meanwhile, government forces had maintained an average of sixty-eight battalion-size operations per week with the number continuing to increase. Over six thousand South Vietnamese soldiers had died in battle during the previous year, but the enemy had incurred twice that number of casualties—a statistic all the more impressive because enemy soldiers who died in artillery and air strikes and of wounds were not included. "Leadership, of course, is a problem in any service," Westmoreland said. "I am impressed in general, however, with the senior officers of the regular forces, particularly at the present time when General Khanh has, I think, appointed his best and most capable officers, and placed them in the key positions. . . . It is absolutely inconceivable to me that the Viet Cong could ever militarily defeat the armed forces of South Vietnam."[13]

Westmoreland's remarks contradicted official assessments that characterized South Vietnamese military operations as unproductive reactions to enemy initia-

[11] "A *Life* Panel," p. 46.

[12] Transcript, *Life* Symposium With Mission Council Members, 14 Nov 64, Westmoreland History, bk. 10, tab 1, p. 44, CMH. The quote was deleted by *Life*'s editors. See also Interv, author with Barry Zorthian, 8 May 76, CMH files.

[13] "A *Life* Panel," p. 46.

tives and that criticized South Vietnamese officers' almost total absorption in politics to the detriment of the war. His views were also considerably more optimistic than those he had set forth in a classified memorandum to Ambassador Taylor. In it he asserted that the South Vietnamese government took U.S. assistance for granted and failed to consult with the United States "prior to making political and military decisions of major impact on governmental operations and pacification. This behavior might be acceptable if the Vietnamese were operating effectively or, at the very minimum, gave evidence of a real desire to do what was required to win the war. The fact is, however, that the conduct of the government is characterized by inefficiency, corruption, disinterest and lack of motivation. The GVN is not winning the war."[14]

If Westmoreland's statements lacked total candor, they nevertheless reflected what he considered his responsibility as chief U.S. adviser to the South Vietnamese armed forces. Charged with encouraging those allies, Westmoreland believed that he stood a better chance of influencing their conduct by praising their accomplishments than by making morale-destroying pessimistic statements likely to be reprinted around the world. He also tended to doubt that the situation was as urgent as many American policy makers believed. Taylor and McNamara agreed that the South Vietnamese government was about to fall apart, but Westmoreland and his officers were convinced time still remained. Khanh's installation of a civilian council, they noted, had given the government a measure of legitimacy. The army was likewise gaining in organization and experience, and the South Vietnamese people appeared to be showing signs of disillusionment with the Viet Cong.[15]

After Westmoreland and the others had concluded their remarks, Ambassador Taylor made certain that no one missed the point of the session. In a statement that could have served as a summary of the Johnson administration's public stance at that time, he stressed that the United States had a vital stake in the war. A great battle had been joined, he said. Although it was too early to say when or how the fighting would end, it was no time to take counsel in fear or to sell the United States short.[16]

The session had the desired effect. *Life* published the interview on 27 November along with articles that viewed the war through the eyes of a USAID adviser in the delta, an Army Special Forces officer in the Central Highlands, and a U.S. Navy team working along the coast. Each piece followed the lines laid down by

[14] Memo, Westmoreland for Ambassador Taylor, 31 Oct 64, sub: The U.S. Posture Toward Emerging GVN, Westmoreland History, bk. 9, tab 24, CMH. See also NSC Working Group on Vietnam, Intelligence Assessment: The Situation in South Vietnam, 13 Nov 64, Chron files, CMH.

[15] All U.S. Army officers in Vietnam were instructed to encourage the South Vietnamese. See Gen Harold K. Johnson, CSA, Report to the Joint Chiefs of Staff, sub: Trip to Vietnam, 8–12 December 1964, MACV Records file 206-02, Historian's Background Papers (1965), copy in CMH files. For Taylor's and McNamara's views, see W. P. Bundy, Notes of an Executive Committee Meeting, 27 Nov 64, and Memo, Westmoreland for Taylor, 24 Nov 64, sub: Assessment of the Military Situation, both in Chron files, CMH. See also Maj William E. LeGro, USA, Notes From a Trip to SEA, Nov 64, Thailand file for 1964, CMH.

[16] "A *Life* Panel," p. 52.

the panel, balancing descriptions of the frustrations and tensions afflicting the American involvement in South Vietnam with portrayals of men who believed strongly in the importance of what they were doing. Although the articles were problem oriented, the Johnson administration could have asked for nothing better. One even quoted a verse from Rudyard Kipling to reemphasize Taylor's request for public patience:

> It is not good for the Christian's health
> to hustle the Asian brown,
> for the Christian riles and the Asian smiles
> and he weareth the Christian down.[17]

Preparing the Public, October–November 1964

Although the interview never directly addressed the question of whether the United States should bomb the North, it did suggest in passing that North Vietnamese infiltration was on the rise. By so doing, it fitted into a whole series of ongoing official revelations designed to suggest that the enemy was becoming more aggressive and that the United States might have to escalate the war.

The process of preparing the public for that possibility was, indeed, already well advanced by the time *Life* published the interview. American policy makers had always reasoned that North Vietnamese efforts to reinforce the Viet Cong constituted an overt act of aggression and had long contemplated using that fact to justify countermeasures. From mid-October, as part of their campaign to ready public opinion, Zorthian and MACV's Office of Information had quietly briefed the Saigon correspondents on the fact that more and more North Vietnamese soldiers were entering South Vietnam. Information officers had released no numbers because hard figures were unavailable, but by 31 October 1964, the Military Assistance Command had completed a study suggesting that infiltration was proceeding at a pace far faster than anyone had suspected.[18]

Ambassador Taylor urged public release of the report almost as soon as the command completed it, but the Johnson administration temporized. Besides fearing that the study might embarrass the president on the eve of the elections by revealing a failure properly to estimate infiltration rates, officials such as William P. Bundy remained unwilling to do anything that might generate further South Vietnamese pressures for a march north. Thus, when questions about the strength of the report's conclusions began to arise in the intelligence community, a more

[17] Marshall Smith, "Junk Navy Has Quietly Perilous Mission," *Life*, 27 Nov 64, p. 38.

[18] Memo, ASD (ISA) for Henry S. Rowen, 25 Jan 64, sub: Interim Report on Vietnam, ISA file 092VN, 20A717, box 64, WNRC. Msg, Saigon 1070 to State, DAIN 428179, 9 Oct 64; Msg, Saigon 1135 to State, DAIN 432186, 14 Oct 64; Memo, JCS 2343/490 for COSA, 13 Nov 64, sub: Analysis of the COMUSMACV Infiltration Study, Viet Cong Forces, RVN, Dated 31 October 1964, VC Infiltration file, CMH.

cautious approach than Taylor's prevailed. While teams of U.S. Intelligence Board analysts proceeded to South Vietnam to verify MACV's facts, Zorthian added the report's major findings to his weekly briefing, allowing them to enter unobtrusively into the public domain.[19]

With the U.S. mission's assistance, Peter Grose published the report's most important details in the *New York Times* on 2 November. They appeared without official confirmation shortly after a spectacular enemy attack at Bien Hoa had claimed four American lives and six B–57 jet bombers. Although the story maintained the low profile the Johnson administration had sought, its publication, in combination with the Bien Hoa attack and an announcement that Ambassador Taylor planned to visit Washington in coming weeks, added to the sense of urgency that was beginning to surround the Vietnam issue. By mid-November, indeed, editors across the United States were calling for an end to procrastination, and Gallup polls were reporting that the U.S. public put the war at the top of the list of problems it wanted solved. Meanwhile, *Life* magazine prefaced its Mission Council interview with the assertion that "President Johnson's first order of business, now that the election is over, is to come to grips with the badly deteriorating situation in South Vietnam. Last month, more Americans were killed there than in any month since the war began. Communist troops, in the highest number ever, infiltrated across the borders."[20]

Taylor left for Washington shortly after *Life*'s statement appeared, prompting intense speculation in the press that he intended to recommend some form of limited escalation. Shortly thereafter, State Department public opinion analysts concluded that Taylor's trip, in combination with the *Life* interview and Peter Grose's revelation, had caused "considerable speculation that the administration is preparing to get the war off dead center."[21]

The American news media were hardly as prepared to support an expansion of the war as the moment made it appear, but barring some event capable of galvanizing U.S. opinion in favor of an outright attack, President Johnson was probably as close as he would ever come to having a public ready for strong action in Southeast Asia. Although newspapers such as the *New York Times* considered escalation foolhardy without a stable South Vietnamese government, others,

[19] Draft Memo, William P. Bundy, 5 Nov 64, sub: Conditions for Action and Key Actions Surrounding Any Decision, doc. 192, *Pentagon Papers*, 3: 593; Memo, ACSI for the Chief of Staff, Army, 13 Nov 64, sub: Analysis of the COMUSMACV Infiltration Study, copy in VC Infiltration file, CMH; JCS J-3 Talking Paper 183–84, 18 Dec 64, Westmoreland History, bk. 11, tab 23, CMH.

[20] Draft Memo, W. P. Bundy, 5 Nov 64, sub: Conditions for Action and Key Actions Surrounding Any Decision, *Pentagon Papers*, 3: 593. Zorthian confirmed that Grose received official help in Interv, author with Zorthian, 13 Apr 76. The article itself appeared as Peter Grose, "Vietnam Outlook Bleaker a Year After Diem's Fall," *New York Times*, 2 Nov 64. For a summary of the editorial mood in the United States, see U.S. Department of State, Bureau of Public Affairs, Vietnam and Related Topics, in American Opinion Summary, 18 Nov 64, FAIM/IR, hereafter cited as U.S. Dept of State, American Opinion Summary. See also George Gallup, "And the View From the Public," *New York Herald-Tribune*, 29 Nov 64; quote from "Alert in Vietnam," *Life*, 27 Nov 64, p. 30.

[21] Quote from U.S. Dept. of State, American Opinion Summary, 25 Nov 64, p. 1. For an example of the speculation, see David Halberstam, "Taylor Expected To Ask Expansion of the War," *New York Times*, 23 Nov 64.

including the *St. Louis Globe-Democrat,* were urging a commitment to victory whatever the consequences. Even those papers that hesitated agreed that a climax seemed imminent. Richard Egan of the *National Observer* thus quoted Senator Richard Russell of Georgia to make the point that the United States appeared to have few options. In Russell's words, "We either have to get out or take some action to help the Vietnamese. They won't help themselves." The *Kansas City Star* commented that no one expected the Johnson administration to bow to its critics and withdraw from South Vietnam.[22]

Despite the favorable public mood, President Johnson declined to take any major new action because he believed that South Vietnamese stability was basic to any course the United States might adopt. Instead, in the hope that the South Vietnamese might rally, he resorted to expedients that added little to measures already in progress. Thus when Ambassador Taylor recommended that the United States link the Saigon government's desire for air strikes against the North to U.S. requests for reform, the president adopted the idea but cautiously refused to authorize more than the first phase of the program. As a warning to the enemy as well as a pledge of American good faith, he agreed to intensify air strikes against enemy infiltration routes and to increase covert South Vietnamese naval attacks along the coast of North Vietnam. Before risking air assaults on North Vietnam itself, however, he insisted that the South Vietnamese move to reform their government. In the same way, while Johnson was willing to plan for joint U.S.-South Vietnamese reprisals against the North in response to spectacular terrorist attacks in the South, he instructed Taylor to tell Khanh that the U.S. government would never risk an expansion of hostilities until there was a regime in Saigon capable of resisting the dangers and exploiting the opportunities that would result.[23]

Planning for expanded naval operations and a widened air war over Laos began almost as soon as Taylor returned to South Vietnam, but neither campaign accomplished much. The naval program never got under way because the monsoon season prevented the small craft operations essential for covert naval attacks. An escalation of the air war over Laos did begin on 14 December, but Johnson again opted for a cautious approach, allowing only two missions of four aircraft each per week. So feeble were the resulting attacks that the North Vietnamese, unable to distinguish the new American sorties from the armed reconnaissance flights that had been occurring since May, missed the point entirely.[24]

The same indecision that hobbled the air war in Laos crippled the Johnson administration's efforts to devise a believable public relations campaign to accompany the new program. With U.S. attacks on the North almost certainly in the

[22] The *New York Times, St. Louis Globe-Democrat, Kansas City Star,* and other U.S. newspapers are quoted liberally by the American Opinion Summary, 25 Nov 64. See also Richard Egan, "Unrest in Saigon Dims Chances of Go North Plan," *National Observer,* 30 Nov 64.

[23] Taylor's suggestions are in Taylor Briefing, 27 Nov 64, sub: The Current Situation in South Vietnam—November 1964, doc. 242, *Pentagon Papers,* 3: 666. See also *Pentagon Papers,* 3: 248-51.

[24] *Pentagon Papers,* 3: 252-54; Msg, State 1394 to Saigon, DAIN 508536 [probably Jan 65].

offing, policy makers became more than ever aware of the need to enlist the support of American and world opinion. The revelation of the enemy's increasing infiltration into South Vietnam once more seemed the best approach, but the president's desire to keep a strong hold on events conflicted with his eagerness to prove North Vietnam the aggressor. Rather than bolster Khanh's call for attacks on the North by revealing the enemy's activities, the administration continued to temporize.

The mood of caution in Washington caused trouble even before Ambassador Taylor returned to South Vietnam. On 1 December Taylor, McNamara, and other policy makers met with President Johnson at the White House. At the conclusion of the meeting, when newsmen entered the oval office for a picture-taking session, one of the reporters overheard the end of a conversation between McNamara and the president about whether Taylor should say anything to the assembled correspondents. "It would be impossible for 'Max' to talk to these people," McNamara told the president softly, "without leaving the impression that the situation is going to hell."[25]

As McNamara had suggested, Taylor slipped out of the White House through a side entrance as soon as the photographers had finished, without saying anything. The news release was equally uncommunicative. "Ambassador Taylor reported that the political situation in Saigon was still difficult," it noted, "but . . . the new government under Prime Minister Huong was making a determined effort to strengthen national unity. . . . Although security problems have increased over the past few months, . . . government forces continue to inflict heavy losses on the Viet Cong." The release said nothing about either President Johnson's instructions to Taylor or the prospective *quid pro quo* agreement with the South Vietnamese. It stated merely that the president had instructed the ambassador to consult urgently with the South Vietnamese government on the measures it had to take to win the war.[26]

Taylor did talk with the press after a second meeting with the president on 3 December, but he said nothing to confirm or deny newsmen's suspicion that the war was taking a new direction. Tantalized but lacking hard facts, the press decided that McNamara's overheard remark represented the only real news they had received that week and used it to denounce the Johnson administration's lack of candor. Charles Ross of the *Chicago Sun-Times* pointed out that by attempting to suppress the facts McNamara had inadvertently served the American people. The *Washington Post* remarked that, if the secretary's words were true, there was no need to put a gloss on them. An informed, mature American public needed to know the truth about Vietnam even if things were "going to hell," if only to know what was wrong.[27]

[25] Charles Mohr, "Johnson Directs Taylor To Press Vietnam on War," *New York Times*, 2 Dec 64.

[26] Department of State *Bulletin*, 21 December 1964, p. 869. See also *Pentagon Papers*, 3: 248–51.

[27] *Pentagon Papers*, 3: 251. The Ross comment and the fact of widespread speculation in the press are noted in Dept of State, American Opinion Summary, 3 and 10 Dec 64. See also "Candor on Vietnam," *Washington Post*, 3 Dec 64.

While some newsmen criticized the Johnson administration, others speculated on the next direction the war would take. A few came uncomfortably close. Working from the premise that any U.S. escalation would take the form of air attacks, Hanson Baldwin suggested potential targets in Laos that might come under fire in a limited air campaign. The *New York Daily News* combined bits and pieces of information already on the public record with a series of shrewd deductions to assert that Taylor had indeed received some sort of conditional authority for air strikes in Laos and North Vietnam.[28]

When less informed reporters picked up the leads supplied by Baldwin and the *Daily News*, Zorthian made little effort to channel or correct the speculative news stories that resulted, on the theory that they worked to the advantage of U.S. forces by confusing the enemy. Ambassador Taylor disagreed. When Secretary Rusk asked whether early release of infiltration statistics might generate pressure for actions beyond what the United States contemplated, Taylor argued strongly for going on the public record as soon as possible. Such action would keep control of the situation and ensure that the enemy received an unexaggerated picture of American intentions.[29]

Taylor proposed "a planned and deliberate method of revealing our . . . program so that we may maintain a measure of control and obtain maximum impact." After the South Vietnamese government issued a formal press release, the United States would describe in broad outline the *quid pro quo* arrangement it was discussing in Saigon. State Department analyst Chester L. Cooper had just completed a white paper on infiltration that revised MACV's earlier statistics and put them in publishable form. It, too, would be released in a series of background briefings—conducted at first by MACV spokesmen but later by the ambassador himself—designed to describe U.S. intentions without discussing specific military steps. By the time those initial moves were complete, the United States would probably have a fair idea of how energetically the South Vietnamese government was prepared to pursue American recommendations. If desired, the ambassador might then make a speech reaffirming U.S. readiness to increase assistance to South Vietnam provided those recommendations were followed.[30]

Taylor concluded his message by recommending publication of Cooper's white paper within the next week to lay the groundwork in public opinion for whatever moves the United States decided to make. He had little concern that release of the document would generate pressure for extreme action; instead, he thought that it would persuade uncommitted nations that American charges of North Vietnamese involvement in the war were well founded.

[28] The news stories are mentioned in Memo, George Ball for the President, 12 Dec 64, sub: Diplomatic Actions Under South Vietnam Program, Chron files, and Msg, Saigon 1775 to State, 10 Dec 64, Westmoreland History, bk. 11, tab 11, both in CMH.

[29] Interv, author with Zorthian, 8 May 76; Msg, State 1231 to Saigon, DAIN 486877, 9 Dec 64. Taylor's response takes up two cables: Msg, Saigon 1775 to State, 10 Dec 64, and Msg, Saigon 1776 to State, 10 Dec 64.

[30] This section is based on Msg, Saigon 1775 to State, 10 Dec 64.

Despite Taylor's strong arguments, President Johnson postponed any decision on the ambassador's proposals until Rusk, McNamara, and other policy makers could decide on the main issue, release of the infiltration study. In the meantime, the old policies prevailed. "Our press handling had been not to interpret . . . White House or Saigon official statements in any way," Acting Secretary of State George Ball told the president, "but to background the more responsible press as fully as possible on a balanced (and thus more favorable) picture of the political and military situations."[31]

On 19 December representatives from the White House and the State and Defense Departments decided against Taylor's program but authorized a few concessions. Public release of the infiltration study would tend, on the one hand, "to create speculation and possibly pressures for greater action than we now have in mind," William Bundy told President Johnson. "On the other hand, it was agreed that the policy of telling the truth on Vietnam—plus the specific pressure from the press in Saigon, which has been promised some form of disclosure for several weeks—made it desirable to give Saigon the authority to indicate the general nature of the evidence and what it shows, on a background basis."[32]

The State Department's cable notifying Taylor of the decision carefully stipulated how much he could reveal. "Our feeling is that the press both here and in Saigon now accepts increased infiltration as fact," State observed, "but that a formal . . . release could be misinterpreted and become the vehicle for speculation." Thus, while general background briefings in Washington and Saigon were to continue, they were to follow established policy and indicate that infiltration was up without referring to any specific numbers. If newsmen pressed for details, Taylor could authorize fuller background briefings to reveal the general nature of the statistics available, but he was also to stress that the picture of enemy infiltration was constantly changing and avoid playing numbers games with the press.[33]

Confrontation: Khanh Versus Taylor, December 1964

In deciding to withhold Cooper's report, the Johnson administration assumed that the public relations initiative remained with the United States, but time was running out. In the late evening of 19 December General Khanh and a group of young South Vietnamese officers dismissed South Vietnam's embryonic legislative body, the High National Council, and the country's civilian premier, Tran Van Huong. A retaliation for the council's refusal to permit the forced retirement of General Minh and other senior generals who blocked the advancement of the

[31] Quote from Memo, Ball for the President, 12 Dec 64, sub: Diplomatic Actions Under South Vietnam Program. The parentheses are part of the quote. See also *Pentagon Papers*, 3: 256.

[32] Notes for the President's Daily Summary, William Bundy, 21 Dec 64, sub: Disclosure of Evidence of North Vietnamese Infiltration Into South Vietnam, Chron files, CMH.

[33] Msg, State to Saigon, 19 Dec 64, FAIM/IR.

younger officers, the act—in effect, a *coup d' etat*—frustrated any hope for Taylor's *quid pro quo* by destroying all semblance of South Vietnamese stability.[34]

Deeply frustrated, Ambassador Taylor summoned four of the young generals to his office to admonish them. Confronting Admiral Chung Tan Cang, Air Vice Marshal Ky, General Thi, and Lt. Gen. Nguyen Van Thieu, he first asked whether all spoke English and then launched into to a lengthy reprimand. "I told you all clearly . . . we Americans were tired of coups. Apparently I wasted my words. Maybe this is because something is wrong with my French because you evidently didn't understand. I made it clear that all the military plans that I know you would like to carry out are dependent on governmental stability. Now you have made a real mess. We cannot carry you forever if you do things like this."[35]

Smarting under Taylor's lash, the generals refused to comply with his demand that they find some way to undo their action. Ky charged that some members of the council were cowards and Communist sympathizers who had obstructed an honest attempt to reinvigorate the armed forces. Cang added, "It seems . . . we are being treated as though we were guilty. What we did was good and we did it only for the good of the country."

Deputy Ambassador U. Alexis Johnson, who saw that the discussion was going nowhere, suggested that if the generals were unable to yield they might at least take no action to impede a later softening of their position. During a news conference slated for that afternoon, instead of announcing the dissolution of the council, official spokesmen might merely affirm that some of that body's members had proved unsatisfactory and had been removed. Seeming neither to accept nor reject Johnson's expedient, the generals responded nebulously that "the door is not closed." But they held their news conference as planned and announced the suspension of the council.[36]

The next day Taylor told General Khanh that the United States could never cooperate with two governments in South Vietnam, one civilian with responsibility and one military with power. Khanh adopted a conciliatory stance, taking full blame for the generals' action and asking whether he should resign as commander in chief of the armed forces. Taylor replied that the situation might indeed improve if Khanh withdrew but that there might also be some merit in his remaining, if the civilian government called for it and the other generals agreed. Later that afternoon Khanh probed further into the question of his resignation by telephoning Taylor to ask whether the United States would be willing to provide travel funds should he and several unspecified generals decide to leave the country.[37]

[34] Msg, Saigon 2230 to State, DAIN 526596, 22 Jan 65, sub: U.S. Mission Monthly Report for December 1964; Msg, State 938 to Bangkok, DAIN, 502564, 29 Dec 64.

[35] This section is based on Msg, Saigon A-493 to State, DAIN 508327, 24 Dec 64, sub: Summary of Conversation, Sunday December 20. Although the message is written in the first person, it is not a stenographic transcript but a summation of full notes taken during the meeting.

[36] Msg, Saigon 1876 to State, DAIN 496842, 21 Dec 64.

[37] Msg, Saigon 1881 to State, DAIN 498127, 21 Dec 64; Msg, State 938 to Bangkok, DAIN 502564, 29 Dec 64.

Taylor and Westmoreland Confer With Reporters

Although Khanh seemed sincere, he quickly proved that he had no intention of resigning and that his overtures were really an attempt to draw Taylor onto weak ground where he could be accused of interfering in South Vietnamese politics. That ploy would make the generals' refusal to reinstate the council a matter of national honor and unify the country's officer corps behind Khanh at a moment when he was losing support to lower-ranking men such as Ky and Thieu.[38]

Khanh broadcast his intentions over Saigon radio on the morning of 22 December. Issuing an Order of the Day to the South Vietnamese armed forces, he announced that it was "better to live poor but proud as free citizens of an independent country than in ease and shame as slaves of the foreigners and communists." Citing as enemies both communism and colonialism in any form, he avowed that the people of South Vietnam would sacrifice to achieve independence but not to carry out the policies of a foreign power.[39]

After setting himself up as the defender of South Vietnamese pride, Khanh summoned *New York Herald-Tribune* correspondent Beverly Deepe to Da Lat for

[38] Msg, Saigon 2230 to State, DAIN 526596, 22 Jan 65, sub: U.S. Mission Monthly Report for December 1964; Msg, State 938 to Bangkok, DAIN 502564, 29 Dec 64.
[39] Msg, Saigon 1896 to State, DAIN 498129, 22 Dec 64.

an exclusive interview. Unless Ambassador Taylor acted "more intelligently," he told the reporter, the United States would lose Southeast Asia "and we will lose our freedom." Taylor's attitude and activities over the previous forty-eight hours had been "beyond imagination as far as an ambassador is concerned. One day I hope to tell the Vietnamese people and the American people about this. . . . It is a pity because Gen. Taylor is not serving his country well." If the United States wanted to solve the Vietnam problem, it would have to be "more practical" and to stop attempting to remake South Vietnam in America's image.[40]

Deepe contacted the U.S. embassy shortly after the interview to ask what Taylor had done to anger Khanh. The embassy replied only that "Ambassador Taylor has undertaken no activities which can be considered improper in any way. . . . All his activities are designed to serve the best interests of both Vietnam and the United States."[41] That evening the State Department added its weight to Taylor's defense, issuing a communique that was almost a slap at Khanh. "Ambassador Taylor has been acting throughout with the full support of the U.S. government" and in recognition of the fact that "a duly constituted government exercising full power . . . without improper interference . . . is the essential condition for the successful prosecution of the effort to defeat the Viet Cong."[42] Secretary Rusk was only slightly less emphatic at a news conference the next morning. Although he had no wish to prejudice ongoing discussions between American and South Vietnamese officials, he stated that without the political unity Taylor was trying to promote, the United States would have to curtail certain unspecified programs of assistance to South Vietnam because they presupposed an effective administrative apparatus.[43]

The American news media contributed a full measure of unfavorable comment. The *New York Post* linked what was happening to McNamara's earlier observation that "the situation is going to hell." Peter Grose of the *New York Times* equated Khanh with the enemy. "It almost seems as if the Viet Cong insurgents and the Saigon government conspired to make the United States feel unwelcome." Noting that the United States had been "mucking about in a serious way in Vietnam for several years," the *Chicago Tribune* charged that the country's generals were "remittance men on the United States' payroll." Without American money and men, "they and the parody of a government which they operate probably would not last a week." Meanwhile, the *New York Herald-Tribune* emphasized the futility of American attempts to pressure the South Vietnamese and the emptiness of Rusk's threats. "We have come full circle from a little over a year ago," the paper said, "when we foolishly allowed ourselves to be induced . . . to help bring down the Diem regime. We were damned for not intervening. We are damned

[40] Beverly Deepe, "Khanh Assails Gen. Taylor," *New York Herald-Tribune*, 23 Dec 64.

[41] Ibid.

[42] Msg, Saigon 1901 to State, DAIN 498128, 22 Dec 64; Msg, State 1328 to Saigon, DAIN 498592, 22 Dec 64.

[43] Bernard Gwertzman, "U.S. Firmly Backs Taylor, Chides Vietnamese Regime," *Washington Star*, 23 Dec 64.

The Scene at the Brink Hotel, 24 December 1964

now for having done so. . . . The issue is not General Khanh versus General Taylor. It is whether the Vietnamese still have the will to survive as an independent state. If they do, they will have to find a way, with or without General Khanh. And we shall have to help them on that way, with or without General Khanh."[44]

Although the State Department supported Taylor fully and most of the U.S. press voiced its hostility to the coup, Taylor was annoyed by Deepe's cooperation in publicizing Khanh's grievances. In retaliation, he gave a detailed account of his conversations with Khanh and the generals at Zorthian's weekly background briefing. He invited all of the reporters in Saigon, except Deepe, to attend.[45]

Taylor undertook the backgrounder because he realized that candor would give the Saigon correspondents perspective on the crisis while countering the charge that he had been unreasonable. Yet since publication of his remarks might only anger the South Vietnamese and worsen his relations with Khanh, he care-

[44] "Vietnam: The Moments of Truth," *New York Post*, 22 Dec 64; Peter Grose, "Ill Will in Vietnam," *New York Times*, 26 Dec 64; "None of Your Sass," *Chicago Tribune*, 24 Dec 64; "Gen. Khanh Vs. Gen. Taylor," *New York Herald-Tribune*, 24 Dec 64.

[45] Msg, Saigon 1930 to State, DAIN 500665, 24 Dec 64; Beverly Deepe, "Taylor Rips Mask Off Khanh," *New York Herald-Tribune*, 25 Dec 64. For confirmation that Deepe was excluded by Taylor's order, see Interv, author with Barry Zorthian, 18 Jun 76, CMH files.

fully stipulated that his comments were off the record. Most of the newsmen complied, but one gave an account of what had transpired to Deepe, who wasted little time taking revenge. Since she had been barred from the briefing, she felt no obligation to respect a rule to which she had never agreed. She published everything the ambassador had said under the lurid title "Taylor Rips Mask Off Khanh." Among the ambassador's more vivid observations, she said, was one in which he had told the generals, "You cannot break the crockery and have others pick it up." He had also ascribed many of the South Vietnamese Army's most recent failures to the fact that the generals stayed in Saigon while mere captains directed the war. If some South Vietnamese commanders were "first class," he had concluded, others bordered on being "nuts."[46]

Deepe's revelations caused an uproar at the U.S. mission because the South Vietnamese appeared on the verge of declaring Taylor *persona non grata*, but in the end the story made little difference. By the time the *Herald-Tribune* published it on 25 December, the enemy had once again demonstrated his ability to strike at will in South Vietnam, giving the United States the leverage it needed to draw Khanh toward compromise.[47]

The United States found a pretext for joint U.S.-South Vietnamese reprisals against the North when enemy sappers bombed the Brink Hotel in Saigon, killing two Americans and wounding fifty-one, but the dispute with Khanh precluded any possibility of a raid. Although Taylor argued strenuously for immediate retaliation to discourage further attacks, the Johnson administration saw no compelling evidence implicating the Viet Cong and feared that the American public might attribute the bombing to malcontents within the South Vietnamese government. The State Department, however, decided to use the incident to pressure Khanh. Launching a two-pronged diplomatic campaign, it instructed the U.S. mission in Saigon to defuse the personality issues underlying the crisis by concentrating on unity and effective government, whatever the formula South Vietnamese leaders found to bring them about. Meanwhile MACV advisers were to exploit their contacts within the armed forces to emphasize to the generals that the Brink bombing would have brought immediate reprisals but for the coup against Huong.[48]

The attempt to cajole the South Vietnamese had no immediate effect. On 26 December Westmoreland's deputy, Lt. Gen. John L. Throckmorton, met with the generals in an attempt to calm them. At no time had Ambassador Taylor intended to disparage anyone, he said, and the United States had never demanded that Khanh resign. The generals responded with varying degrees of hostility. Although Ky noted that Khanh appeared to have manipulated the whole crisis

[46] Interv, author with Zorthian, 18 Jun 76; Deepe, "Taylor Rips Mask Off Khanh."

[47] Interv, author with Zorthian, 18 Jun 76; Msg, State 1347 to Saigon, DAIN 500637, 24 Dec 64.

[48] *Pentagon Papers*, 3: 262. The State Department's instructions are contained in two consecutive messages: Msg, State 1346 to Saigon, DAIN 500537, 24 Dec 64, and Msg, State 1347 to Saigon, DAIN 500637, 24 Dec 64.

for his own ends, Thieu implied vaguely that someone was lying, and Cang remained adamantly unmoved.[49]

The first sign that the generals were relenting came on 30 December, when Taylor won their agreement to the establishment of an ad hoc committee. Composed of representatives of all parties to the dispute, it provided a forum for debating the issues on their own merits in an atmosphere above personalities. The next step came shortly thereafter, when General Khanh told an American observer with close connections to the U.S. mission that he was willing to resolve his difficulties with Taylor. Admitting that South Vietnam could never win the war without U.S. aid, he called for concessions on both sides and repudiated his interview with Deepe. The reporter had attributed statements to him that he had never made, he declared. All he had ever said was that Taylor's conduct had been "unimaginable."[50]

American pressure on Khanh nevertheless continued over the next week. When the general proposed establishing a military "organ of control" to oversee future civilian governments, Taylor killed the idea, informing several important South Vietnamese officers through Huong that the United States would never support another government imposed by the military. Two days later, Khanh called Westmoreland to his office to inquire about future joint attacks on the North, only to be told politely that the uncertainties arising from the coup had all but sidelined the idea. Meanwhile, MACV advisers in the field discreetly informed their South Vietnamese counterparts that Khanh's failure to resolve the crisis was blocking vigorous prosecution of the war.[51]

By 6 January 1965, the ad hoc committee had announced tentative agreement on a formula for ending the crisis. Under the new arrangement there would be no High National Council. Instead, the army would restore full control to a civilian government under Huong, which in turn would begin planning for the immediate election of a truly representative national assembly. Stressing that all the United States had wanted was an effective working relationship between the military and civilian arms of the South Vietnamese government, Taylor accepted the compromise. Taylor and Khanh signed a joint communique on 9 January sanctioning the agreements.[52]

The situation returned to normal for only ten days. Buddhist leaders had already informed the U.S. mission that they were implacably opposed to any government headed by Prime Minister Huong. When Huong returned to power,

[49] Msg, Saigon 1980 to State, DAIN 503047, 29 Dec 64; Msg, Saigon 1955 to State, DAIN 501529, 26 Dec 64.

[50] Msg, Saigon 1980 to State, DAIN 503047, 29 Dec 64; Msg, Saigon 1989 to State, DAIN 503703, 30 Dec 64; Msg, Saigon 1999 to State, DAIN 504161, 30 Dec 64; CIA Field Rpt, DAIN 504177, 30 Dec 64, sub: Nguyen Khanh's Wish To Settle Existing Problems.

[51] Msg, Saigon 2014 to State, DAIN 506202, 2 Jan 65; Msg, Saigon 2023 to State, DAIN 506455, 3 Jan 65; Msg, Saigon 2041 to State, DAIN 507830, 5 Jan 65; Msg, Saigon 2053 to State, DAIN 509063, 6 Jan 65.

[52] Msg, Saigon 2054 to State, DAIN 508645, 6 Jan 65; Msg, Saigon 2068 to State, DAIN 508841; Msg, MACV 003 to NMCC, 9 Jan 65, December Crisis file, CMH.

they began looking for an excuse to riot. Seizing upon one of the new regime's first official acts, a decree of 17 January that enlarged the army's draft calls, Buddhist agitators joined with disaffected students in a series of antigovernment, anti-American riots designed to bring Huong down. Khanh's spokesmen attempted to reason with the movement's leaders, but to no avail. While looters sacked the U.S. Information Agency library in Hue and disturbances spread from one city to another, Buddhist haranguers denounced Huong as an American lackey and called for armed resistance to the United States. Order returned only on 27 January, when the generals finally yielded to their antagonists, ousting Huong.[53]

Questions Arise, January 1965

The continuing turmoil sparked a debate in the U.S. Congress over further American aid to South Vietnam. A few antiwar senators such as Frank Church of Idaho and Albert Gore of Tennessee called for complete American disengagement, but even such proadministration stalwarts as Richard Russell of Georgia and Michael "Mike" Monroney of Oklahoma began to voice doubts and to advocate full hearings on the conduct of America's Vietnam policy. Tabulating congressional opinion at the height of the December crisis, the Associated Press found the Senate deeply divided. While only 3 of the 63 legislators who responded wanted immediate withdrawal, 31 recommended a negotiated settlement after further improvement of the U.S.-South Vietnamese position, 10 favored prompt negotiations, 8 sought commitment of U.S. forces against North Vietnam, and 11 said they had no opinion.[54] "Do we forsake what we have done?" Senate Republican Leader Everett M. Dirksen of Illinois intoned plaintively. "Do we go further and venture north and invite possible complications with Red China? Or do we just play along?"[55] *New York Herald-Tribune* correspondent Laurence Barrett supplied the answer. "Those who favor sterner action on one extreme or a ceasefire on the other are becoming more restive. The majority in between meanwhile clings to the policy of more-of-the-same, not with conviction or hope, but in the grip of an inertia born of not knowing what else to do."[56]

American newspapers were as perplexed as Congress but relatively unified in their condemnation of the U.S. government's lack of leadership on the Vietnam question. Although conservatively oriented journals such as the *Seattle Times* made spirited assaults on critics of government policy, most of the press failed to see any positive direction in the course events were taking. Syndicated columnists Rowland Evans and Robert Novak asserted that "If the United States does

[53] Msg, Saigon 2016 to State, DAIN 506572, 2 Jan 65. For a summary of the crisis itself, see *Pentagon Papers*, 3: 261-62.
[54] "Significant Rumblings," *Newsweek*, 18 Jan 65, p. 13; *Pentagon Papers*, 3: 263.
[55] "Debating Vietnam," *New Orleans Times Picayune*, 9 Jan 65.
[56] Laurence Barrett, "And in Washington, Pressing Decisions," *New York Herald-Tribune*, 17 Jan 65.

not soon add a new dimension to the war, our diplomats may find themselves the victims of a humiliating 'peace conference.'" *Life* magazine lamented that "Red intentions are becoming more credible while U.S. accomplishments have been clouded by our lack of clear intentions." The *New York Times* avowed, "Apathy is . . . not a policy. The United States has been stalling for time, but time has been working against us. The policy of drift is getting more and more dangerous, carrying with it . . . the possibility of falling by inadvertence and indirection into a major war."[57]

Underlying the concern of the press was the conviction of many newsmen that the U.S. government had purposely lied about its involvements in South Vietnam. Keyes Beech of the *Chicago Daily News* scored the Johnson administration for suppressing much of the North Vietnamese infiltration story. The *New York Times* published an article by Senator Wayne Morse of Oregon charging that the United States had fraudulently claimed to support free government in Southeast Asia when it had in fact been maintaining an American beachhead. The issue gained momentum when Arthur Dommen of UPI for the first time revealed the full extent of American air operations in Laos. That disclosure prompted the State Department to respond to queries from the press with an avowal that whatever the United States had done had been justified by Communist aggression.[58]

If Congress and the news media were troubled, the U.S. public was hardly less dissatisfied. No one had taken to the streets in protest and the majority of the American people paid little attention to groups advocating peace, but a Gallup poll taken toward the end of January revealed that almost everyone favored some sort of action to resolve the problem. Four out of five of those who said they followed the war closely believed that South Vietnam was losing to the Viet Cong. Two out of three agreed that the country would never form a stable government, but few wanted a unilateral American withdrawal. Instead, claiming that U.S. leaders had been right in entering the war, 50 percent believed that the United States was obliged to defend independent nations from Communist aggression. By a score of four to three, those interviewed even asserted that the United States should commit American troops if the danger of a military crisis arose. If there was to be no backing down, however, Americans were still willing to accept an honorable compromise. Eighty-one percent said they would support a peace conference that included the leaders of Southeast Asia and mainland China.[59]

For its part, the Johnson administration equated any negotiated U.S. withdrawal with "surrender on the installment plan."[60] On 14 January the Army Chief

[57] "Reply to Defeatism," *Seattle Times*, 7 Jan 65; Rowland Evans and Robert Novak, "Vietnam's Continuing Crisis," *Washington Post*, 8 Jan 65; "To LBJ: What Is Our Aim in Vietnam?" *Life*, 8 Jan 65, p. 5; "The Latest Coup in Vietnam," *New York Times*, 28 Jan 65.

[58] Msg, Joint State/Defense 1513 to Saigon, 22 Jan 65, Chron files, CMH; Wayne Morse, "We Must Leave Vietnam," *New York Times Magazine*, 17 Jan 65; *Pentagon Papers*, 3: 264; Department of State *Bulletin*, 8 February 1965, p. 167.

[59] "Significant Rumblings," *Newsweek*, 18 Jan 65, p. 13; George Gallup, "Americans Believe S. Vietnamese Are Losing War to Communists," *Washington Post*, 31 Jan 65.

[60] Memo, McGeorge Bundy for the President, 7 Feb 65, sub: The Situation in Vietnam, Key Materials file for February 1965, CMH.

of Staff, General Harold K. Johnson, rebuffed those who were calling for immediate action by telling a Los Angeles meeting of the National Security Industrial Association that the United States was in South Vietnam to defend freedom and that the American people should demonstrate the maturity that had long been theirs by exhibiting "patience, persistence and determination."[61] Three days later, Senator Morse's article appeared in the *New York Times Magazine*, but next to a piece by former Ambassador Lodge that disputed many of its arguments. "Pulling out of Vietnam," Lodge asserted, "is exactly the same as turning Vietnam over to the communists."[62] Assistant Secretary of State for Southeast Asian Affairs William P. Bundy was more conciliatory. On 23 January 1965, he told the Columbia, Missouri, Chamber of Commerce that "apropos of the headlines, . . . I think we are doing our job and that the media are doing theirs. The picture that you, as thoughtful citizens, get is in fact the picture that we have on all essential points. If that picture is complex or not entirely clear, believe me our picture is the same, for that is the nature of the situation."[63]

As those efforts proceeded, Rusk and McNamara began a quiet effort to relieve some of the pressure surrounding the question of North Vietnamese infiltration into the South. First they held a briefing on 21 January to inform congressional leaders of the basic facts contained in the State Department's white paper on the subject. To cut off leaks and to refute allegations appearing in the press that the Johnson administration was suppressing the facts, they then instructed Zorthian to proceed with a backgrounder detailing the same information.[64]

Although prepared to reveal most of the infiltration story, the Johnson administration remained unwilling to release a formal white paper on the subject. The State Department's instructions to Zorthian thus warned against releasing any of the backup documents used to prepare briefings for the press. In the same way, U.S. mission representatives were to inform the South Vietnamese government that a backgrounder on the subject would occur, but they were to avoid encouraging South Vietnamese participation because that might "create pressures and impact beyond what we desire." If correspondents began to question the delay in releasing a formal paper, Zorthian was to stress that sources had become much more numerous in recent weeks and that careful checking and compilation were required before the mission could release the facts.

The State Department's instructions had the desired effect. By 26 January the Saigon correspondents had begun writing low-keyed articles on MACV's revision of infiltration statistics, to the accompaniment of little adverse editorial comment. Seymour Topping of the *New York Times* noted in passing that the new estimates were part of a recent survey submitted to congressional leaders in con-

[61] Harold K. Johnson, "The Defense of Freedom in Vietnam," Address Before the National Security Industrial Association, 14 Jan 65, Department of State *Bulletin*, 8 February 1965, p. 176.

[62] Henry Cabot Lodge, "We Can Win in Vietnam," *New York Times Magazine*, 17 Jan 65.

[63] William P. Bundy, "American Policy in South Vietnam and Southeast Asia," an Address Before the Columbia, Missouri, Chamber of Commerce, 23 Jan 65, Department of State *Bulletin*, 8 February 1965, p. 168.

[64] This section is based on Msg, Joint State/Defense 1513 to Saigon, 22 Jan 65.

junction with an analysis of the value of air attacks on North Vietnam to isolate the Viet Cong from their source of leadership and supplies.[65]

Despite the effort to keep a firm grip on events, U.S. policy makers nevertheless recognized that whatever initiative they had in South Vietnam was fast slipping away and that they could no longer avoid a decision on widening the war. During the last week of December, as the Khanh-Taylor imbroglio reached its height, enemy forces occupied the village of Binh Gia in Phuoc Tuy Province of the III Corps Tactical Zone in an unprecedented multiregiment operation. In the four-day contest that ensued, South Vietnamese commanders directed more of their attention to the power struggle in Saigon than to the battle, in which 177 of their men were counted as killed, 181 wounded, and 104 missing. American casualties for the same engagement were 6 killed, 9 wounded, and 3 missing. During the first week of January the Communist Chinese *People's Daily* described the engagement as "a smartly conducted, tough pitched battle of annihilation" that proved the Viet Cong had "grown into a formidable liberation army."[66]

Shortly after the battle ended, William Bundy told Secretary Rusk that even if the overall impact of the political crisis and the defeat at Binh Gia were difficult to assess, there were ample indications that the morale of the Saigon government was "very shaky indeed." To many Asian and European nations, he said, the United States appeared to be linking additional action against the North to the attainment of a more perfect government in the South than could reasonably be expected. Reprisal raids against North Vietnam as soon as the enemy provided an excuse seemed the answer. "They might not save South Vietnam," Bundy said, but "we would still have appeared to Asians to have done a lot more about it."[67]

Ambassador Taylor shared most of Bundy's conclusions. On 6 January he cabled President Johnson that "we are faced with a seriously deteriorating situation characterized by continued political turmoil, irresponsibility and division within the armed forces, lethargy . . . and signs of mounting terrorism by the Viet Cong directly at U.S. personnel." Adding that the United States was on a losing track and had to risk a change, he asserted that reprisals and air operations against North Vietnam should begin just as soon as South Vietnam attained a minimal government. In the meantime, President Johnson should set the stage for action by informing the American public about enemy infiltration and by initiating aggressive naval patrolling along the enemy coast. When the United States decided to act, Taylor concluded, it would be able to justify the decision on the basis of infiltration, of Viet Cong terrorism, of attacks on the patrols, or some combination of the three.[68]

[65] Seymour Topping, "Hanoi's Troop Aid Now Held Bigger," *New York Times*, 27 Jan 65.

[66] Airgram, Hong Kong A–473 to State, 15 Jan 65, sub: Communist China Joint Week No. 2, Binh Gia file, CMH. See also Msg, MACV to NMCC, DAIN 507420, 5 Jan 65; Msg, MACV J312 0686 to CINCPAC, 8 Jan 65, sub: Analysis of Situation in Phuoc Tuy Province, CMH files.

[67] Memo, William P. Bundy for Secretary Rusk, 6 Jan 65, sub: Notes on South Vietnamese Situation and Alternatives, doc. 248, *Pentagon Papers*, 3: 684.

[68] Msg, Saigon 2052 to State, 6 Jan 65, for the President from the Ambassador, Westmoreland History, bk. 12, tab 14, CMH.

Although Johnson told Taylor that he was "inclined to adopt a policy of prompt and clear reprisal," he sought again to avoid any line of action that might later restrict his freedom of movement. Besides the question of South Vietnamese political stability, he had American and world public opinion to consider. Before embarking on a major escalation, he would have to evacuate American women and children from South Vietnam. How, he asked, could he communicate that to the press without appearing to be running away?[69]

At Taylor's suggestion, the president sent White House Adviser McGeorge Bundy to South Vietnam to determine firsthand both the condition of the South Vietnamese government and the types of pressures best applied against Hanoi and the Viet Cong. Bundy and his assistants arrived in Saigon on 3 February; within several days they had tentatively concluded that the war was going just about as badly as it had seemed from Washington.[70]

Justifying Escalation, February–March 1965

If any doubts remained, they vanished abruptly on the morning of 7 February, Saigon time, when enemy mortarmen and sappers killed 9 Americans and wounded 108 in a brazen attack on the U.S. barracks and airstrip at Pleiku. Already predisposed to strong action and convinced that the United States could no longer allow the enemy to attack Americans with impunity, President Johnson responded by authorizing a series of joint U.S.-South Vietnamese air strikes against targets in North Vietnam and by ordering all American dependents to leave South Vietnam.[71] White House spokesmen linked the president's actions to the enemy's whole posture of aggression in the South, citing as justification not only the Pleiku raid but also Viet Cong attacks on South Vietnamese airfields and villages. "These attacks were only made possible," they noted, "by the continuing infiltration of personnel and equipment from North Vietnam." President Johnson meanwhile hinted publicly that further American action might be in the offing. "We have no choice now but to clear the decks," he said, "and make absolutely clear our continued determination to back South Vietnam in its fight to maintain its independence."[72]

While the president made his decisions, McGeorge Bundy returned from Saigon with recommendations that once more stressed the need for a program of sustained reprisals against the North. Time was running out, Bundy told the presi-

[69] Msg, State 1419 to Saigon, 8 Jan 65, for Ambassador from the President, FAIM/IR.

[70] Msg, Saigon 2057 to State, 6 Jan 65, Westmoreland History, bk. 12, tab 14, CMH; Msg, State 1581 to Saigon, 2 Feb 65, for Ambassador from McGeorge Bundy, FAIM/IR; Msg, Saigon 2420 to State, 7 Feb 65, for McNamara and Vance from McNaughton, Special files, 1965, CMH.

[71] Memo, MAC J00 for Ambassador Taylor, 8 Feb 65, sub: Weekly Assessment of Military Activity for the Period 31 January to 7 February 1965, Westmoreland History, bk. 13, tab 27, CMH.

[72] "United States and South Vietnamese Forces Launch Retaliatory Attacks Against North Vietnam," Department of State *Bulletin*, 22 February 1965, p. 238.

dent. Since the South Vietnamese were incapable of successfully prosecuting the war and negotiations with the Communists held no serious promise, only a program of continuing, gradually escalating attacks appeared to offer any hope. Establishing that the United States had "the will and force and patience and determination to take the necessary action," the program would compensate for the American failure to motivate the South Vietnamese. It would also give the president the leverage he needed "to speak on Vietnam . . . with growing force and effectiveness." If the United States nevertheless failed, the attacks would also weaken the charge that the Johnson administration had failed to do all it could to avert defeat.[73]

Bundy stressed that if the enemy was to believe that further aggression in the South was self-destructive, the United States could not give the impression that it was responding to enemy initiatives. The attacks therefore would have to continue to be linked to the enemy's total conduct in South Vietnam. Highly visible incidents such as the assassination of a province chief might have to be cited at the outset, but once the reprisals were clearly under way, a white paper establishing the full extent of Communist infiltration from the North and weekly lists of enemy atrocities should be enough to justify anything the president wanted to do. Announcements would have to state that the United States had no designs upon North Vietnamese territory and to specify that the severity of the raids would fluctuate according to the tempo of Viet Cong activity in the South, but, for the rest, the United States should execute its policy with the least possible discussion. Taking care to avoid an appearance of boasting that might make it hard for Hanoi to shift ground, U.S. officials would instead use every forum to point out that the true cause of the problem was aggression from North Vietnam.

Ambassador Taylor said many of the same things in a cable to the White House. Admiral Sharp also concurred. "While it may be politically desirable to speak publicly in terms of a 'graduated reprisal program,' I would hope that we are thinking, and will act, in terms of a . . . steady, relentless movement toward our objective."[74] Backed by his advisers, President Johnson on 8 February approved the program in outline, noting that "I am now prepared to go forward with the best government we can get." Three days later a Viet Cong attack on a U.S. enlisted men's barracks at Qui Nhon gave the White House the pretext it needed to announce a second air strike against North Vietnamese targets. As Bundy had recommended, official spokesmen justified the attack with a catalog of enemy atrocities, mentioning Qui Nhon only in passing.[75]

The American news media accepted the necessity of responding to the Communists' attacks, for the most part agreeing with a comment by the *New York Daily News* that President Johnson's air strikes were "a good crackback." Yet while

[73] This section is based on Memo, McGeorge Bundy for the President, 7 Feb 65, sub: The Situation in Vietnam, Key Materials file for February 1965, CMH.

[74] For Taylor's and Sharp's recommendations, see *Pentagon Papers*, 3: 315.

[75] Msg, State 1653 to Saigon, 8 Feb 65, for Taylor from the President, FAIM/IR; Department of State *Bulletin*, 1 March 1965, p. 290.

Bundy and Westmoreland at Pleiku

the conservative *Daily News* went on to advocate striking Communist China's nuclear arms facilities in addition to further attacks on the North, many reporters and editors saw little purpose in further escalating the war. The *St. Louis Post-Dispatch* avowed that "the new exchange of strikes simply emphasizes the bankruptcy of American policy." James Reston of the *New York Times* argued that America's "crooked course" in South Vietnam meant the United States was digging deeper into "the accustomed military rut." Conservative columnist David Lawrence warned against an unthinking slide into all-out war. The editors of the *New York Times* asserted that "the only sane way out is diplomatic, international, political, economic—not military." Arthur Krock of the *New York Times* meanwhile traced the roots of the news media's uneasiness to the Johnson administration's failure to prove that the guerrillas killing Americans were predominantly North Vietnamese, and columnist Max Lerner observed that those who said the war was "a futile folly" would go unchallenged until President Johnson clarified what he was doing in South Vietnam.[76]

[76] "A Good Crackback—But," *New York Daily News*, 8 Feb 65; David Lawrence, "Mounting Crisis in the Vietnam War," *New York Herald-Tribune*, 8 Feb 65; James Reston, "Washington: The Undeclared and Unexplained War, *New York Times*, 14 Feb 65; "The Dangers in Vietnam," *New York Times*, 9 Feb 65; Arthur Krock, "In the Nation: The Bombs of February," *New York Times*, 14 Feb 65. Max Lerner

Well aware that clarifications were necessary, the Johnson administration prepared to defend its policies. As early as 9 February, the Joint Chiefs of Staff had finished reviewing MACV's October study of North Vietnamese infiltration but had decided that the more recent Cooper report was better suited to information officers' needs. By the twelfth, Defense and State Department analysts were hard at work incorporating new intelligence into the Cooper draft and had begun compiling a data base of continuing enemy atrocities for possible later use. At the State Department, Acting Legal Adviser Leonard Meeker composed a memorandum defining the basis for U.S. attacks on North Vietnam in international law, and unidentified sources began leaking analyses to the press suggesting that North Vietnam could avoid further destruction by terminating its support for the Viet Cong.[77]

As those efforts proceeded, President Johnson and his staff attempted to endow the American involvement in South Vietnam with high moral purpose. At a 12 February luncheon celebrating Abraham Lincoln's birthday, Johnson compared the war in South Vietnam to Lincoln's preservation of the Union. Lincoln had proved that democracy could work, the president noted. The United States thus became a "city on the hill" charged with carrying that example to the world. "History and our own achievements have . . . thrust upon us the principal responsibility for the protection of freedom on earth," Johnson continued. "We do not ask for this task. But we welcome it."[78]

Several days later Vice President Hubert H. Humphrey took up the theme at an international symposium at the United Nations on Pope John XXIII's 1963 encyclical, "Peace on Earth." Acknowledging with Saint Augustine of Hippo that war would persist until the end of time not because men loved peace the less but because they loved their own version of peace the more, Humphrey described America's role in South Vietnam as a form of peace-keeping. "Today in Vietnam," he said, " . . . freedom is endangered by the systematic attempt of foreign-backed subversives to win control of the country. Our policy is clear. . . . We will resist aggression. We will be faithful to a friend. We seek no wider war. We seek no domination. Our goal in Southeast Asia is today what it was in 1954 . . . peace and freedom for the people of Vietnam."[79]

The administration's campaign became more specific on 25 February, when Secretary Rusk announced at a State Department press conference that a white paper on North Vietnamese infiltration would shortly become available. Two days later the paper appeared. Contending that fully 75 percent of the 4,400 Viet Cong

and the *St. Louis Post-Dispatch* are quoted in an extensive survey of February press coverage in "Sizing-up Vietnam," *Time*, 19 Feb 65.

[77] JCSM 155-65, 9 Feb 65, sub: MACV Infiltration Study, ISA 381VN, 70A3717, box 46, WNRC; Msg, Wheeler JCS 0531-65 to Westmoreland, 11 Feb 65, Westmoreland Papers, CMH; Draft Msg, William Bundy to Taylor, 14 Feb 65, and Memo, Leonard C. Meeker for William Bundy, 11 Feb 65, sub: Attached Legal Memorandum on United States Actions in North Vietnam, both in Chron files, CMH. For mention of the leaks, see *Pentagon Papers*, 3: 330.

[78] Department of State *Bulletin*, 8 March 1965, p. 334.

[79] United States/United Nations Press Release 4500, in ibid., p. 326.

known to have entered South Vietnam in the first eight months of 1964 were ethnic North Vietnamese, the 65-page document cited twenty-five interviews with captured infiltrators as proof. It contained an extensive list of captured weapons and ammunition of obviously Communist manufacture to demonstrate that the North Vietnamese were the Viet Cong's main supplier and detailed the organization Hanoi had developed to control the war in the South. If peace could be restored to South Vietnam, the paper concluded, the United States would be ready at once to reduce its military involvement. The choice between peace and an increasingly destructive conflict was up to Hanoi.[80]

Presented with much fanfare, the white paper had little of the effect policy makers expected. Instead, it produced a vehement backlash. Although the *Baltimore Sun* and the *Washington Daily News* saw in the document "overwhelming evidence" that the struggle in South Vietnam was more than a civil war, many journals agreed with the *St. Louis Post-Dispatch*, which characterized the study as an obvious attempt to justify further attacks on the North, and a weak one at that. Conservative papers such as the *Omaha World-Herald* scored the report's failure to mention that Hanoi's actions were really part of a "global communist conspiracy" and wondered whether U.S. efforts against North Vietnam were merely "a kind of public relations war" in which the United States went to extremes to keep from antagonizing its true opponents, Moscow and Peking. "Since the Cuban Missile Crisis of 1962, when the administration manipulated the news and claimed the right to do so," the *World-Herald* charged, "no American can be sure whether his government is giving him the whole story." The liberal *New Republic* was just as critical. Of the 25 infiltrators named in the study, it noted, 16 were native to South Vietnam and only 8 had been positively identified as North Vietnamese. The *Providence Journal* meanwhile observed that if the Communists had broken the Geneva Agreements of 1954, the United States had done likewise, if only by supporting Diem in his refusal to hold elections in 1956 to reunify the country. The freedom that the United States insisted it was trying to defend in South Vietnam had never existed. In fact, if the North Vietnamese were infiltrating men and supplies, the bulk of the resistance was still indigenous, a struggle in which many South Vietnamese sincerely believed they were fighting for their nation's independence after generations of foreign rule.[81]

In the end, after all the critics had registered their complaints, Chester Cooper could only lament that no official publication could have accomplished the ambitious objectives the white paper's proponents had hoped to achieve. It was impossible, he said, to provide sufficient documentary evidence of Hanoi's direction and support of the war in the South. Although captured documents and interro-

[80] Department of State *Bulletin*, 15 March 1965, p. 362. See also State Department Publication 7839, *Aggression From the North: The Record of North Vietnam's Campaign To Conquer South Vietnam* (Washington, D.C.: Government Printing Office, 1965).

[81] "White Paper on Vietnam," *Washington Daily News*, 1 Mar 65; "Vietnam Basics," *Baltimore Sun*, 25 Feb 65; "Weak Reed To Lean On," *St. Louis Post-Dispatch*, 2 Mar 65; "White Paper," *Omaha World-Herald*, 2 Mar 65; "The White Paper," *New Republic*, 13 Mar 65, p. 10; "Flaws in Our Case for Vietnam Support," *Providence Journal*, 28 Feb 65.

gation records were declassified and incorporated into the white paper, the most important sources were too sensitive to use.[82]

Cooper was only partially correct; for if the report lacked force, it was also the victim of months of procrastination in which the United States had progressively surrendered whatever initiative it possessed to a vain hope that the South Vietnamese government might stabilize. Instead, coup followed coup, with Khanh himself falling just days before the white paper appeared. When the long-awaited bombing campaign finally began on 2 March, three out of four Americans, according to pollster Louis Harris, no longer believed victory was possible in South Vietnam. Although 83 percent of the U.S. public rallied to the president's side when the bombing began, it did so more out of duty than enthusiasm. ''I'm in favor of anything to prevent war,'' a Florida man told Harris pollsters, ''but as conditions are now, I see no alternative but to stay on and do what has to be done to end this thing.'' The reaction of the press to the white paper reflected that mood, making the study, in the eyes of many, not the justification of a new and victorious initiative, but, as the *New York Times* avowed, ''a tacit admission of failure.''[83]

[82] Chester L. Cooper, *The Last Crusade* (New York: Dodd, Mead & Co., 1970), p. 264.
[83] Harris, *Anguish of Change*, p. 58; ''The One-Way Street,'' *New York Times*, 7 Mar 65.

6

Censorship Considered

Although Zorthian considered the U.S. mission's relations with the press as good as possible under the circumstances, the sensitivity of Washington agencies to news stories from Saigon continued to grow. President Johnson's decision in November 1964 to begin outright bombing attacks in Laos (BARREL ROLL) caused it. Johnson wanted to avoid appearing to escalate the war, but the press continued to emphasize the widening nature of American involvement, creating a host of problems. For while the attacks had done little to decrease enemy infiltration, they had forced the Communists to decentralize operations in Laos, an effect North Vietnam's Russian allies appeared willing to tolerate, but only if the United States avoided excessive publicity. There were also possible Chinese reactions and world public opinion to consider. If the United States appeared too bellicose, the Chinese might misread the situation and decide to intervene. At the very least, so the reasoning went, too much publicity would increase public awareness of the escalation and fuel agitation for a negotiated settlement.[1]

The president's reluctant commitment to BARREL ROLL at first precluded public relations problems. Only a few planes went out on each mission, and their attacks so closely resembled the armed reconnaissance flights of previous months that the Communists failed to take much notice. No enemy propaganda statements appeared to draw the attention of the press to the program.[2]

That situation changed on 12 January 1965, when President Johnson, intent upon demonstrating the American will to remain in Southeast Asia and recognizing that the raids were having little effect, authorized a heavy attack on an important bridge in northern Laos. The strike gave the desired signal but had

[1] U.S. Information Agency, Report of Far East Public Affairs Conference at Baguio, PI, 11–15 January 1965, ISA 092 VN, 70A371, box 31, WNRC; Msg, Saigon 2077 to State, 7 Jan 65, Air Ops-Laos file, CMH; Msg, Wheeler JCS 739-65 to Westmoreland, 1 Mar 65, Westmoreland Papers, CMH; Msg, State 1881 to Saigon, 3 Mar 65, Westmoreland History, bk. 14, tab 17, CMH.

[2] Msg, Saigon 2073 to State, DAIN 509865, 7 Jan 65, Army Staff Communications Center files, Army War College.

an unwanted side effect. Shortly after it ended, Radio Hanoi and Radio Peking went on the air to claim that Communist gunners had shot down several American aircraft and that the United States had once more escalated the war.[3]

In the past, American information officers had countered enemy announcements of this sort by saying that an incident had indeed occurred but that the plane had been on a legitimate reconnaissance mission at Laotian request and had gone down while suppressing enemy fire. In this case Communist news agencies had announced a major American attack. If official spokesmen stayed with the old formula, the Saigon correspondents would inevitably discover the deception and begin a new round of denunciations.[4]

Trapped between the president's desire for a low profile and the need to preserve official credibility, State and Defense Department public affairs officers finally settled upon a tactic that they hoped would avoid untruth while stifling speculation in the press. Shortly after Communist announcements began to circulate, they issued a deliberately unclear statement that admitted the loss of two planes but avoided stating that the aircraft had been on a reconnaissance mission. Pressed on the point, they fell back on the standard rejoinder that "this is an operational matter upon which we cannot comment."[5]

Instead of reducing conjecture, the carefully worded release had the opposite effect. Confronted by an official refusal to clarify what had happened, reporters linked the information Communist broadcasters had provided with the U.S. government's failure to mention reconnaissance to deduce that a change in policy had occurred. Their first dispatches concentrated on whether the attack had been an escalation and whether strikes of that sort could be either effective or moral, but their comments soon broadened. Laurence Barrett of the *New York Herald-Tribune* argued that the U.S. public's right to know the nature and extent of the war overrode tactical and political reasons for keeping quiet about the raid. United Press International repeated the charge that U.S. fighter bombers had been attacking Communist supply routes in Laos for the past seven months. A Reuters dispatch from Saigon noted that the aircraft used in the raid could well have come from American bases in Thailand, and *Time* magazine surmised that the North Vietnamese were circumventing U.S. attacks in Laos by funneling more and more of their aid to the Viet Cong through Cambodian ports.[6]

Toward the end of January, President Johnson's friend Senator Monroney visited Saigon, where he met with both Zorthian and Westmoreland. He told the two that there was, as he put it, "general unhappiness in Washington" with the character of the news reporting coming out of South Vietnam. Convinced

[3] Memo, ASD PA for the SECDEF, 19 Jan 65, sub: News Media Treatment of BARREL ROLL NINE, DDI Air Incidents/Policy file.

[4] Ibid.

[5] Ibid. See also Msg, State to Vientiane, DAIN 516308, 13 Jan 65.

[6] Msg, State 592 to Vientiane, 13 Jan 65, FAIM/IR; Memo, ASD PA for SECDEF 19 Jan 65, sub: News Media Treatment of BARREL ROLL NINE. For a summary of press coverage, see Msg, State 596 to Vientiane, 14 Jan 65, FAIM/IR; Laurence Barrett, "Secrecy in Southeast Asia," *New York Herald-Tribune*, 19 Jan 65; "The Quiet Escalation," *Time*, 22 Jan 65, p. 22.

that Monroney was Johnson's personal emissary, General Westmoreland had few doubts about the senator's meaning: the president himself was becoming increasingly concerned about the U.S. mission's failure to keep the Saigon correspondents under control.[7]

Guidelines for the Press Evolve, February–March 1965

Air strikes on 6 and 11 February in response to the Viet Cong attacks at Pleiku and Qui Nhon further strained the situation. Believing that "the more public the challenge we present to the DRV [Democratic Republic of Vietnam] the more difficult it becomes for them and their friends to back down," Ambassador Taylor recommended that official spokesmen say as little as possible to newsmen after the reprisals.[8] Information officers agreed in principle but still felt a need to maintain the Johnson administration's credibility. As a compromise, they drafted an announcement that drew attention away from the reprisals and toward North Vietnamese aggression.[9]

The approach worked well after the 6 February attack because the Saigon correspondents were concentrating on the destruction at Pleiku and received little advance warning that U.S. reprisals were imminent. That was not the case on 11 February. Aware that the United States might retaliate if another major enemy attack occurred, more than twenty newsmen congregated at Da Nang, the main base for air strikes against North Vietnam. When the enemy struck at Qui Nhon and American fighter-bombers again took off for targets in the North, the reporters counted the departing aircraft and wired their home offices of the event even before the planes had reached their targets. The reporters then contacted Air Marshal Ky, who confirmed that attacks on North Vietnam were indeed in progress.[10]

Premature news stories possibly forewarning the enemy of the strikes were only the first of the problems to arise on 11 February. At a series of post-strike press conferences, MACV information officers announced the types of bombs dropped, released aerial photographs of strike results, and introduced two pilots who had participated in the attacks. General Wheeler promptly cabled Westmoreland to note that the briefers had announced erroneously that the jets had hit a target 160 miles north of the Demilitarized Zone, a slip, he said, that had informed the Communists of a potential American target, giving them time to prepare defenses. In the same way, one of the pilots had responded to a reporter's question by noting that although enemy MiG fighters had failed to challenge the

[7] Westmoreland quotes Monroney in Msg, Westmoreland MAC 309 to Gen George V. Underwood, Chief of Army Information, 21 Jan 65, Westmoreland Papers, CMH.

[8] Msg, Saigon 2186 to State, 18 Jan 65, Westmoreland History, bk. 12, tab 25, CMH.

[9] Msg, State to Saigon, 6 Feb 65, FAIM/IR.

[10] Msg, State 1702 to Saigon, 11 Feb 65, FAIM/IR; Msg, Saigon 2538 to State, 13 Feb 65, DDI News from Vietnam file; Msg, Westmoreland MAC 831 to Wheeler, 17 Feb 65, Westmoreland Papers, CMH; [Reuters], "Saigon's Planes Attack North," *New York Times*, 11 Feb 65.

Aerial View of Da Nang Air Base. *Note proximity of populated areas.*

attack, they well might have. That kind of "idle speculation," Wheeler insisted, injected more menace into the situation than the facts warranted.[11]

Sensitive to the president's mounting displeasure with official press policies, the Office of the Assistant Secretary of Defense for Public Affairs drafted guidance to restrict information on future air strikes against North Vietnam. The rules cabled to Saigon on 12 February permitted release of the times of attack, the locations and general categories of targets, the participation of either South Vietnamese or U.S. aircraft, the names of American killed and wounded after search and rescue operations were completed, and "very general characterizations" of mission success. At the same time, they restricted the access of the press to information that might either embarrass the military services, help the enemy, or increase discussion of the war.[12]

Under those restrictions, military spokesmen were to release only photographs approved by the Department of Defense lest newsmen inadvertently publish pictures that revealed the use of such weapons as napalm and antipersonnel cluster

[11] Msg, Joint State/Defense 1701 to Saigon, 11 Feb 65, DDI News from Vietnam file; Msg, Wheeler JCS 553-65 to Sharp and Westmoreland, 13 Feb 65, Westmoreland Papers, CMH; Interv, author with Barry Zorthian, 1 Dec 76, CMH files.

[12] Msg, Defense 5083 to CINCPAC, 12 Feb 65, DDI Air Incidents/Policy file.

bombs or learn that pilots had mistakenly attacked a nonmilitary target. For the same reason, nothing was to be said about the kinds of bombs dropped other than that they were "conventional." Since some aircraft striking North Vietnam flew from bases in Thailand and since the Thais wanted to avoid any indication that their territory was involved, information officers might reveal the names of aircraft carriers launching strikes but were to describe all attacks originating from South Vietnam and Thailand only as "land based." They were also to refrain from telling how many aircraft took part in a raid because an announcement of that number might give the enemy a means for gauging the effectiveness of his radar. If reporters noted a big strike against a target previously hit, moreover, they would undoubtedly question the effectiveness of the earlier attack.[13]

Shortly after dispatch of those rules, both General Wheeler and Admiral Sharp advised Westmoreland to limit the access of the press to potentially damaging information. He might begin by ending the practice of having airmen involved in an attack brief the Saigon correspondents after the action had ended. In the same way, the first announcement in all cases where air attacks involved planes based outside of South Vietnam might come from the commander in chief, Pacific, in Hawaii, where, Sharp said, "the total picture is brought together most quickly and most accurately."[14]

Although they agreed that the expanding air war required new public affairs guidance, information officers in Saigon found serious fault with the rules. In a 13 February message to the Department of Defense, they pointed out that the enemy could hardly avoid concluding that all targets of military significance in North Vietnam were on President Johnson's list, and anyone familiar with jet aircraft recognized that enemy fighters, given sufficient warning, were capable of intercepting any attack. Because the Communists were well aware that only three airfields in South Vietnam—Bien Hoa, Tan Son Nhut, and Da Nang—were capable of handling jet fighter-bombers, refusal to mention those bases in news releases contributed little to military security while compelling official spokesmen to conceal the obvious. In the same way, reporters were aware of the size of attacks on North Vietnam and could recite the types of bombs and ammunition being used. An unwillingness to provide at least round figures on the number of planes involved in a raid and the resort to the word *conventional* to describe well-known types of armament would achieve little beyond the destruction of official credibility. Reluctance to allow newsmen access to pilots would have a similar effect. Correspondents needed the colorful details returning airmen could provide and would interpret the silencing of those officers as a cover-up. The information officers concluded that the United States would have to coordinate with the South Vietnamese any effort to restrict the press unless it wanted a repetition of the problems caused by Marshal Ky's premature remarks on the 11 Febru-

[13] Ibid.; Msg, MACOI 4511 to ASD PA, 13 Feb 65, CMH files; Msg, Defense 6726 to MACOI, 10 Mar 65, DDI Air Incidents/Policy file.

[14] Quote from Msg, Sharp to Westmoreland, 14 Feb 65, Westmoreland History, bk. 11, tab 40, CMH. See also Msg, Wheeler JCS 553-65 to Sharp and Westmoreland, 13 Feb 65, Westmoreland Papers, CMH.

ary attacks. The most advisable course would be to convene a conference of all the senior information officers involved with the war to discuss the direction future policies should take.[15]

When the Defense and State Departments made no immediate response to those suggestions, Zorthian took action on his own. He knew that South Vietnam was so open and news sources so abundant that restrictions on the press would fail unless newsmen voluntarily agreed to cooperate. He therefore contacted the most important Saigon correspondents to obtain their promise to hold stories on air strikes until the planes returned. "This does not necessarily commit us to any announcement or briefing," he told his superiors in Washington, but simply to issuing a "go ahead on information the news media may have obtained by personal observation and other sources." When the State Department raised no objections, Zorthian distributed a memorandum to the Saigon correspondents incorporating many of the rules covered in the Defense Department's 12 February guidelines. He requested voluntary compliance.[16]

General Westmoreland shared Zorthian's views. He told the Joint Chiefs of Staff that a high-level conference to update information policy was imperative. "Since the rules of the game are changing rapidly," he said, "it seems to me that we should consider arrangements similar to those exercised during the Korean conflict. This would involve providing for accredited war correspondents (we might want to give them another name) and censorship in some limited form."[17]

Although the State and Defense Departments approved a conference, scheduling it for Honolulu in mid-March, Westmoreland's suggestion of censorship produced no formal comment. General Wheeler merely noted in his response that the Joint Chiefs of Staff wanted to make sure that "media representatives do not receive operational information of assistance to the enemy, or, should they receive such information unavoidably, that they do not dispatch it in time to be of value."[18] Censorship nevertheless remained in the back of everyone's mind, to surface as an open issue early in March, after the Saigon correspondents made a series of revelations that threatened both operational security and American relations with the South Vietnamese. The breach occurred following a decision by President Johnson on 26 February to send two battalions of U.S. marines to protect Da Nang Air Base, a major change in the nature of American involvement. American officials considered the base vital to attacks against North Vietnam and believed that Marine units were essential to protect it. The South Vietnamese, however, remained sensitive to an influx of foreign troops. Only reluctantly sanctioning the move, they stipulated that the marines come ashore

[15] Msg, MACOI 4511 to ASD PA, 13 Feb 65. See also Msg, Saigon 2538 to State, 13 Feb 65; Msg, Westmoreland MAC 792 to Wheeler, 15 Feb 65, Westmoreland Papers, CMH; Interv, author with Zorthian, 1 Dec 76.

[16] Msg, Saigon 2560 to State, 13 Feb 65, DDI Air Incidents/Policy file; Interv, author with Zorthian, 1 Dec 76.

[17] Msg, Westmoreland MAC 831 to Wheeler, 17 Feb 65.

[18] Msg, Wheeler JCS 641-65 to Westmoreland, 19 Feb 65, Miles Policy/Strategy files, CMH; Interv, author with Zorthian, 1 Dec 76.

with as little publicity as possible.[19] "The concern of the Vietnamese," Westmoreland later told Wheeler, "is that the arrival of this large contingent of Americans could trigger demonstrations with overtones of cessation of hostilities and peace by negotiation."[20]

In compliance with South Vietnamese wishes, the State and Defense Departments ordered the U.S. mission in Saigon to prevent premature disclosures of the landing. Reporters at Da Nang could nevertheless see that the base was preparing for the arrival of American troops. On 2 March they filed dispatches to that effect. Although the revelation in the end caused few if any problems with either the South Vietnamese or the enemy, it startled official Washington.[21] "Rash of stories under Saigon dateline on forthcoming landing of Marines," McNamara cabled Westmoreland, "has seriously compromised policy and decision making here. The irresponsible, if not insubordinate actions which have led to these leaks must be stopped. We are taking action to prevent similar occurrences in the future at this end, including the application of severe disciplinary penalties. Please do the same in Saigon."[22]

Westmoreland responded that there had been no leak to the press and that newsmen had once more built their stories on readily observable circumstantial evidence. "I am sure you appreciate we have been operating here under 'maximum candor' policy which has been encouraged by Washington agencies." Underscoring his earlier call for an examination of censorship, he added, "I have felt for some time and have so expressed myself to General Wheeler and Admiral Sharp that this policy must be modified in view of changed nature of military activities."[23]

The need for a better way of protecting sensitive information became even more apparent shortly after Westmoreland's cable reached Washington. On 2 March, following several postponements brought on by continuing administrative chaos within the South Vietnamese government, the United States finally began a program of regular, gradually intensifying air attacks against North Vietnam (ROLLING THUNDER). In the process a loophole appeared in Zorthian's agreement with the Saigon correspondents. Until then reporters had been able to discern U.S. attacks in Laos only from leaks, casualty announcements, and Communist propaganda. Lacking details, they wrote infrequently on the subject. With the start of routinely announced operations against North Vietnam, however, they had little difficulty deducing when Laos was the target. On 3 March, for example, the United States launched a large air strike from Da Nang but refused to confirm that the raid was going into North Vietnam. Reporters knew that there

[19] Msg, Wheeler JCS 736–65 to Westmoreland, 27 Feb 65, Westmoreland History, bk. 13, tab 75; *Pentagon Papers*, 3: 423; Msg, MAC J00 6394 to CINCPAC, DAIN 564036, 2 Mar 65.

[20] Msg, MAC J-3 to JCS, DAIN 551733, 18 Feb 65.

[21] Msg, Joint State/Defense 6068 to Saigon, 27 Feb 65, Miles Policy/Strategy files; Msg, Westmoreland MAC 1110 to McNamara, 3 Mar 65, Westmoreland Papers. Both in CMH. For an example of the news stories, see Tad Szulc, "More Marines Due for Vietnam Duty," *New York Times*, 2 Mar 65.

[22] Msg, McNamara OSD 754 to Westmoreland, 2 Mar 65, Westmoreland History, bk. 14, tab 9, CMH.

[23] Msg, Westmoreland MAC 1110 to McNamara, 3 Mar 65.

was only one other place that the aircraft could go and cabled word to their editors that Laos had been bombed just as soon as the planes returned to base.[24]

The front page stories that followed brought a quick, vehement reaction from President Johnson. "Highest authority continues to be gravely concerned by speed and completeness of discussion of operational details of military missions in Laos and North Vietnam," the State Department told Ambassador Taylor.

> Latest example is reporting of number of planes in most recent BARREL ROLL operation, apparently based on Danang observation. All such numbers and quantities redouble international pressure against these operations and serve no useful purpose. In the case of Laos, they also complicate Souvanna's problem. . . . We recognize that there are Vietnamese as well as U.S. sources for much of this information. We also recognize that Danang and perhaps other air bases are open for observation, but we believe that if U.S. sources sternly refused details, few reporters will seek out accurate facts by themselves. Should they persist, we believe you should consider placing the environs of airfields including even city of Danang off limits to unauthorized U.S. citizens. These are military operations and their size and shape should be kept firmly classified.[25]

The Director of the United States Information Agency, Carl Rowan, arrived in Saigon a few days later. Remarking that the press corps in Saigon had begun to grow and was becoming unwieldy—by the end of the year more than 250 reporters would be present—he reported to the State Department that correspondents were competing strenuously for what news there was and that more irresponsible revelations were bound to be the result. Control was impossible under non-wartime conditions, but some arrangement to reduce current difficulties seemed imperative. At the very least, contingency planning should begin for the "stringent measures" that would become necessary if the war escalated much further.[26]

Officials at the U.S. mission shared Rowan's concern but remained convinced that neither censorship nor involuntary restraints on the press would do much good. The United States was operating from a sovereign country, Ambassador Taylor observed, in which newsmen were free to travel by other than U.S. military means and to file dispatches through cables and telephones operated by the South Vietnamese. Under those circumstances, the South Vietnamese government would have to impose censorship on foreign correspondents, and there was no guarantee it would confine its supervision to military matters. Any attempt to close the city of Da Nang to American citizens would be equally impractical, and even denying access to the Da Nang Air Base posed difficulties. South Vietnam's commercial airline occupied a terminal at one end of the field, where newsmen could easily observe departing and arriving aircraft if excluded from the rest of the base.[27]

[24] Msg, Saigon 2876 to State, 6 Mar 65, Westmoreland History, bk. 14, tab 21, CMH; [AP], "U.S. Bombing Raid on Red Aid Route in Laos Reported," *New York Times*, 4 Mar 65.

[25] Msg, State 1881 to Saigon, 3 Mar 65.

[26] Msg, Saigon 2873 to State, 6 Mar 65, DDI Public Affairs Operations/Origins file.

[27] Msg, Saigon 2876 to State, 6 Mar 65.

USIA's Carl Rowan *(center) tours South Vietnam.*

Taylor decided nevertheless that he had to respond in some manner to the mounting pressure for restrictions. He outlined the steps he was taking and made a number of suggestions in a message to the State Department on 6 March. The U.S. mission, he said, had already approached the South Vietnamese government to arrange for the closing of the Da Nang and Bien Hoa Air Bases to all but escorted newsmen. That step would at least reduce the access of the Saigon correspondents to South Vietnamese and American airmen and limit the amount of specific information they could acquire. For the rest, since newsmen were about to confirm that American planes based in Thailand were participating in attacks on North Vietnam, Taylor suggested that official spokesmen make verification more difficult by limiting strike announcements to a simple statement that American and South Vietnamese planes had attacked North Vietnam, giving no indication of the size of the operation or whether the aircraft had been land or sea based. For the same reason, information officers might stop announcing the number of sorties flown in support of ground operations and revealing when planes went down in Laos. Although announcements were unavoidable when a pilot was killed, wounded, or captured in Laos, official spokesmen might dampen speculation in the press by stating blandly that an airman had been lost in connection with operations in Southeast Asia.[28]

The Deputy Assistant Secretary of State for Public Affairs, James L. Greenfield, believed Taylor's suggestions might cause more problems than they solved. In passing them to the State Department's Far Eastern Affairs Division, he noted that Communist wire services and radio programs would continue to announce the times and locations of strikes in Laos, whatever the United States did to cut off discussion of the subject. If further limitations were imposed, correspondents would verify the announcements by paying civilian South Vietnamese to count the aircraft leaving Da Nang and would surely write stories condemning the new information policy as an attempted cover-up. Since Zorthian's arrangements with the Saigon correspondents had already solved the most pressing security problems at Da Nang, Greenfield concluded, playing down the raids in Laos appeared to be Taylor's main concern. Perhaps that problem could be eliminated by launching those raids only from aircraft carriers and fully secure bases outside of South Vietnam.[29]

The State Department sent Greenfield's suggestion to the Defense Department but with the notation that there were no fully secure bases in Southeast Asia. Defense Department officials concluded, in any case, that events themselves would shortly remedy whatever problems remained. "There will be so many airstrikes," Admiral Francis J. Blouin of the Office of International Security Affairs avowed, ". . . the press will find it difficult to tie in communist announcements with specific missions."[30]

Ambassador Taylor himself reconsidered the suggestion that the Military Assistance Command restrict information on the number of planes involved in an air strike, settling instead for an earlier recommendation that official spokesmen give the size of attacks on North Vietnam in round numbers. The Defense Department accepted the modification. It continued to insist that the command withhold all information on the bases from which strikes originated, but did sanction occasional, carefully coordinated pilot briefings and routine mention of the most common types of bombs.[31]

In the meantime nothing was done about Taylor's decision to close Da Nang Air Base to all but escorted reporters. South Vietnamese security officers accordingly announced on the morning of 15 March that newsmen could visit the installation only in the company of an information officer. Shortly thereafter local U.S.

[28] Ibid.; Interv, author with Barry Zorthian, 6 Dec 76, CMH files. Zorthian noted that Taylor always drafted his own messages to Washington.

[29] Memo, James L. Greenfield for Leonard Unger, Deputy Assistant Secretary of State for Far Eastern Affairs, 10 Mar 65, sub: Suggested Steps To Reduce Press Coverage of Air Strikes From Danang and Bien Hoa, DDI Air Incidents/Policy file.

[30] Memo, Leonard Unger for Rear Admiral Francis J. Blouin, 12 Mar 65, sub: BARREL ROLL and the Press, DDI Air Incidents/Policy file. Quote from Memo, Blouin for Unger, 16 Mar 65, sub: BARREL ROLL and the Press, ISA 381 1965, 70A3717, box 46, WNRC.

[31] Msg, Saigon 2950 to State, 13 Mar 65; Msg, Joint State/Defense/USIA 7890 to Saigon, 26 Mar 65; and Msg, Defense 6826 to CINCPAC, 10 Mar 65, all in DDI Air Incidents/Policy file.

commanders notified correspondents that they were no longer welcome at American service clubs and canteens at the base.[32]

As Greenfield had predicted, the move sparked a vehement reaction in the press. Berating the entire information program, the Associated Press pointed out that MACV's rules required a military escort for each correspondent visiting Da Nang Air Base but that only two information officers were available to serve the more than thirty newsmen present in the area. In a widely quoted statement, AP's managing editor, Wes Gallagher, charged that the Pentagon's news restrictions threw into doubt the American people's ability to obtain a true picture of the war. "Barring correspondents from free access to air bases and other military installations and providing an 'escort' for every correspondent," he said, "is clearly aimed not at security matters but at controlling what American fighting men might say." Terming the closing of Da Nang Air Base the sort of "totalitarian abuse" that had "never occurred in the darkest days of World War II," Richard Starnes of the *New York World-Telegram* took up Gallagher's theme. In past conflicts American reporters had always been able to collect war news at its source by flying on combat missions and by sharing the fighting man's lot at the front. That was no longer the case, he avowed; many raids against North Vietnam had been revealed only by Radio Hanoi.[33]

Arthur Sylvester responded to the mounting criticism by announcing that the restrictions at Da Nang had been imposed by the South Vietnamese without notice to the United States and that the Defense Department continued its policy of "complete candor with newsmen." The editors of the *Chicago Tribune* interpreted the statement to mean that the Pentagon allowed correspondents to report "anything they see in a dark closet after the door has been closed." The *New York Herald-Tribune* compared the Da Nang imbroglio to incidents that had occurred during the Diem era. The exclusion of correspondents from messes and clubs at the base put the U.S. government in the incongruous position of trusting the South Vietnamese employees of those establishments over some of its own citizens.[34]

The Honolulu Information Conference, March 1965

The controversy was at its height when the information conference Zorthian had requested convened in Honolulu between 18 and 20 March to consider whether censorship of the press in Vietnam would produce the results the John-

[32] ASD PA, Report of the Honolulu Information Conference, 18–20 March 65, tab D, DDI Information Conference folder, hereafter cited as Honolulu Conference Report; Memo 23d Air Base Group, Danang, for Office of Information, 15 Mar 65, sub: Security, Honolulu Conference Report, tab E.

[33] [AP], "Vietnam Curbs Hit by Newsmen," *Baltimore Sun*, 18 Mar 65; Richard Starnes, "Pentagon Has Viet Story—But It's Kept Secret," *New York World-Telegram*, 19 Mar 65.

[34] "New Restrictions Disavowed by U.S.," *New York Times*, 19 Mar 65; "Managed News From Both Ends," *Chicago Tribune*, 20 Mar 65; "Pentagon Candor on Viet War," *New York Herald-Tribune*, 21 Mar 65.

Zorthian (*left*) Meets With Newsmen

son administration sought. Recognizing that the Da Nang policy had been a sharp blow to official relations with the news media, the assembled delegates—Zorthian; Greenfield; Bankson; Baker; the new MACV Chief of Information, Col. Benjamin W. Legare, U.S. Army; and representatives of all U.S. government agencies concerned with the war in South Vietnam—concluded that the uproar prefigured what was likely to happen if the Johnson administration decided to impose a restrictive press policy. American success in South Vietnam depended upon the support of the public, they noted in their final report, and that support was likely to waver if "any significant number of our people believe . . . they are being misled."[35]

Working from that premise, the group rejected any form of field press censorship, opting firmly for the system of voluntary cooperation Zorthian had already put into effect. Censorship would require the legal underpinnings of a declaration of war as well as an enormous logistical and administrative effort. The censors would need jurisdiction over all communications and transportation facilities connecting South Vietnam with the rest of the world and parallel authority over civilian mail. That would necessitate a large number of multilingual military per-

[35] This section is based on the Honolulu Conference Report.

sonnel to do the censoring and expanded, U.S.-controlled teletype and radio circuits in South Vietnam to move the censored material. Even if the United States could meet those conditions, the South Vietnamese remained an unknown quantity. Since they were responsible for their own internal affairs, they would necessarily play an important part in any censorship program. Yet lacking a concept of American-style freedom of the press, they would undoubtedly exercise their prerogatives with a heavy hand. In any case, many of the Saigon correspondents were foreigners beyond the reach of American military regulations and likely to resist any attempt to bring them under control.[36]

Voluntary cooperation, as opposed to censorship, provided a number of advantages. Besides retaining the policy of maximum candor that had all along contributed to what the conferees termed "accurate and constructive news coverage," the approach allowed for a measure of control over the Saigon correspondents. In return for accreditation to cover the war, military transportation around South Vietnam, and access to important briefings and interviews, reporters would have to agree to abide by certain rules designed to protect military security. Those who accepted that obligation would, in addition to their other benefits, gain entree to candid, sometimes classified information, while those who refused would find their privileges at an end. Since the Saigon correspondents had already behaved responsibly in agreeing to withhold reports on air strikes in progress, the procedure had an additional merit. It recognized the good faith and honor of news media representatives, "a recognition well-deserved by most."

The guidelines for release of information on the air war that the information officers recommended were basically the same as those already in effect. Although the policy of "no answer, no lies" would have to continue with regard to air strikes in Laos, the sooner those strikes could be announced, the better. "All information about strikes should be released except that which has legitimate . . . security implications or . . . which must be withheld in the national interest. When it is necessary to withhold information, there should be sound reasons that can be given to media representatives on an off-the-record or background basis."

Only slightly less important than what was released was the way in which the U.S. government made the announcement. In the past, when an attack on North Vietnam had been imminent, the U.S. mission and South Vietnamese officials had drafted a joint communique for initial release in Saigon, transmitting it to Washington for revision and final clearance. As soon as the aircraft returned, the Saigon correspondents had received the document—a statement of grievances against North Vietnam designed to justify the attack before world public opinion—and then had attended a briefing on the strike by the MACV air operations officer. Meanwhile, backup briefings had been conducted at the White House, the Pentagon, and even the State Department. Although intended to corroborate the

[36] The Honolulu Conference Report fails to make much mention of possible problems with the South Vietnamese, but the question definitely figured into the discussion, as noted in Interv, author with Zorthian, 1 Dec 76.

information the Military Assistance Command and the South Vietnamese were releasing in Saigon, those sessions in fact had often undercut the efforts of the U.S. mission by being more elaborate and informative. Taking place in Washington, the briefings had also tended to draw attention away from South Vietnam's contribution to the war. The inconsistencies and shifting emphases that had resulted from so many release points were eroding the policy of maximum candor and diminishing Zorthian's authority as overall coordinator of information policy.

The whole complicated process might have been necessary when the attacks on North Vietnam had begun, especially the practice of justifying the strikes by linking them to enemy provocations. Now the procedure had become unwieldy. In addition, the coupling of strictly military matters to obvious propaganda had done nothing to improve official relations with the press. For that reason, public statements justifying the attacks should be separated from strike announcements and find expression either in special press releases on enemy terrorism or in briefings dealing exclusively with North Vietnamese aggression. If stronger emphasis seemed necessary, the South Vietnamese could publish the reports on enemy activities that they made to the International Control Commission. In the same way, since the South Vietnamese remained sensitive to anything that made them seem inferior to the United States, Washington agencies had to make certain that the first announcement of air strikes took the form of a joint U.S.-South Vietnamese communique that always originated in Saigon. With a straightforward approach free of political taint, the conferees concluded, ROLLING THUNDER attacks might at last cease to be the subject of sensational or critical stories, becoming instead a routine part of the daily MACV communique.

When the conference ended on 20 March, Rodger Bankson submitted a final report to the White House and other concerned agencies. After listing all the recommendations suggested in Honolulu, he detailed a number of strictly organizational matters: the need for better information officers, ways to improve the South Vietnamese government's relations with the press, recommendations for an interagency coordinating committee to control the large number of urgent cables from Washington agencies that appeared in Saigon every time the press said something unusual about the war. Bankson sent a preliminary draft of the report to the Joint Chiefs of Staff for comment.[37]

General Wheeler himself replied, making several suggestions that required only minor adjustments of wording but rejecting outright one of the conference's main proposals. "I believe the United States would weaken its position in world opinion by removing statements justifying attacks on NVN from strike announcements," Wheeler said. "In my opinion, constant repetition of such justifications is necessary, using all possible occasions and means to do so." Formulas adaptable to many circumstances might substitute for statements tailored to specific situations in advance, but the conference's recommendations should be revised to

[37] Honolulu Conference Report.

read that all strike announcements would contain a statement of the basic justification for air attacks against North Vietnam.[38]

Unwilling to delay approval of the report for the sake of one point, Arthur Sylvester substituted Wheeler's exact wording for the original recommendation. The Defense Department approved the amendment on 3 April, along with the rest of the report. Four days later Zorthian briefed the Saigon correspondents on the rules, requesting their voluntary cooperation.[39]

Shortly thereafter, the MACV Office of Information modified Taylor's restrictions on the access of newsmen to Da Nang Air Base. Correspondents would have to obtain from the South Vietnamese National Press Center in Saigon a renewable identification card, valid for one month. They would have to be escorted while crossing South Vietnamese areas of the base, but once they had agreed to abide by security regulations and had received clearance from a newly established press center in Da Nang, they were to have free access without escort to all unclassified sectors within American portions of the base. If they had any doubts about security aspects of the stories they wrote, they had only to submit their copy to the MACV Office of Information for review.[40]

The new guidelines met with immediate criticism from the press, much of it centering upon the fact that Zorthian would be administering the rules. Newsmen had little objection to Zorthian himself or to protecting information of value to the enemy, but the minister-counselor for public affairs had just been appointed the head of a new agency, the Joint U.S. Public Affairs Office (JUSPAO), that was to take charge of both relations with the news media and propaganda operations against the enemy. Fearing that propaganda might somehow taint the news, syndicated columnist David Lawrence and Associated Press president George Beebe objected.[41]

Aware of the mood of the press and concerned lest the arrangement diminish the credibility of the military information program, Rodger Bankson on 28 April recommended that the U.S. mission exclude the MACV Office of Information from the Joint U.S. Public Affairs Office. On 7 May the State Department concurred; it left Zorthian in charge of both propaganda and press relations in Saigon but deleted the MACV Office of Information from the organizational chart describing the new agency.[42]

Satisfied by the arrangement, the Saigon correspondents said little on the subject, but the wire services continued to argue that propagandists had gained con-

[38] Chief of Staff Memorandum (CSM)-518-65, Wheeler for ASD PA, 29 Mar 65, sub: Information Policy in Vietnam, DDI Information Conference folder.

[39] Msg, Joint State/Defense/USIA 8389 to CINCPAC, COMUSMACV, 3 Apr 65, DDI Information Conference folder; Msg, Saigon 3266 to State, 7 Apr 65, DDI Air Incidents/Policy file.

[40] Msg, Saigon 3322 to State, 10 Apr 65, DDI Air Incidents/Policy file.

[41] [AP], "Editors Criticize USIA in Vietnam," *New York Times*, 21 Apr 65; "Truth or Propaganda," *New York Times*, 23 Apr 65; David Lawrence, "U.S. Censorship Policy in Viet Assailed," *New York Herald-Tribune*, 27 Apr 65.

[42] Memo, OASD PA for USIA, 28 Apr 65, sub: JUSPAO, DDI PA OPNS ORGNS file; Msg, State 578 to Saigon, 7 May 65, FAIM/IR.

trol of news from South Vietnam. The publisher and editor of the *Denver Post* and a member of the USIA Advisory Committee, Palmer Hoyt, thus advised President Johnson in June that "a major damaging flap" was inevitable unless the administration took action. Hoyt's suggestion led to a move by the State Department to make Zorthian minister-counselor for press relations only.[43]

Ambassador Taylor objected to the change. The separation of Zorthian's responsibility for propaganda from his work as minister for public affairs, he said, would frustrate the purpose of the Joint U.S. Public Affairs Office, which had been designed to combine the complementary operations of propaganda and public affairs under one roof and one director in order to cut down on needless duplication and waste. The U.S. Information Service was hardly in a position to censor anything. Although Zorthian advised both the U.S. mission and Westmoreland on public relations, Westmoreland had command responsibility for military news and in fact made most of the decisions on the subject. The issue thus went beyond the U.S. Information Service to the principle of civilian control over military news. Beebe, Lawrence, and the others were harking back to World War II and the Korean War, sentimentally hoping to reinstate the practices of a simpler day when military authority over news had been relatively unencumbered by political considerations. To give in to those concerns in a context as complicated as the one in South Vietnam could only make matters worse.[44]

Taylor's argument prevailed. For the rest, the basic apparatus for dealing with the news media in South Vietnam was in place. The men who formulated the policy had failed to separate military information from politics, but they had probably been naive in thinking they could. At the very least, they had created a system capable of giving the American people a reasonably accurate accounting of the war without at the same time helping the enemy. Whether the government of the United States would use it in that manner remained to be seen.

[43] Msg, State 3035 to Saigon, 24 Jun 65, FAIM/IR.

[44] Msg, Saigon 68 to State, 7 Jul 65, FAIM/IR.

7

Counterinsurgency Combat Operations

The policies agreed to at the Honolulu Conference in mid-March 1965 addressed immediate issues: censorship and the air war. Although the assembled public relations experts discussed the possibility that American ground forces might shortly begin fighting as units in South Vietnam, they deferred recommendations on the subject. Since classified documents on sensitive topics seemed to surface in the press with embarrassing regularity, they reasoned that a leak on so controversial a question might only provoke public relations problems.[1]

That a large infusion of American troops might become necessary to forestall almost certain defeat in South Vietnam was nevertheless apparent at the time to most senior American military planners. As early as 12 February, more than a month before the Honolulu Conference, the Joint Chiefs of Staff had suggested that Secretary of Defense McNamara consider the possibility of major troop commitments. On 20 February they had renewed their request, recommending the deployment of a full U.S. Army division in South Vietnam's Central Highlands.[2]

On 25 March General Westmoreland submitted an estimate of the situation in South Vietnam that supported the Joint Chiefs' call for American units. The South Vietnamese government was becoming increasingly feeble, Westmoreland warned, while its armed forces—led by men more concerned with political intrigue than with fighting the enemy—showed incipient signs of collapse. American troops were the only solution: 33,000 men immediately—a U.S. Army division, an Army airborne brigade, and an additional Marine battalion—and, if the bombing of North Vietnam produced no weakening of enemy will, more by mid-year.[3]

Despite Westmoreland's and the Joint Chiefs' recommendations, President

[1] Interv, author with Barry Zorthian, 19 Sep 77, CMH files.
[2] Msg, JCS 5147 to CINCPAC, 12 Feb 65, for Sharp from Wheeler, Plans and Policy file; Msg, JCS 1008-65 to CINCPAC, 20 Feb 65, for Sharp from Wheeler, Westmoreland Papers. Both in CMH.
[3] MACV, Commander's Estimate of the Situation in South Vietnam, 25 Mar 65, Westmoreland History, bk. 14, tab 38.

Marines Come Ashore at Da Nang, March 1965

Johnson and his civilian advisers remained unconvinced of a need for divisions. Although willing to have two battalions of marines guard the Da Nang Air Base, they were uncertain how the American public would react to a massive American buildup and feared that the presence of large numbers of foreign troops in South Vietnam might foster South Vietnamese xenophobia. They had visions of American soldiers, trapped in the Central Highlands, having to fight their way to the sea through both the enemy and mutinous South Vietnamese Army units engaged in a civil war. When the Viet Cong exploded a bomb near the entrance of the U.S. embassy in Saigon on 30 March, killing two Americans and fifteen South Vietnamese and wounding a large number of passersby, President Johnson thus refused to inaugurate another round of reprisals against North Vietnam. Unsure of his next steps and preferring a cautious stance that would preserve maximum freedom of action, he attempted instead to justify any decision he might make by publicly contrasting American self-restraint with North Vietnam's continuing provocations.[4]

[4] For mention of civilian concerns, see MFR, 4 Apr 65, sub: Summary of Meeting at Honolulu Between Taylor, Sharp, Westmoreland et al., Westmoreland History, bk. 15, tab 9, CMH. *Pentagon Papers,* 3: 374; William C. Westmoreland, *A Soldier Reports* (Garden City, N.Y.: Doubleday, 1976), p. 130.

The Mission of the Marines

In the end, Johnson settled upon an approach that he hoped would prepare for U.S. troop commitments while postponing the kinds of actions that might increase political tensions surrounding the Vietnam issue. Declining on 3 April to send all the fighting men the generals had requested, he instead ordered a Marine battalion landing team and a jet aircraft squadron to Da Nang and a second Marine landing team to Hue—in all, a mere 3,000 men. He also authorized the Army to send a full logistical command—20,000 men—to South Vietnam to begin improving the country's port and supply facilities. The marines already at Da Nang were to assume a more aggressive posture, patrolling in force and engaging the enemy in what Johnson called "counter-insurgency combat operations."[5]

Although the term Johnson chose was vague, Admiral Sharp had no doubts about the president's intentions. When the Military Assistance Command notified the marines at Da Nang of their new mission, stressing their continuing duty to guard U.S. installations rather than to take the offensive, Sharp told Westmoreland that "this is not what our superiors intend" and that the marines should begin seeking out the enemy aggressively as soon as possible.[6] Sharp went on to associate the new role of the marines with the defense of Da Nang, but MACV Chief of Operations, Brig. Gen. William E. DePuy, told Westmoreland that the distinction was a formality. The admiral was attempting to comply with a request from Secretary of State Rusk that the command for the time being depict the mission of all American troops in South Vietnam as defensive.[7]

During June, when Westmoreland was about to take the offensive in earnest and requested Sharp's assurance that Johnson's 3 April directive gave him the authority to do so, Sharp reiterated that it did and that clearer language would only have reduced the command's flexibility. "This phrase [counterinsurgency combat operations] under-went close and careful study," Sharp avowed, "and it stands today as the direct order from the highest authority through the Secretary of Defense, the Joint Chiefs of Staff, and me."[8]

At a 3 April meeting with Ambassador Taylor shortly after Johnson announced his decision to his top policy-making staff, Secretary Rusk characterized the president's caution as the product of domestic political considerations. "The president felt that he must not force the pace too fast," Rusk observed, "or congress and public opinion, which had been held in line up to now through the president's strenuous efforts, would no longer support our actions in Vietnam."[9]

The press guidance to the U.S. mission in Saigon that accompanied word of

[5] Msg, State 2184 to Saigon, 3 Apr 65, FAIM/IR; MFR, 3 Apr 65, sub: Meeting Between Secretary Rusk, Taylor, McGeorge Bundy et al., in Washington, D.C., Westmoreland History, bk. 15, tab 5, CMH.
[6] Msg, CINCPAC to MACV, 14 Apr 65, Westmoreland Papers, CMH.
[7] Memo, MACV J-3 for Westmoreland, 17 Apr 65, sub: Mission for the 9th MEB, in COMUSMACV's Notebook, Honolulu Trip, 18 Apr 65, tab 11, file 1797-66, 2/80, 67A4604, box 2, WNRC.
[8] Msg, Sharp to Westmoreland, 13 Jun 65, Westmoreland History, bk. 16, tab 26, CMH.
[9] MFR, 3 Apr 65, sub: Meeting Between Rusk, Taylor, McGeorge Bundy et al., in Washington, D.C.

Johnson's decision reflected the president's concern. "The pacing of deployments is of critical import," the State Department declared. "We do not desire [to] give [the] impression of a rapid, massive build up." Although the new Marine units were to deploy at the earliest possible moment after the ambassador gained South Vietnamese concurrence, all other deployments were to be "spaced over a period [of] time with publicity . . . kept at the lowest possible key." The Office of the Assistant Secretary of Defense for Public Affairs was to authorize all statements on the subject. If reporters began to ask questions, official spokesmen were to deflect them by describing the movement of troop transports toward South Vietnam as routine Pacific Fleet maneuvers.[10]

Little further press guidance on the subject went to Saigon. On the matter of the marines' new mission, information officers tailored their announcements to the three-phase program Westmoreland had devised to move his troops in easy stages toward full combat. During the first phase, when the additional Marine units began patrolling around Da Nang Air Base, MACV spokesmen repeated the undramatic language that had accompanied the arrival of the earlier Marine contingents on 7 March: "The limited mission of . . . [the] marines will be to relieve government of Vietnam forces now engaged in security duties for action in the pacification program and in an offensive role against communist guerrilla forces."[11] Quietly and off the record, information officers affirmed that the marines would fight if attacked and would deepen Da Nang's defenses by patrolling out from the base's perimeter, but they also stipulated that those activities were necessary to any defensive mission. Several weeks later, when Westmoreland prepared to inaugurate his second phase, moving the marines further onto the offensive as a mobile reaction force within a fifty-mile radius of Da Nang, MACV spokesmen observed routinely that U.S. forces had "a combat support role in addition to their defensive mission."[12] Only during May, when Westmoreland contemplated beginning the third phase, using the marines to support South Vietnamese units anywhere in the I Corps Tactical Zone, did the question of fully revealing the offensive mission of U.S. forces arise. Then the Johnson administration once more temporized, postponing a decision.

From the moment the first marines arrived in early March the Saigon correspondents appeared to have few doubts about the direction the United States was taking. Accepting the development as the natural outcome of U.S. deployments, they paid considerable attention to the arrival of American troops but wrote only a few stories on the marines' shift to the offensive. By the end of May, information officers in Saigon had come to suspect that the news media had given the subject so little coverage that the American public was at best only partially aware of what was happening in South Vietnam.[13]

[10] Msg, State 2184 to Saigon, 3 Apr 65; Msg, JCS 8387 to CINCPAC, 3 Apr 65, FAIM/IR.

[11] Msg, Saigon 3539 to State, DAIN 625622, 26 Apr 65. For an outline of Westmoreland's plan, see Msg, MACV J-3 11535 to CINCPAC, 10 Apr 65, Plans and Policy file, CMH.

[12] Msg, Saigon 3820 to State, 20 May 65, FAIM/IR.

[13] Msg, Saigon 3539 to State, DAIN 625622, 26 Apr 65; Msg, Saigon 3820 to State, 20 May 65.

Pulitzer Prize Winners Malcolm Browne, Peter Arnett, and Horst Faas

The Tear Gas Controversy

If reporters accepted the evolving nature of American ground combat in South Vietnam, there was no easing of tension between newsmen and officials. Trouble with the press became acute, indeed, shortly after the first Marine units arrived in South Vietnam, while President Johnson was considering whether to involve American ground forces in the war. It centered on the South Vietnamese Army's intention to employ tear and nausea-producing gases when those agents might assist in the rescue of prisoners of war.

Information officers in the field in South Vietnam recognized as early as December 1964 that the tactic might cause an outcry in the press. They suggested that the MACV Office of Information brief the Saigon correspondents on the subject to make the point that the gases in question were standard riot control agents in use around the world. That step would lessen the surprise and confusion bound to result if newsmen learned of the story on their own and would allow the command to differentiate clearly between tear gas and the lethal chemicals employed during World War I.[14]

[14] Interv, author with Zorthian, 19 Sep 77; Interv, author with Col Ralph Ropp (U.S. Army Support

Sensitive to any question involving American prisoners of war and unwilling to give the enemy even the slightest propaganda advantage, the State and Defense Departments refused to permit a prior briefing. As the information officers had predicted, when South Vietnamese troops employed tear gas twice during December 1964 and once again during January 1965, newsmen began to hear rumors. Associated Press reporter Peter Arnett contacted the MACV information liaison officer in the III Corps Tactical Zone, Capt. Richard Bryan, to learn the details. Holding to orders despite his own preference for a full disclosure, Bryan refused to respond. Aware that he had angered Arnett, the officer cabled the command to warn that the story was out and that Arnett would write about it as soon as he could tie the information he had to a specific incident. The command never responded.[15]

On 20 March Arnett's associate, photographer Horst Faas, learned that a South Vietnamese division was planning an operation in Hau Ngia Province and that one regiment was armed with tear gas. Because he knew of Arnett's fruitless encounter with Bryan, Faas joined the unit in the field without contacting the MACV Office of Information. Seeing firsthand that the troops were indeed in possession of gas canisters, he returned to Saigon to tell Arnett, who put the news on the Associated Press wire.[16]

Arnett began his dispatch by quoting a Radio Hanoi report that the United States and South Vietnam were using "poisonous chemicals" against the Viet Cong and that a twelve-year-old girl had suffered a swollen face in a recent gas attack. "By tacit agreement," he continued, "gas was not used in World War II or Korea." After relating Faas' experience, the reporter quoted a comment by a U.S. adviser to the effect that although the chemicals involved were nonlethal and of no lasting effect, they were still difficult to justify to an American public that remembered the mustard gas of World War I. Only toward the end of the piece did Arnett remark that American officers believed tear and nausea gases to be the most humane way to clear areas where the enemy was holding women and children hostage. Even then he implied that the South Vietnamese lacked the sophistication to use the tactic properly. "One case in which an experiment fizzled was reported in the II Corps area," he said. "Vietnamese troops moved in after a gas attack which they believed had put a Viet Cong unit out of action. But firing erupted from the gas-filled area. The troops were reported to have fled in disorder."[17]

Except for the quotation from Radio Hanoi with its unverifiable assertion that a girl had been injured, much of what Arnett said was true. The South Vietnamese did lack experience in using gas, and the tactic had thus far achieved few if any concrete results. Yet the image of the girl and the references to experiments, mus-

Command IO, 1965) and Lt Col Richard Bryan (MACV IO in the III Corps Tactical Zone), 24 Jan 77, CMH files.

[15] Interv, author with Ropp and Bryan, 24 Jan 77.

[16] Msg, Westmoreland MAC 1678 to Wheeler, 28 Mar 65, Westmoreland Papers, CMH.

[17] [AP Dispatch], untitled, datelined Tokyo, 22 Mar 65, DDI Gas file.

tard gas, and tacit agreements set a tone that obscured the reference to the possibility that tear gas might limit civilian casualties. "To the uninformed, all gas is poisonous," General Wheeler later told Westmoreland, "and an experiment is something conducted by a mad doctor in a secret laboratory."[18] Arnett had written a deliberately negative story, Barry Zorthian added, one whose worldwide anti-American impact "even a neophyte journalist would have known."[19]

The article circulated around the world for several hours before Zorthian and the MACV Office of Information learned of its existence. They never caught up. Information officers in Saigon were still attempting to determine the facts when Admiral Sharp's press spokesman in Hawaii telephoned to inform them that the secretary of defense wanted the Military Assistance Command to issue a statement within the hour. Thirty minutes later briefers told the Saigon correspondents that "In tactical situations in which Viet Cong intermingle with or take refuge among non-combatants, rather than use artillery or aerial bombardment, Vietnamese troops have and use a type of tear gas in the area. It is a non-lethal type of gas which disables temporarily, making the enemy incapable of fighting. Its use in such situations is no different than the use of disabling gases in riot control."[20]

Westmoreland later admitted that the communique was less than perfect. "We were somewhat stampeded by the OSD directive," he told Wheeler, "and while the spokesman's wording is defensible, it might have been more facile and lower keyed. He used the right words, *type of tear gas* and *riot control*, and the wrong, *non-lethal* and *disabling*." The result was that newspapers around the world paid little attention to MACV's explanation, preferring to interpret the statement as confirmation of Arnett's story.[21]

Further attempts to clarify the situation met with little more success. At a formal press conference in Saigon the next day, the briefer, a senior colonel from MACV's Operations Division, underscored the harmlessness of the chemical agents involved and declared that the United States and South Vietnam had never experimented with gas. Then he contradicted himself by saying that both countries were merely "interested in developing techniques and tactics of employment under varying situations."[22] Shortly thereafter Rusk and McNamara issued statements emphasizing that the gases in question were standard riot control agents in use in many countries, but by then war critics in the United States and elsewhere had begun to charge that the Johnson administration had confirmed it was experimenting with toxic chemicals. "The argument that the non-toxic gas is more merciful than anti-personnel weapons has some merit," the *Washington Post* declared, "but not much. The trouble is that although the gas may not be

[18] Msg, JCS 1071 to Westmoreland, 25 Mar 65, Westmoreland Papers, CMH. See also Msg, MACV J-3 9171 to Defense, 23 Mar 65, DDI Gas file.

[19] Msg, Saigon 3124 to State, 28 Mar 65, for Greenfield from Zorthian, DDI Gas file.

[20] Ibid.; Msg, Saigon 3053 to State, 23 Mar 65, DDI Gas file.

[21] Msg, Westmoreland MAC 1678 to Wheeler, 28 Mar 65.

[22] Msg, MACOI 9168 to OSD PA, 23 Mar 65, sub: Supplemental Press Trends No. 075A-65, DDI Gas file.

poison, the word is, and all the propaganda resources in the world cannot explain away its employment as an act of Christian charity.'' Even gas that only temporarily disabled could kill the very young, the very old, and those with heart and lung ailments, the *New York Times* asserted. ''In Vietnam, gas was supplied and sanctioned by white men against Asians. This is something that no Asian, communist or not, will forget.'' The Federation of American Scientists meanwhile avowed, ''We find it morally repugnant that the United States should find itself the party to the use of weapons of indiscriminate effect, with principal effectiveness against civilian populations. . . . The characterization of such applications as 'humane' is incomprehensible, to say the least.''[23]

Newspapers and governments throughout the world echoed those themes. Radio Moscow charged that the United States was using increasingly barbarous methods in South Vietnam. The Canadian Broadcasting System compared the South Vietnamese use of gas on the Viet Cong to the German use of gas on Canadians during World War I at the Battle of Ypres. Other Canadian commentators decried American readiness to use unconventional arms against Asians, a circumstance they likened to the atomic bombing of Hiroshima. In England Prime Minister Harold Wilson was jeered by members of his own party when he observed in the House of Commons that he wanted to check the facts before making any comments on the subject.[24]

The reaction in Asia was just as vehement. The Tokyo newspaper *Asahi* charged that ''the U.S. bombing of North Vietnam and particularly the use of poison gas has given the impression that Asians are guinea pigs in chemical warfare tests, and this has lost America many friends.'' Japanese delegates to the United Nations in New York told American diplomats that the comment in *Asahi* applied to the rest of Asia as well. United Nations delegates from Burma, Ceylon, India, and Afghanistan were responding emotionally to the news, they said, at the very moment when their governments had seemed likely to accept the justice of U.S. air strikes against North Vietnam. The only thing that the United States could do to resolve the crisis, the diplomats counseled, was ''keep quiet.''[25]

The reasons underlying the outcry were difficult to ascertain at the time because the agents involved were of a variety different from the lethal gases used during World War I and subsequently banned by the Geneva Conventions. In the end, the cause may have been less the gases themselves than a combination of the dread inspired by the very word *gas* and a growing concern on the part of the world community that a major war of international significance was brewing in

[23] OSD PA, News Release 183-65, 23 Mar 65; Statement by Secretary of Defense Robert S. McNamara, and Dept of State, News Release 59, 24 Mar 65, Secretary Rusk's News Conference of 24 March 1965, all in DDI Gas file; ''Blackening Our Name,'' *Washington Post*, 24 Mar 65; ''Gas (Non-lethal) in Vietnam,'' *New York Times*, 24 Mar 65; Federation of American Scientists, News Release, 25 Mar 65, sub: Scientists Denounce U.S. Use of Gas Weapons, DDI Gas file.

[24] ''Vietnam Gas Use Draws Protests,'' *Baltimore Sun*, 23 Mar 65; Msg, Ottawa 1197 to State, 25 Mar 65, FAIM/IR; Clyde H. Farnsworth, ''War-Gas Debate Stirs Commons,'' *New York Times*, 24 Mar 65.

[25] Memo, USIA for the President, 25 Mar 65, sub: Daily Reaction Report; Msg, USUN, New York, 5768 to State, 24 Mar 65. Both in DDI Gas file.

Southeast Asia. As Philip Geyelin of the *Wall Street Journal* observed at the height of the outcry,

> the real significance of the gas "crisis" is that it should ever have stirred such an uproar at all. . . . What this suggests, at the least, is a very large gap indeed between the public pledges of support and the true feelings of a good number of allied nations. It also suggests that grudging public support would give way quite quickly to open opposition should the U.S. war effort in Vietnam really give the world something more than the use of riot gas to worry about. . . . The world is a lot more nervous than might have been suspected about just what the U.S. is up to, a lot more skeptical about the U.S. side of the argument, and a lot more eager for any kind of solution, however injurious to Western interests, so long as it heads off the danger of a wider war.[26]

Whatever the controversy's cause, its effects endured for months. In Saigon Ambassador Taylor almost immediately withdrew General Westmoreland's authority to employ tear gas. Secretary McNamara tied the general's hands further in July when he told correspondents at a Saigon backgrounder that "If by itself it would save the situation, I wouldn't [use gas]. . . . My God, I don't want to go through that again. . . . We cannot explain the difference between a riot control agent and a lethal one."[27]

As a corollary to the flurry over gas, a project designed to destroy enemy crops in remote areas of South Vietnam also came under review. In mid-March 1965, just before the gas controversy arose, MACV intelligence detected four Viet Cong battalions and several independent companies hiding in Binh Dinh Province, a heavily populated region 250 miles northeast of Saigon. The command requested permission to destroy crops in those portions of the province fully controlled by the Communists in order to restrict the enemy's ability to live off the land, a request approved by Ambassador Taylor. Arnett's story appeared shortly thereafter. Although newsmen had known of herbicide operations almost from the time they began in 1962 and had written scarcely a critical comment, State Department officials suddenly saw potential embarrassment in the program and requested that the Joint Chiefs of Staff put an end to it.[28]

Although the Joint Chiefs refused, they were unable to avoid a compromise that limited Westmoreland's flexibility. "While 'gas crisis' is running its course," that agreement stipulated," . . . it would be preferable that major crop destruction programs . . . be stretched out or otherwise reduced in visibility, provided this can be done without publicity and without serious problems [with the South Vietnamese]." The Military Assistance Command was to take "maximum measures to reduce publicity" and to prepare "to meet any inquiries with a full ration-

[26] Philip Geyelin, "Vietnam Vexation: Outcry Over Use of Gas Points Up U.S. Aloneness There," *Wall Street Journal*, 26 Mar 65.

[27] MFR, OSD PA [Jul 65], sub: Excerpts of SECDEF Background Briefing—July Visit to Saigon; Msg, MAC J311 to CINCPAC, 9 Sep 65. Both in DDI Gas file.

[28] Msg, Saigon 3004 to State, 18 Mar 65, and Msg, State A-4 to Saigon, 11 Jul 64, sub: Response to Press Queries on Use of Chemical Weed Killers for Defoliation, both in FAIM/IR. Msg, Wheeler JCS 1071 to Westmoreland, 25 Mar 65; Msg, State 2110 to Saigon, 27 Mar 65, FAIM/IR.

ale, including measures taken to provide for inhabitants of the area.''[29]

The U.S. mission immediately postponed spraying in Binh Dinh Province and notified the State Department that it would expand the program to populated regions only after it had studied the public relations impact of those operations in remote areas. Should an adverse reaction develop, the program would cease.[30]

A U.S. Air Force Defoliation Mission

In response to questions from newsmen, U.S. spokesmen in Saigon emphasized that crop destruction operations were South Vietnamese in origin and referred all further questions to South Vietnamese government spokesmen who had been instructed to make low-keyed announcements every time an herbicide operation ended. Whenever the opportunity arose, MACV information officers also pointed out that the chemicals involved were standard weed-control agents in use throughout the United States, harmless to humans, animals, water, and soil.[31]

Although the U.S. mission's careful handling of the issue kept the Saigon correspondents from criticizing crop destruction until late in the war, officials in Washington remained sensitive to the potentially explosive nature of the program long after the controversy over Arnett's revelations had ended. In July 1965, when the Military Assistance Command petitioned for permission to expand herbicide operations, the State Department concurred only reluctantly, warning that the whole question was still a matter of serious political concern. Because of that concern and of questions about how the program would affect the loyalty of South Vietnam's peasants to the Saigon government, the department instructed Westmoreland to obtain the permission of both the ambassador and a senior South Vietnamese official before approving each operation. In general, crop destruction was to concentrate on remote areas where the enemy had difficulty finding food and where early reestablishment of the government's control appeared doubtful. If the command decided military advantages outweighed political and psychological drawbacks in some particular instance, a spraying operation might occur in a heavily populated area, but only with prior authorization from Washing-

[29] Quotes from Msg, Wheeler JCS 1071 to Westmoreland, 25 Mar 65. See also Msg, State 2084 to Saigon, 24 Mar 65, FAIM/IR.

[30] Msg, Saigon 3089 to State, 25 Mar 65, FAIM/IR.

[31] Msg, State 2128 to Saigon, 30 Mar 65, and Msg, Saigon 203 to State, 20 Jul 65, both in FAIM/IR.

ton. Under all circumstances, the people living in target regions were to be warned that spraying was imminent. At the same time, to lessen hostility toward the South Vietnamese government, all damage that resulted from the spraying was to be laid to the Viet Cong, who had refused to leave the target area.[32]

Censorship Reconsidered and Rejected

As the controversy over gas subsided, the Johnson administration began to explore ways both to avoid similar outcries in the future and to ensure a climate of opinion in the United States conducive to the hard decisions it was having to make. In an effort to demonstrate to the world that the United States had allies in its fight to defeat Communist aggression, the State Department began talks with Australia, South Korea, and other Asian nations to obtain troops and military assistance for South Vietnam. President Johnson also prepared a major policy statement to emphasize America's desire for peace and the Communists' preference for war. Speaking at Johns Hopkins University in Baltimore on 7 April, he declared the willingness of the United States to negotiate with North Vietnam "without preconditions" and proposed massive American regional development of Southeast Asia as an alternative to continued war. The president had made those offers, the State Department later informed Ambassador Taylor, with no expectation of a cease-fire or of a halt to American activity in South Vietnam prior to a North Vietnamese withdrawal. He had merely sought to show that the United States was "prepared to do anything reasonable to explore the way to peace" and to throw the burden of any future escalation upon the enemy by making it clear that the Communists were the ones impeding a peaceful settlement.[33]

Although the president's comments attracted favorable reviews in the American news media and throughout the rest of the world, Johnson was disturbed to learn after delivering them that news reports from Saigon were counteracting the conciliatory image he had tried to create. On the day of the speech, U.S. Ambassador to India Chester Bowles reported that wire-service stories out of Saigon were building an impression among Indian newspaper readers of large-scale aerial warfare in Southeast Asia. North Vietnamese assertions that the United States was hitting nonmilitary targets and using napalm, Bowles said, were poorly balanced in those stories by American accounts of enemy sabotage and atrocities. Several days later, as if to confirm the ambassador's cable, MACV briefers routinely announced that three thousand more U.S. marines would shortly arrive in South Vietnam, prompting more articles in the press on the expansion of the war. In a dispatch that emphasized growing U.S. involvement in South Vietnam, one reporter asserted that "American war planes, swarming against North Vietnam in unprecedented numbers, wrecked three bridges and scored for the first

[32] Msg, State 294 to Saigon, 29 Jul 65, and Msg, State 370 to Saigon, 7 Aug 65, both in FAIM/IR.
[33] Msg, State 2217 to Saigon, 6 Apr 65, FAIM/IR; *Pentagon Papers*, 2: 355.

time against Mig fighters. . . . South of the border, fresh landings of U.S. Marines—about 3,000 men and a jet squadron—were in the offing at Danang and Hue. The Navy and Air Force launched 220 planes laden with 245 tons of bombs and rockets for this 20th and most massive of the air strikes that started two months ago."[34]

The president's displeasure took little time reaching General Westmoreland. "Highest authority," General Wheeler cabled, "is increasingly unhappy at press releases which forecast impending U.S. reinforcements to South Vietnam, discuss U.S. military actions to include targets and extension of target system, and represent magnitude and weight of effort of U.S. strikes against [North Vietnam]." The president was concerned that prospective deployments might become public knowledge before the administration had consulted congressional leaders. He also believed that, besides being of possible value to the enemy, emphasis upon the weight of American attacks tended "to discount and overshadow his speeches and statements establishing the moderation and restraint of U.S. actions in the face of provocations" against South Vietnam. "It is a fact," Wheeler said, "that the situation in the U.S. is exacerbated and pressures upon highest authority increased by press coverage of items such as those cited above."[35] Wheeler added in a subsequent message that "we recognize . . . the press release on landing of marines and press briefings on operations have been in accord with agreed procedures. . . . My intent was to acquaint you with the situation . . . in Washington and solicit thoughts from the field which might help in these regards. It may well be that nothing short of press censorship will serve this end."[36]

General Westmoreland replied that censorship might indeed be the only answer to the problem but that "practical considerations" made it impossible.[37] Admiral Sharp agreed. "In view of the increasing tempo of air strikes and proposed deployments to South Vietnam, I expect press coverage to move into an even higher key. As we escalate, so will the reporting of the press. I doubt that even with field press censorship this could be avoided, and it is quite likely that censorship would even have an inflammatory effect." Sharp added that the sensationalized reporting and premature revelations worrying the president had done nothing to affect the security of American forces in South Vietnam but that he would still instruct subordinate commanders to "avoid statements which add fuel to the already burning fire." While news reporting from South Vietnam indeed appeared "overheated and troublesome," it was hardly as unfavorable as it had been in the past.[38]

Despite Sharp's and Westmoreland's views on censorship, the Johnson administration declined to rule it out. The Deputy Assistant Secretary of Defense for Public Affairs, Philip Goulding, instructed Special Assistant for Southeast Asia

[34] Msg, New Delhi 2849 to Saigon, 7 Apr 65, FAIM/IR; Msg, Wheeler JCS 1271-65 to Westmoreland, 10 Apr 65, Westmoreland Papers, CMH.

[35] Msg, Wheeler JCS 1271-65 to Westmoreland, 10 Apr 65.

[36] Msg, Wheeler JCS 1272-65 to Westmoreland, 10 Apr 65, Westmoreland Papers, CMH.

[37] Msg, Westmoreland MAC 1985 to Wheeler, 11 Apr 65, Westmoreland Papers, CMH.

[38] Msg, CINCPAC to JCS, 10 Apr 65, for Wheeler from Sharp, Westmoreland Papers, CMH.

Rodger Bankson to investigate the issue again. On 8 May Bankson submitted a report that resurrected all the arguments against censorship formulated at the Honolulu Conference and again emphasized that the South Vietnamese were of necessity central to the program but would never handle newsmen fairly. Reputable reporters would exercise restraint if given good reasons for doing so, Bankson concluded, and "a mixture of experience, understanding, patience, and mutual effort" would best achieve the common goal newsmen shared with information officers: "the maximum release of information without endangering military security."[39]

In the end the only change in policy resulting from the president's concern was a ruling from the Office of the Assistant Secretary of Defense for Public Affairs on 10 April forbidding prior announcement of troop movements. That step took care of premature disclosures, but did nothing to remedy the basic problem: that the United States was moving deeper and deeper into war with only the most tenuous consent of the American public and Congress, a process that was bound to involve severe public relations problems.[40]

The Stir Over Escalation

By mid-April 1965 President Johnson had finally decided on the direction he should take in South Vietnam. With his program of increasingly severe air attacks against North Vietnam failing to produce an appreciable weakening of enemy will, he concluded that something more was needed. "President's belief," Secretary Rusk cabled Taylor, "is that current situation requires use of all practicable means of strengthening [U.S.] position in South Vietnam and that additional U.S. troops are important if not decisive reinforcement."[41]

With that decision, officials in Washington and Saigon began to discuss how many combat units Westmoreland needed at once and how many more he would require in the near future. The Defense Department immediately approved in principle a brigade-size force as security for the Bien Hoa–Vung Tau region near Saigon and another multibattalion force to conduct "counter-insurgency combat operations" in enclaves along the South Vietnamese coast. After McNamara, Taylor, Wheeler, Sharp, and Westmoreland met at Honolulu on 20 April, McNamara recommended raising American strength by 44,000 men to bring the total to 75,000. A request would then go to South Korea and Australia asking them to add another 7,250 men.[42]

[39] Memo, Rodger Bankson for Philip Goulding, 8 May 65, sub: Censorship, DDI Censorship file.

[40] Msg, Joint State/Defense/USIA 8876 to All Military Commands, 10 Apr 65, Ground Rules file, CMH.

[41] Msg, State 2322 to Saigon, 15 Apr 65, FAIM/IR.

[42] *Pentagon Papers*, 3: 410; Msg, Defense to Saigon, 15 Apr 65, Westmoreland History, bk. 15, tab 23; John T. McNaughton, Minutes of the April 20, 1965, Honolulu Meeting, 23 Apr 65, Miles Policy/Strategy files. Both in CMH.

Questions about whether the Army should call up its reserves and a crisis in the Dominican Republic that resulted in American intervention delayed a final decision on those recommendations, although President Johnson approved deploying the U.S. Army's 173d Airborne Brigade to Bien Hoa and three Marine battalion landing teams to Chu Lai. Meanwhile the American news media continued to speculate. An Associated Press news analysis on 22 April noted the possibility that American units might soon be fighting in South Vietnam and asked, if they did, whether a joint U.S.–South Vietnamese command would be wise. Another report a week later by United Press International revealed that Australia was planning to send a battalion to South Vietnam.[43]

Lacking permission to announce the mission of U.S. troops and therefore unable to allay that kind of speculation, information officers in Saigon adhered to the press guidance they had received on 3 April, affirming developments as they occurred in as routine a manner as possible. Officials in Washington, for their part, recognized that it was impossible to conceal America's growing involvement in the war, but rather than settle for a matter-of-fact approach they once more attempted to set up a diversion.[44]

In early May the State and Defense Departments told Zorthian that a review of MACV news releases and briefings had revealed an overemphasis on the American effort. Information officers should begin to stress South Vietnamese accomplishments and the fact that the American role was still only advice and support. Although news stories about the effort to eradicate malaria, the number of tons of rice supplied to refugees, and the new schools being built were of little interest to newsmen, "We would hope . . . eventually to turn reporters, who consider themselves 'war correspondents,' into 'counter-insurgency correspondents,' fully knowledgeable on all aspects of the Free World effort in Vietnam."[45]

Zorthian responded that while MACV's daily briefings did indeed give preponderant attention to military aspects of the war, there was little he could do to change them. Experience had shown that any attempt to shorten briefings, lump ground and air actions together, or in any way play down the military effort produced only hard questioning for details. He nevertheless proposed two steps: that the U.S. embassy redouble its efforts to convince the South Vietnamese to shoulder briefing responsibilities for their portion of the war and that the Military Assistance Command change the format of its daily news conference to allow time for embassy spokesmen to brief the press on nonmilitary developments. Although concern about missing an occasional news break on civilian matters would probably keep most newsmen in their seats for the added briefing, Zorthian said, the State Department should have few illusions about how much copy those changes would generate. Most nonmilitary events were simply too undramatic.[46]

[43] *Pentagon Papers*, 3: 410. Msg, Saigon 3539 to State, 26 Apr 65, and Msg, State 2451 to Saigon, 28 Apr 65, both in FAIM/IR.

[44] Msg, Saigon 3539 to State, 26 Apr 65; Msg, Joint State/Defense 2498 to Saigon, 4 May 65, FAIM/IR.

[45] Msg, Joint State/Defense 2498 to Saigon, 4 May 65.

[46] Msg, Saigon 3688 to State, DAIN 643985, 8 May 65.

Troops of the 173d Airborne Brigade Near Bien Hoa

Officials in Washington, however, continued the attempt to play down the American role in the war. "High levels as well as I," Arthur Sylvester cabled the command two days after receiving Zorthian's message, "have noted with concern what appears to be a gradual departure from the policy established by the Honolulu public affairs conference, particularly with reference to the amount of detailed information released about air strikes against North Vietnam." It appeared in Washington, he said, "that each U.S. military element is seeking maximum visibility. While this is natural, and credit for a tough job well done is a definite factor in morale, we must not take the war away from the Vietnamese. As more U.S. military personnel are introduced, this trend, unless stopped, could be harmful to our efforts."[47]

So directed, MACV information officers cut back on the details they released and inserted statistics on Viet Cong terrorism into their nightly communiques. In hopes of attracting the attention of the press, Ambassador Taylor also made special visits to nonmilitary projects such as the Saigon waterworks. Zorthian meanwhile inaugurated the nonmilitary briefings he had outlined in his message and redoubled his attempts to get the South Vietnamese to hold news confer-

[47] Msg, Defense 1897 to CINCPAC, MACV, 11 May 65, DDI Air Incidents/Policy file.

ences on their operations. When the State and Defense Departments decided in May to reveal that U.S. Navy warships standing off the coast of South Vietnam had on several occasions fired in support of ground operations, he likewise urged that the release be part of the regular Saigon briefing, where the news would appear to involve only the use of a new type of armament and would be less likely to seem a major escalation of the war.[48]

Zorthian, Westmoreland, and the MACV Chief of Information, Colonel Legare, were nevertheless convinced that any attempt to downplay the American role in the war would fail. "Reporters could see for themselves that the marines and the 173d's paratroopers were not sitting tight in their foxholes waiting for the enemy to come to them," Westmoreland would later recall. "They could easily see that American units were patrolling in some depth and sometimes engaging in full-scale offensive operations."[49] On 11 May Legare telephoned the Office of the Assistant Secretary of Defense for Public Affairs to tell Bankson that a comprehensive background briefing for selected newsmen on the concept of American combat operations in South Vietnam was imperative and that, unless directed otherwise, General Westmoreland intended to give it on 13 May.[50]

The verdict against the briefing was unanimous. When Arthur Sylvester checked with the State Department, the U.S. Information Agency, and the White House, all opposed the idea. The Johnson administration, Bankson informed Legare, was unprepared for the "impact" of such a session. Westmoreland would have to withhold comment on the subject for the time being.[51]

Although Westmoreland complied, both he and Zorthian continued to press for an official announcement on the role of U.S. troops. Several days after Bankson's call, they drafted a brief statement for release to newsmen as soon as American units first undertook an outright combat operation of major proportions. In submitting it to the State Department for approval, they noted that various correspondents had discussed the prospect of an expanded mission for American forces and that a spokesman for the U.S. mission had pointed out that both the marines and the airborne brigade had combat support roles. Nevertheless, they observed, "We do not believe impact of decision has fully sunk home and we expect questions will be raised when and if . . . first actual combat support mission is undertaken."[52]

A few days later the American news media gave the U.S. mission in Saigon an opportunity to renew its request. When newspapers in Washington revealed that the United States would shortly have 75,000 men in South Vietnam and the Saigon correspondents began discussing command relationships between those

[48] Memo, Benjamin H. Read, Ex Sec, State Dept, for Honorable Horace H. Busby, White House, 22 May 65, sub: Viet Cong Terrorism, FAIM/IR; Mission Council Action Memo 98, 18 May 65, MCAM No. 93 thru 102 file, 67A4662, box 5, WNRC; Msg, Saigon 3889 to State, 26 May 65, FAIM/IR.

[49] Westmoreland, *A Soldier Reports*, p. 135.

[50] MFR, Rodger Bankson for Arthur Sylvester, 14 May 65, sub: Background Briefing on Concept for U.S./Allied Combat Operations in Support of RVNAF, DDI Rules of Engagement file.

[51] Ibid.

[52] Msg, Saigon 3820 to State, DAIN 658756, 20 May 65.

troops and the South Vietnamese Army, Ambassador Taylor promptly sought authority for a background briefing to keep speculative stories on the subject from embarrassing the South Vietnamese government.[53]

The State Department postponed any comments on command relationships while it studied the conflicting arguments surrounding the issue, but it reconsidered its earlier ruling against a statement of mission. When it learned that the Viet Cong, after a two-month lull, had begun to mount attacks and ambushes with units of up to regimental size and that Westmoreland might soon have to use American troops to prevent a major South Vietnamese defeat, the agency told Taylor to begin planning to announce the combat mission of U.S. troops. The handling of such a development with the press would be far more effective, it reasoned, if official spokesmen anticipated the event rather than having to reply to reporters' questions as they had during the tear gas controversy.[54]

The Johnson administration suggested that the South Vietnamese make the announcement in order to give the subject as little emphasis as possible, but Zorthian disagreed. With South Vietnamese spokesmen yet to hold regular news conferences, the Saigon correspondents would almost certainly interpret the approach as an attempt to minimize the American role in the war. Since U.S. troops were the subject, American briefers had to make the announcement.[55]

The State Department rejected Zorthian's reasoning. The South Vietnamese should incorporate the announcement into one of their periodic press communiques, it said, leaving any questions that might arise for American briefers to answer quietly at the regular 5 o'clock briefing. There was "great advantage in getting reporters accustomed to having the government of Vietnam . . . become the source of information on decisions concerning the U.S. role in South Vietnam."[56]

The exchange continued into the first week in June with the State Department seeking ways to emphasize South Vietnamese control over the war and Zorthian objecting that the nature and scope of the American involvement precluded any possibility of doing so. In the end the Viet Cong settled the matter. On 31 May they mauled two South Vietnamese Army battalions at the village of Ba Gia in Quang Ngai Province and shortly thereafter forced government units to abandon several district capitals in the Central Highlands. At the height of the battle the South Vietnamese asked Westmoreland for help. Although they subsequently withdrew the request, the incident convinced the State Department that Westmoreland might in the near future have to commit U.S. troops to a major battle.

[53] Msg, Saigon 3858 to State, 24 May 65, and Msg, State 2688 to Saigon, 25 May 65, both in FAIM/IR.

[54] Msg, State 2702 to Saigon, 26 May 65, FAIM/IR; Msg, Saigon 4074 to State, 5 Jun 65, sub: Estimate of Political-Military Situation, Westmoreland History, bk. 16, tab 21, CMH.

[55] Msg, State to Saigon, DAIN 666406, 26 May 65; Msg, Saigon 4014 to State, 2 Jun 65, FAIM/IR. Although Zorthian's rejoinder, Saigon 3912 to State, is missing from official files, it is mentioned in Msg, State 2747 to Saigon, DAIN 671033, 29 May 65. Zorthian confirmed his views in Interv, author with Barry Zorthian, 29 Jun 77, CMH files.

[56] Msg, State 2747 to Saigon, DAIN 671033, 29 May 65.

Yielding to Zorthian's wishes, the department authorized him to issue the low-keyed announcement he had sought.[57]

The MACV operations officer, General DePuy, made the announcement on 4 June at the regular MACV briefing, in response to a question planted by Colonel Legare with a trusted correspondent. "When U.S. troops were brought in several weeks ago," the reporter noted, "we were told one of the possibilities was that when the enemy was sighted and fixed, there would be American troops engaged. They were fixed [at Ba Gia] for two or three days. What is the combat mission of the American troops here? When are they going to go into action?" DePuy responded that while the defensive mission of U.S. forces remained the same, American units would supply combat support to the South Vietnamese Army when necessary. After observing that the details were still to be worked out, he changed the subject so deftly that other correspondents failed to follow through with questions.[58]

Although the decision had been made to reveal the mission of U.S. troops, the Johnson administration remained unwilling to let go completely. The communique it released the day after DePuy's briefing was therefore less than candid.

> As you know, American troops have been sent to South Vietnam recently with the mission of protecting key installations there. In establishing and patrolling their defense perimeters they come into contact with the Viet Cong and at times are fired upon. Our troops, naturally, return the fire. It should come as no surprise . . . that our troops engage in combat in these and similar circumstances. Let me emphasize that the Vietnamese government forces are carrying the brunt of combat operations. Those U.S. forces assigned as advisers to the armed forces of Vietnam remain in that capacity.[59]

Noting the discrepancy between the two statements—DePuy had mentioned that U.S. forces might undertake missions as circumstances required while the State Department had said all combat was strictly defensive—Zorthian told the department that he would anticipate reporters' questions by repeating DePuy's statement as soon as the occasion arose and that he would have the South Vietnamese do the same. A slip by a State Department briefing officer in Washington nevertheless brought the matter into the open before Zorthian could act. At the regular afternoon briefing on 7 June, a newsman asked, "Has any request been received from the South Vietnamese government for combat assistance from our troops beyond that which is already being supplied?" The briefing officer, Robert McCloskey, responded, "Apparently there was one under circumstances that I can't precisely recall but the request was later withdrawn." The reporter then wanted to know whether the incident had occurred during the battle at Ba Gia, "when the marine battalion was alerted." McCloskey responded that "it

[57] Msg, State 2746 to Saigon, 29 May 65; Msg, Saigon 4014 to State, 2 Jun 65; and Msg, MACV 18608 to OSD PA, 2 Jun 65, all in FAIM/IR; Brigadier General Edwin H. Simmons, "Marine Corps Operations in Vietnam, 1965-1966," *U.S. Naval Institute Proceedings*, April 1968, p. 14; Msg, Saigon 4058 to State, 5 Jun 65, FAIM/IR.

[58] Msg, MACV 18896 to OSD PA, 4 Jun 65, sub: Press Trends No. 148A-65, FAIM/IR; Msg, Saigon 4058 to State, 5 Jun 65.

[59] Msg, State 2810 to Saigon, 5 Jun 65, FAIM/IR.

was certainly recently and I would presume it was the occasion to which you refer."

Coming at a time when public attention in the United States was focused on the first walk in space by *Gemini IV* astronauts, McCloskey's slip attracted little attention, but a session with the press the next afternoon made a deeper impression. Pressed by newsmen, McCloskey admitted that American forces were available for offensive combat. "What you are saying," a reporter then asked,

> means that the decision has been made in Washington as a matter of policy that if Westmoreland receives a request for U.S. forces in Vietnam to give combat support to Vietnamese forces he has the power to make the decision?
> *A.* That is correct.
> *Q.* Could you give us any understanding, Bob, as to when Westmoreland got this additional authority?
> *A.* I couldn't be specific but it is something that has developed over the past several weeks.
> *Q.* Could you tell us whether there has been any instance yet where General Westmoreland has made a decision to use American forces for such combat?
> *A.* So far as I know, the answer is no.
> *Q.* Bob, you said this has been stated publicly out there in Vietnam. Can you give us a reference to that public statement?
> *A.* It was at least the subject of some background comment by American military authorities in Saigon some time within the past eight to ten days.
> *Q.* You are not aware of any on-the-record announcement of policy?
> *A.* I cannot account for anything specifically.[60]

McCloskey's remarks triggered complaints by legislators and the press that the new combat mission of American forces represented a dangerous and reckless departure from accepted policy. Senator Jacob K. Javits of New York charged that "we have been moving in the direction of a massive, bogdown land struggle in Asia without any specific consent by congress or the people for that kind of war." Senator Ernest Gruening of Alaska warned that what the president was doing was unconstitutional.[61] The *New York Times* observed that "The American people were told by a minor State Department official yesterday that, in effect, they were in a land war on the continent of Asia. . . . The nation is informed about it not by the president, not by a cabinet member, not even by a sub-cabinet official, but by a public relations officer. There is still no official explanation."[62]

The White House attempted to dilute the impact of McCloskey's statement by announcing that there had been no change in the mission of U.S. ground combat units and that the primary mission of those troops remained the same: the safeguarding of important military installations such as the air base at Da Nang. "If help is requested by appropriate Vietnamese commanders," the statement nevertheless revealed, "General Westmoreland also has authority within the assigned mission to employ these troops in support of Vietnamese forces faced with aggressive attack when other effective reserves are not available. . . . If

[60] Msg, State 2832 to Saigon, 8 Jun 65, FAIM/IR.
[61] John W. Finney, "U.S. Denies Shift on Troop Policy in Vietnam War," *New York Times*, 10 Jun 65.
[62] "Ground War in Asia," *New York Times*, 9 Jun 65.

General Westmoreland did not have this discretionary authority, a situation might occur and great advantage might be won by the Viet Cong because of delays in communications."[63]

The statement, and another by Secretary Rusk on national television to the effect that "obviously, we don't expect these men to sit there like hypnotized rabbits waiting for the Viet Cong to strike," did nothing to quell the criticism that began to rise in the press. "The Johnson administration's decision authorizing a combat role for American troops in South Vietnam is only confirmed by yesterday's White House statement," the *New York Times* charged, "yet the statement is carefully drafted to give the impression that the United States is not embarking on a radical new course." There could be only one interpretation, *Times* reporter John W. Finney observed. "The White House was disturbed by the conclusion, drawn from yesterday's statement on a wide scale, that the administration was deepening the commitment in Vietnam by undertaking open combat against the guerrillas."[64]

Making much the same point that information officers in Saigon had stressed for weeks, Kenneth Crawford of *Newsweek* noted that to anyone who followed daily casualty reports from South Vietnam, the mission of U.S. troops had been obvious for some time. The State Department's announcement made that mission appear to be "some new and momentous step toward further escalation of the war," he said, and the attempt by the White House to clarify that department's explanation only intensified the impression that the administration was "trying to cover up some sinister innovation."[65]

Columnist Arthur Krock had the last word. The only reason for the stir created by the State Department's announcement, he declared, was

> the administration's evasive rhetoric on every occasion when our military role in Vietnam is expanded. . . . The self-evident purpose of the White House statement was to modify the . . . public conclusion that the primary mission of United States troops in South Vietnam has been fundamentally changed. "Not recently or at any other time," said the White House, has the president given any such order to General Westmoreland. The general has always been authorized to do "whatever is necessary." . . . Thus the administration reverted to a semantic quibble. . . . For there is certainly fundamental "change" in a "mission" which begins as strategic counsel and technical assistance within a government territory, proceeds to bombing outside that territory, . . . moves onward to "perimeter defense" that inescapably leads to ground combat, and finally is given authority for expansion into formal ground warfare.

Krock ended by quoting from Lewis Carroll's *Alice in Wonderland*: " 'The question is,' said Alice to Humpty Dumpty, 'whether you *can* make words mean so many different things.' "[66]

[63] Cir 2470, State Dept to All Diplomatic Posts, 9 Jun 65, FAIM/IR.

[64] "Ground War in Washington," *New York Times*, 10 Jun 65; Finney, "U.S. Denies Shift on Troop Policy in Vietnam War."

[65] Kenneth Crawford, "On Taking the Heat," *Newsweek*, 21 Jun 65, p. 36.

[66] Arthur Krock, "By Any Other Name, It's Still War," *New York Times*, 10 Jun 65.

8

The Ground War

While accusations continued in the news media that the Johnson administration had lied about the mission of U.S. troops, the Viet Cong increased their pressure in South Vietnam, showing improved training and discipline and stronger fire power. By 7 June 1965, they had mounted damaging regimental assaults on targets in Phuoc Long and Quang Ngai Provinces, and General Westmoreland had concluded that they possessed the resources to attack at will in all four of South Vietnam's corps tactical zones.[1]

That the enemy might indeed launch a series of nationwide attacks had disturbing implications for the United States. As General Westmoreland and Ambassador Taylor noted in messages to the president, many units of the South Vietnamese Army had become so demoralized during the enemy offensive that their will to keep fighting would probably collapse in the face of continued significant enemy victories. To guard against that possibility, the United States would probably have to commit American units to combat.[2]

The situation worsened on the night of 9 June, when the Viet Cong attacked a U.S.-advised South Vietnamese Special Forces camp near Dong Xoai, a district capital located some ninety kilometers north of Saigon. After penetrating the camp's perimeter and surrounding its defenders, the Communists ambushed and destroyed two waves of South Vietnamese reinforcements, killing or wounding some 650 government soldiers. So resounding was their victory that the deputy chief of the Viet Cong's Political Affairs Division, Brig. Gen. Tran Quoc Vinh, later termed the battle the most glorious Communist achievement of 1965.[3]

Alarmed by the extent of the debacle and by the Viet Cong's apparent deci-

[1] Msg, Saigon 4074 to State, 5 Jun 65, sub: Estimate of Political-Military Situation, Westmoreland History, bk. 16, tab 21, CMH; Msg, MACV 19118 to CINCPAC, 7 Jun 65, FAIM/IR.

[2] Ibid.

[3] Msg, MACV 2008 to CINCPAC, 13 Jun 65, FAIM/IR files; MACV History, 1965, p. 222, CMH files; MACV J261, CDEC Log no. 01-3398-67, 7 May 67, sub: Speech by Brig Gen Tran Do, Westmoreland History, bk. 17, tab A-4.

sion to seize a base from which it could dominate all of Phuoc Long Province, Westmoreland cabled Admiral Sharp at the height of the battle on 13 June to inform him that the situation at Dong Xoai had become critical. Against an estimated five regular enemy battalions supported by artillery, local guerrillas, and civilian porters, the South Vietnamese Army could muster no more than one understrength ranger battalion, two understrength infantry battalions, and the remains of the battalions mauled on the first day of the attack. Since bad weather had for the time being precluded air strikes and since the already depleted South Vietnamese general reserve was needed for the defense of Saigon, he could see only one course of action: if the battle failed to turn in favor of government forces within the next twelve hours, the Military Assistance Command would have to move battalions of the 173d Airborne Brigade north to Dong Xoai to dislodge the Viet Cong.[4]

The prospect of committing American units to combat while newspapers in the United States were accusing the Johnson administration of lying about the mission of U.S. troops appalled Admiral Sharp. In a telephone call and in a later cable, he reminded Westmoreland that if U.S. forces entered the battle and suffered defeat, ''particularly in the immediate wake of adverse publicity on this subject, the political consequences could be embarrassing and might even jeopardize or change the course of our present plans regarding the use of U.S. forces in Vietnam.'' Sharp told Westmoreland to balance carefully the benefits of using American troops against the limited value of the camp at Dong Xoai and to attempt to achieve his ends by the massive application of air power. If he nevertheless decided to commit ground forces to combat, he was to notify the ambassador, the commander in chief, Pacific, and the Joint Chiefs of Staff before issuing any orders.[5]

Westmoreland yielded to Sharp's wishes. Although he moved a battalion of the 173d to Phuoc Vinh, an old French base within striking distance of Dong Xoai, he took advantage of a break in the weather to drive the enemy back with air attacks and never committed ground units to combat.[6]

Guidelines for Reporting the Ground War

The decision kept the outcry Sharp had feared from developing but uncovered a new problem. For weeks the Saigon correspondents had been watching for any sign that U.S. troops might take the offensive. When they saw units

[4] Msg, Westmoreland MAC 3072 to CINCPAC, 13 Jun 65, Westmoreland Papers, CMH; Msg, MACV 20024 to CINCPAC, 13 Jun 65, FAIM/IR.

[5] Memo, MACV for Gen Westmoreland, 13 Jun 65, sub: Telephone Call From Adm. Sharp, Miles Policy/Strategy files, CMH; Msg, Sharp to Westmoreland, 13 Jun 65, Westmoreland History, bk. 16. tab 26, CMH.

[6] Msg, Westmoreland MAC 3077 to Sharp, 13 Jun 65, Westmoreland Papers, CMH.

Admiral Sharp

of the 173d going on alert and moving to Phuoc Vinh, they began transmitting the news to their editors. By the time the MACV Office of Information learned what was happening and requested that the correspondents refrain from discussing the operation, news bulletins of potential value to the enemy were leaving Saigon over commercial lines almost certainly subject to Viet Cong monitoring.[7]

Aware that similar breaches of security would occur if the 173d actually went into combat, the Military Assistance Command and the U.S. mission moved quickly to devise rules to keep the Saigon correspondents under control. If the enemy attacked the unit while it was at Phuoc Vinh, Zorthian told the State Department, information officers would confirm the time and place of the incident at the regular briefing but would ask newsmen to comply voluntarily with MACV's refusal to reveal casualty figures until the engagement was over. That step would keep the Communists from learning how effective their tactics had been. In the same way, if the 173d entered combat, MACV spokesmen would confirm the move only when the news was of no further use to the Viet Cong and discuss casualties only after the fighting had ended. In either case, briefers in Saigon would adhere to the lines of the 9 June White House statement on the mission of U.S. troops. They would confirm that the battalion had gone to Phuoc Vinh at the request of the South Vietnamese government and stipulate that the deployment was well within General Westmoreland's authority.[8]

Endorsing the guidelines, the State Department directed Zorthian to reconsider the whole question of restraints on the press in light of the fact that American troops would shortly begin genuine offensive operations. That inquiry was to cover all the topics originally surveyed by the Honolulu information conference in March, including the wisdom of inaugurating field press censorship, the practicality of a system that would merely "advise" newsmen on security matters without actually censoring their dispatches, and the possibility of adopting a set of voluntary restrictions similar to those already in effect for the air war.[9] Before drafting a reply, Zorthian consulted a number of the Saigon correspondents. He

[7] Msg, Saigon to NMCC, 13 Jun 65, FAIM/IR.
[8] Ibid.
[9] Msg, State 2891 to Saigon, 14 Jun 65, FAIM/IR.

found them as concerned about preserving military security as he and amenable to further restriction, provided unscrupulous newsmen gained no advantage over those who cooperated.

In view of the reporters' obvious sympathy for MACV's need to restrict sensitive information, Zorthian proposed to State and Defense a set of guidelines that he believed would meet official requirements without doing violence to the news media.[10] In the past, he noted, official briefers had routinely announced troop movements, the units participating in a battle, and the number of killed and wounded in specific engagements. That practice would no longer be possible. If State and Defense approved, the command would announce future deployments only when the news was obviously in enemy hands, and briefers would describe the magnitude of particular operations only in general terms without revealing participating units. To preserve official credibility, the command would release weekly totals of Americans killed, wounded, and missing, but it would never associate those figures with individual battles or skirmishes. Instead, it would announce day-to-day casualties only as light, heavy, or moderate.

Although he believed that most newsmen would abide by the rules, Zorthian nevertheless stressed that problems might arise. A few reporters might attempt to circumvent the new guidelines, in which case the command would have to threaten to exclude offenders from government briefings and facilities and deny them the right to accompany troops into the field. Washington agencies could, of course, foster greater acceptance of the restrictions by promoting the program with top editors and publishers in the United States, but in case that effort failed, they ought also to draft a contingency plan for full field press censorship.

The State Department took Zorthian's proposals under consideration, debating whether to announce the program formally or to tell the press about it on a background basis. A formal news conference on the subject, the department told Zorthian, "might . . . make news of a particularly unfortunate kind."[11]

Zorthian agreed that the backgrounder approach was attractive but recommended against it. The news media would make a story of the change no matter how discreetly officials handled it, he told the State Department. A formal announcement would at least have the virtue of containing "phraseology of our own selection . . . as against . . . an oral briefing which can be misquoted and presented selectively."[12]

The State and Defense Departments delayed approving Zorthian's guidelines until they could consult with the South Vietnamese, but the Saigon correspondents knew some sort of restrictions were in the offing and began to speculate. John Maffre of the *Washington Post* believed censorship probable but questioned its wisdom. The South Vietnamese would necessarily become involved, he told his readers, and might find themselves unable to resist the temptation to "put the screws" to the press. Other correspondents disagreed. In private conferences

[10] This section is based on Msg, Saigon 4205 to State, 15 Jun 65, DDI PA OPNS ORGNS file.
[11] Msg, State 2911 to Saigon, 15 Jun 65, FAIM/IR.
[12] Msg, Saigon 4231 to State, 16 Jun 65, FAIM/IR.

with Zorthian they actively recommended censorship of military information as the best means of maintaining fair competition among newsmen.[13]

Whatever their position on the issue, all of the newsmen were finely attuned to any sign of official news management and resented even the appearance of manipulation. How sensitive they were was demonstrated on 18 June, shortly after Westmoreland launched the first B–52 raid of the war. Westmoreland had been experimenting with air power since at least mid-April, when the Military Assistance Command had conducted an unprecedented 400-plane tactical air attack against an enemy base in War Zone C, eighty-five kilometers northwest of Saigon. Code-named BLACK VIRGIN and designed to ascertain whether fighter-bombers attacking in waves could destroy large enemy ground installations, the operation had taken twelve hours to complete and had done little more than disrupt Communist movement in the area. Dust and rising smoke had quickly obscured the target, making accurate bombing impossible.[14]

Determined to eliminate enemy sanctuaries and to boost the morale of the South Vietnamese Army by demonstrating U.S. support, Westmoreland gained permission to try again on 18 June with B–52's, the most powerful bombers in the American arsenal. He hoped that the B–52's, armed with racks of heavy bombs and striking from altitudes that made them invisible from the ground, would deliver the sort of punch smaller aircraft lacked.[15]

Westmoreland advised against revealing the operation to the press, but both the State and Defense Departments disagreed. U.S. Special Forces officers accompanied by teams of South Vietnamese irregular troops would enter the target zone shortly after the attack to assess the bombers' effectiveness, they said, giving a large number of people knowledge of what had happened. If someone leaked the story to the press before an official statement appeared, reporters might decide that the United States was trying to hide something and start an outcry similar to the one that had occurred over the use of gas.[16]

Washington agencies were so sensitive to this possibility that the Defense Department devised elaborate procedures to allay any adverse reaction in the press. While the Joint Chiefs of Staff drafted a complicated communications plan to keep track of the minute-by-minute progress of the bombers toward the target, the Office of the Assistant Secretary of Defense for Public Affairs composed an announcement for delivery to the press by Secretary McNamara as soon as the Joint Chiefs learned the bombs were falling. Playing down the newness of the tactic and emphasizing Communist provocations, the statement asserted that the innovation was an attempt to save civilian lives by forestalling enemy assaults upon population centers such as the village of Dong Xoai.[17]

[13] John Maffre, "U.S.-Viet Cong Conflict, Censorship Held Likely," *Washington Post*, 15 Jun 65; Msg, Saigon 4238 to State, 19 Jun 65, FAIM/IR.

[14] Briefing, sub: Facts on First B–52 Strike, Westmoreland History, bk. 1, tab A–1.

[15] Ibid.

[16] Msg, MACV 20217 to JCS, 15 Jun 65, Miles Policy/Strategy files, CMH; Msg, State 2933 to Saigon, 16 Jun 65, FAIM/IR.

[17] Msg, JCS to CINCPAC, 17 Jun 65, Miles Policy/Strategy files, CMH. Msg, JCS 4100 to CINCPAC, 17 Jun 65, and Msg, State 2939 to Saigon, 17 Jun 65, both in FAIM/IR.

When he heard of the plan, Zorthian cabled the State Department to point out that a Washington announcement would conflict with the well-established policy of revealing all combat developments in Saigon first. The Defense Department responded by approving the release of a joint U.S.-South Vietnamese communique in Saigon before the Washington statement, but the concession proved at best a token. Soon after the bombs began to fall on the morning of the attack, Rodger Bankson at the Pentagon telephoned the MACV Office of Information to tell Colonel Legare to proceed with his news release. Soon thereafter, while Legare was still assembling the Saigon correspondents, the Defense Department made its announcement, preempting the news from Saigon.[18]

Word of the Defense Department's haste took some time to reach South Vietnam. The delay precluded an immediate outcry by the Saigon correspondents but left room for more problems to develop. For although newsmen pressed the U.S. mission for details, the command refused to give any until its scouts returned, creating a news vacuum which both the press and Washington agencies rushed to fill.[19]

The imbroglio that developed began when two of the B-52s collided in midair and crashed en route to Vietnam. Learning of the accident, news commentators in the United States weighed the cost of the raid and the loss of the planes against the value of the target and questioned whether the U.S. Air Force should have been destroying jungle huts with expensive intercontinental bombers. Since one purpose of the raid had been to set a precedent to justify further B-52 strikes in South Vietnam, those criticisms prompted an immediate response from the Defense Department. Drawing upon preliminary reports from the Military Assistance Command, the Office of the Assistant Secretary of Defense for Public Affairs issued a news release calling the mission a success and citing many of the details the command had thus far refused to give.[20]

Despite the Defense Department's claims, the attack had done little actual damage to the enemy, a fact readily apparent to any diligent reporter. Official spokesmen, for example, asserted in their announcement that the raid had probably caused "numerous Viet Cong casualties" but could enumerate only three—all killed by the ground survey teams. In the same way, Defense Department spokesmen contended that the raid had destroyed 2,500 pounds of enemy rice, a large communications center, and from 20 to 30 buildings; yet the communique suggested that the survey teams had once more caused all of the damage. Thus, when information officers avowed that "it is extremely significant that this Viet Cong headquarters has been overrun and destroyed, that an area which had been considered unassailable has been entered, and that [South Vietnamese] troops could follow up the bombing to accomplish their ground objective," reporters dis-

[18] Msg, Saigon 4240 to State, 17 Jun 65, Miles Policy/Strategy files, CMH; Msg, State 2930 to Saigon, 17 Jun 65, FAIM/IR; Msg, Saigon 4297 to State, 20 Jun 65, DDI War Zone D file.

[19] Msg, Saigon 4297 to State, 20 Jun 65.

[20] Ibid.; Msg, NMCC to CINCPAC, 18 Jun 65, for IO from Sylvester, DDI War Zone D file; Msg, Saigon 3761 to State, 14 May 65, Arc Light file, CMH.

A B–52 Bombing Raid

counted all of those claims and concluded that the Johnson administration was trying to hide the failure of its raid behind another public relations ploy.[21]

Radio and television commentators were among the most vocal of the critics. Walter Cronkite of CBS News charged that the Pentagon was "attempting to put the best possible light on what . . . appeared to be a mission that failed." Peter Jennings of ABC News said that in light of the raid's high cost "this [has been] . . . the most spectacular disappointment of the Vietnam War." Syndicated radio columnist Paul Harvey exaggerated the whole affair. "Our biggest bombers went into combat for the first time. Our mighty, globe-girdling B–52s, designed to obliterate whole cities, have been thrown into the jungle in Vietnam—and the results have been unimpressive. . . . Twenty-seven of the giants rained napalm, rockets, and bombs on three square miles. . . . [Yet] when . . . ground troops rushed in where we'd just bombed they found no dead communists, in fact collided with a bunch of live ones. So I don't know."[22]

By the morning of 19 June, editors in the United States had begun to spread

[21] Msg, NMCC to CINCPAC, 18 Jun 65, for IO from Sylvester.

[22] Radio-TV Reports, Inc., Dialog: Detailed Broadcast Log, 18 Jun 65, DDI B–52 file, hereafter cited as Radio-TV-Defense Dialog.

the controversy to South Vietnam by cabling their correspondents to inform them that news of the raid had been more available in Washington than in Saigon. The correspondents turned to the MACV Office of Information to complain bitterly that the U.S. mission's maladroit handling of the affair had put them in an impossible position. Their employers maintained them in South Vietnam at great expense so that they could make the earliest possible reports, yet when the most important air strike of the war occurred, Pentagon correspondents received the story first. Obviously, the reporters charged, since the Military Assistance Command had failed to reveal details of the strike itself, all impetus on the matter had come from Washington, where officials had decided to hide the failure of the attack behind a mass of distorted facts.[23]

The command responded by reading Sylvester's Washington news release to the correspondents. The reporters found the information meager and continued to charge cover-up. Westmoreland's chief of staff, General Stilwell, attempted to quiet the clamor by holding a special briefing to explain the communique, but the reporters continued to demand evidence that the bombing itself had done any damage to the Viet Cong.[24]

Stilwell tried again at a noon briefing on 20 June, where he introduced three of the U.S. Army advisers who had accompanied the search parties and who had been in the target area on previous occasions. In the past, the advisers said, War Zone D had always been a dangerous place, where even well-trained battalions encountered insurmountable enemy resistance. Yet the B–52 raid had so dazed the Communists that three small reconnaissance teams of forty-one men each landed with ease and began destroying enemy installations. The Communists regrouped quickly, but they were obviously off balance and must have suffered many casualties. "If I ever have to go into an area like that again," one officer asserted, "I hope those B–52s . . . are there."[25]

As soon as the advisers finished their briefing, the correspondents pressed to the attack. "How effective do you think the B–52 bombing was in covering and destroying this area?" one reporter wanted to know. "Damn effective" came the reply. "We were told yesterday no huts were destroyed or bunkers caved in by bombs," the reporter countered. "Here's the point," the adviser responded. "Three teams covered [only] ten percent of the area up there. . . . " That was just the problem, the newsman interrupted. "If you multiply no damage by ten you still have no damage."[26]

The briefing ended a few moments later, but the controversy simmered on into the evening. At the regular 5 o'clock MACV briefing Joseph Fried of the *New York Daily News* observed that neither Stilwell's statement nor those of the advisers squared with one made by the State Department the day before to the effect that

[23] Msg, Saigon 4297 to State, 20 Jun 65; Msg, Saigon 4293 to State, 19 Jun 65, DDI War Zone D file.
[24] Msg, Saigon 4297 to State, 20 Jun 65; Msg, MACV 21020 to OASD PA, 19 Jun 65, sub: Press Trends 163A65, FAIM/IR.
[25] Msg, MACOI to NMCC, 20 Jun 65, DDI B–52 file.
[26] Ibid.

the raid had done "substantial damage in itself." Would the command, he asked sarcastically, care to confirm that contention? Although the briefer refused to be drawn into an exchange and dismissed the question, the U.S. mission notified the State Department that "For your information, . . . as you will see from press trends reports, . . . no concrete evidence was obtained of either damage or casualties."[27]

A few days later, as part of an official critique of the operation, the Military Assistance Command strongly recommended that the U.S. mission in Saigon alone announce all future B-52 raids. If a simultaneous Washington communique became necessary, the Saigon correspondents should be told about it so that they could avoid wasting money cabling news that was already common knowledge in the United States. The Defense Department concurred, responding that all information on U.S. activities in South Vietnam would normally come from Saigon.[28]

Both General Westmoreland and Admiral Sharp agreed that the raid had been a success. If evidence of actual damage was lacking, they reasoned, the strike appeared nevertheless to have thrown Communist forces in War Zone D off balance. It had certainly helped make the B-52 an adjunct to the American arsenal in South Vietnam. "What is now important," Sharp told Westmoreland the day after the attack, "is to get off a request for another, or perhaps a series of . . . missions before the political climate changes. . . . The main thing is to establish a pattern."[29]

Planning for a second operation began almost immediately, with a large survey force of South Vietnamese paratroopers and elements of the U.S. 173d Airborne Brigade taking the place of Special Forces teams in order to ensure adequate penetration of the target and a thorough search. President Johnson and some members of the National Security Council questioned using the 173d out of concern that the enemy might seek a propaganda coup by attempting to defeat the unit, but General Wheeler assured them that the brigade would have the advantage of all the air and artillery support the Military Assistance Command could muster. In addition, the MACV Office of Information intended to keep newsmen on a tight leash during the attack in order to eliminate any security problem that might arise from that quarter.[30]

MACV's public relations plan for the operation was a test of Zorthian's principle of voluntary cooperation. Since the ground portion of the raid would involve using American troops in the already controversial role of supporting the South Vietnamese in combat, information officers decided the best way to avoid criticism would be to allow correspondents to see for themselves what was going

[27] Msg, Saigon 4299 to State, 20 Jun 65, FAIM/IR; MFR, DDI, 19 Jun 65, sub: Statement by Dept of State, DDI War Zone D file.

[28] Msg, MACV 21538 to CINCPAC, 24 Jun 65, DDI War Zone D file; Msg, Defense 5597 to CINCPAC, 12 Jul 65, DDI Releasing Authority in RVN file.

[29] Msg, Sharp to Westmoreland, 19 Jun 65, Miles Policy/Strategy files, CMH.

[30] Msg, MACV 21358 to CINCPAC, 23 Jun 65, DDI War Zone D file; Msg, Wheeler JCS 2330-65 to Sharp, 24 Jun 65, Westmoreland Papers, CMH; Msg, Saigon 4414 to State, 28 Jun 65, FAIM/IR.

on. To that end, they made room for six reporters to accompany the 173d into combat and told them they could write what they wanted.[31]

Although the press was to have a free hand in reporting the operation, the South Vietnamese had yet to approve Zorthian's voluntary guidelines, so the Military Assistance Command and the U.S. mission imposed interim rules for the press to follow. The command would not release casualty statistics until after the operation had ended, and everyone using the pool's reports would have to refrain from cabling the story to the United States until the command gave the word. Information officers reasoned that most reporters would abide by the arrangement and that the few who went off on their own would find their dispatches "dying from lack of nourishment."[32]

Although the Military Assistance Command canceled the B–52 portion of the plan after a Special Forces team reconnoitered the target and found it vacant, General Westmoreland decided to go ahead with the ground assault because the area housed a number of enemy bases and posed a threat to the air base at Bien Hoa. That decision turned the mission into a standard ground operation and prompted the command to replace the indefinite period reporters would have to wait before announcing the attack with a simple 36-hour rule.[33]

Commencing on 28 June and lasting four days, the operation encountered light resistance from the enemy and only a few complaints from newsmen. Apparently nettled, for example, by MACV's refusal a day earlier to supply details of the crash of a C–123 transport, on a secret mission, which had gone down within sight of a UPI photographer, Peter Arnett of AP charged that MACV's 36-hour rule embargoed information obviously in enemy hands. U.S. mission public affairs officers denied the contention, asserting that except for AP almost the entire press corps in Saigon accepted the need for the rule. "Hell hath no fury," Zorthian added in derision, "like a wire service scooped."[34]

Yet if most of the Saigon correspondents went along with MACV's rules, they were still uneasy. NBC News correspondent Sid White cataloged their doubts. Referring to Zorthian's continuing conversations with newsmen on the possibility of future restraints, he told his audience,

> In recent weeks there have been several instances, mostly at the connivance of officials in Washington, of attempts to withhold or distort the facts. The most recent effort in this direction was the disclosure that military officials in Saigon will soon change their method of reporting ground combat casualties by providing weekly . . . rather than day-to-day summaries. . . . It appears that this is a move that is being taken to offset expected casualties as Americans are committed more frequently to combat. . . . That is to say . . . [officials] won't give a true picture which might make us look bad.[35]

White's opinion was hardly that of the majority, but similar concerns surfaced

[31] Msg, Saigon 4372 to State, 25 Jun 65, FAIM/IR; Msg, Saigon 4414 to State, 28 Jun 65.
[32] Ibid.
[33] Msg, MACV 22042 to CINCPAC, 27 Jun 65, FAIM/IR; Msg, Saigon 4414 to State, 28 Jun 65.
[34] Msg, Saigon 4416 to State, 28 Jun 65, and Msg, Saigon 4430 to State, 29 Jun 65, both in FAIM/IR.
[35] Msg, State 40 to Saigon, 3 Jul 65, FAIM/IR.

regularly during June and July at the nightly MACV briefings. On one occasion, referring to MACV's refusal to give casualty statistics for an engagement involving the 173d Airborne Brigade, a reporter asked pointedly whether the policy was local in origin or the result of directions from Washington. The briefer responded that the idea had been the command's and attempted to allay concern that officials were trying to hide something by promising to have participants in the engagement brief the press as soon as the troops returned. Expedients of that sort were nevertheless at best temporary. Until the Military Assistance Command established a consistent set of rules for reporting the ground war, reporters would continue to suspect every shift in policy.[36]

Recognizing the need, the State and Defense Departments decided during the first week in July to go ahead with Zorthian's proposed guidelines and began to press the U.S. mission to brief correspondents on the regulations. They could then start simultaneous consultations with managing editors in the United States. Zorthian was prepared to comply but still had to contend with the South Vietnamese Ministries of Defense and War. Although professing general agreement with the proposed guidelines, those agencies insisted upon studying every detail carefully and at length.[37]

Receiving approval at last, Zorthian read the rules to the Saigon correspondents on 12 July. The Military Assistance Command would announce casualties by number once a week, he told the reporters, but would describe losses for particular engagements only as *light, moderate,* or *heavy*. Official spokesmen would refuse to confirm troop movements until the information was clearly in enemy hands and would never identify units participating in specific combat operations by number or type. If reporters uncovered information of that sort on their own, they were to consider it classified and refrain from using it.[38]

During the question-and-answer period that followed, newsmen concentrated upon MACV's policy for announcing casualties. To a request for the criteria that officials would use to determine whether battle losses were light, moderate, or heavy, Zorthian replied that no exact measurement was possible and that correspondents would have to rely upon his judgment. He had no objection to telling newsmen the actual numbers off the record so that they could gauge their descriptions accordingly, but he warned that the first time a reporter published the figures the practice would cease. Zorthian added, "I certainly would not deny any correspondent the right, if he's on an operation, to say that he saw a casualty or he saw a man die, providing he doesn't give close to the statistical rundown. . . . If he wants to give the evaluation that it was a very hard fight [with] . . . a lot of casualties . . . he would have the right to report that. It's when [the numbers] . . . begin to get finite . . . [that] it gets to be of importance or aid to the other side."

[36] Msg, Saigon 89 to State, 8 Jul 65, FAIM/IR.
[37] Msg, State 54 to Saigon, 6 Jul 65, and Msg, Saigon 52 to State, 6 Jul 65, both in FAIM/IR.
[38] This section is based on Msg, MACV 24135 to OASD PA, 12 Jul 65, sub: Supplemental Press Trends 186A-65, FAIM/IR.

President Johnson Confers With His Advisers. *From left, George Ball, Robert McNamara, Robert Komer, and Dean Rusk.*

The answer did little to satisfy some of the correspondents. Referring to a recent operation in War Zone D, they asked how the Military Assistance Command would have handled that engagement's ten killed and forty-six wounded had the new system been in effect. The MACV representative assisting Zorthian replied that he would have considered those casualties light because the operation had lasted four days.

''Everything is relative,'' the newsmen rejoined. ''If we understand that there are twenty casualties in an engagement, off the record, it would make an enormous difference to us whether [they involved] . . . a platoon or a battalion; but since we're not allowed to say it's a platoon or a battalion, if it were a platoon we'd have to say . . . heavy.'' Neither Zorthian nor the other information officers present took much notice of the comment, but in the months to come the issue would grow into a major problem for the command.

The backgrounder ended shortly after that exchange, but not before Zorthian had decided that the guidelines would need further clarification. As it stood, he cabled the State Department, although the Military Assistance Command had asked reporters orally to observe the rules, the whole process had been handled informally. Something more definite was needed, a statement in writing that could be available for new correspondents arriving in Saigon. The State Department agreed, authorizing Zorthian to put the guidelines into effect on the morning of 15 July and to add a paragraph emphasizing the voluntary nature of the rules and the fact that certain categories of information were to remain restricted until the command itself decided otherwise.[39]

The news media reacted calmly when Zorthian issued the final version of the guidelines. A few journals such as the *Chicago Tribune* warned that the Johnson administration was attempting to limit free discussion of the war, but the comments of Keyes Beech of the *Chicago Daily News* were more characteristic of what the press had to say. "The consensus among responsible newsmen and others," Beech wrote from Saigon, "is that some restriction of information useful to the enemy is long over due."[40]

A Favorable Public Mood, June–July 1965

The cooperation of the press in accepting the MACV guidelines reflected an attitude growing in the United States that American objectives in Southeast Asia deserved support. Although the antiwar movement conducted a number of "teach-ins" at American colleges during June and July and the campus organization Students for a Democratic Society sponsored a protest march on Washington that attracted several thousand demonstrators, Harris polls revealed that 62 percent of the American people approved of President Johnson's handling of the war and that 79 percent believed South Vietnam would fall to the Communists unless the United States stood firm. The polls also revealed doubts—a substantial 32 percent believed that the United States might lose the war if it developed into a major conflict—but the overall figures so impressed the State Department that it ordered U.S. missions throughout the world to cite them whenever the American public's support for the war came into question.[41]

Although public opinion polls represented at best an imprecise measure of support for the war, there was no mistaking the attitude of the CBS television network, which decided to produce a series of "Vietnam Perspective" programs for airing during August. Inviting the participation of a number of administration spokesmen—Rusk, McNamara, McGeorge Bundy, Wheeler, Taylor, and U.S. Ambassador to the United Nations Arthur Goldberg—and none of the critics of the war, the network allowed those officials to review and edit tapes of their comments prior to broadcast. It then published transcripts of the programs in book

[39] Msg, Saigon 119 to State, 12 Jul 65, and Msg, State 111 to Saigon, 12 Jul 65, both in FAIM/IR.

[40] "Creeping Censorship," *Chicago Tribune*, 17 Jul 65; Keyes Beech, "U.S. Considers Tighter Rein on Security," *Chicago Daily News*, 14 Jul 65.

[41] For a summary of antiwar movement activities during 1965, see Msg, State 376 to Wellington, 24 Nov 65, FAIM/IR. For the poll data, see Msg, State Circular to All Diplomatic Posts, 13 Jul 65, FAIM/IR.

form at its own expense. The introduction by Walter Cronkite stated that the work was "an important historical document, commended to our reading by history itself."[42]

Although the public mood appeared favorable, the Johnson administration continued to put little faith in American public opinion. Speaking in London during July at a meeting of a Joint United States–United Kingdom Information Working Group, the Assistant Secretary of State for Public Affairs, James Greenfield, admitted candidly that the high public regard for the war was based on low U.S. casualty rates and that criticism would necessarily rise as casualties increased. General Westmoreland was of the same mind. Seeing no likelihood of victory and almost certain defeat unless U.S. ground forces entered combat in large numbers, he told General Wheeler that the United States should prepare U.S. and world opinion for the rigors ahead by "airing an objective and complete analysis of the problem we face and what we must do about it." Since the news from South Vietnam was already bad, additional information would make little difference. The approach might, indeed, put the Johnson administration "in a position to counter-attack in good faith the distorted reporting of the crepe-hangers."[43]

A highly publicized trip to South Vietnam by Secretary McNamara during mid-July served part of the purpose Westmoreland sought. Before McNamara left, President Johnson told a news conference that "it will be necessary to resist . . . aggression and therefore to have substantially larger increments of troops."[44] That remark prompted a spate of news stories speculating on the possibility that Johnson would increase draft calls and muster the reserves. By the time McNamara returned to Washington with word that the situation was "worse than a year ago (when it was worse than the year before that)," the American public appeared to have become reconciled to the idea that the United States would begin taking a more active part in the war.[45] Thus, when McNamara recommended increasing the number of U.S. troops in South Vietnam to 175,000 by 1 November and William Bundy proposed a broad public relations campaign to prepare the public and generate international support, the Assistant Secretary of State for International Organizations, Harlan Cleveland, advised against any promotional buildup. "As a result of Secretary McNamara's trip and the press reports coming out of Saigon," he told Rusk, "I have the very definite impression . . . that public opinion is already substantially conditioned to expect an increase in our force level to 200,000 men. . . . If the president decides to move

[42] CBS News, *Vietnam Perspective* (New York: Pocket Books, 1965), p. xvi. See also Memo, Sylvester for the Secretary of Defense, 6 Aug 65, sub: CBS TV Show on Vietnam, file 091.122, 1965, 70A3717, box 43, WNRC.

[43] U.S. Dept of State, Record of the U.S./U.K. Information Working Group Meeting, London, 20–21 Jul 65, FAIM/IR. Westmoreland's comment is in Msg, Westmoreland MAC 3240 to Wheeler, 24 Jun 65, Westmoreland Papers, CMH.

[44] Fact Sheet, U.S. Army, 16 Jul 65, sub: Presidential Press Conference, DDI Build Up of U.S. Forces file.

[45] Memo, McNamara for the President, 20 Jul 65, sub: Recommendations of Additional Deployments to Vietnam, FAIM/IR.

Zorthian and Sylvester in Saigon

along these lines, it will not come as much of a surprise. The surprise would be if he decided not to act. . . . A build up of the contemplated size cannot be played in 'low key'; but we can certainly avoid all the paraphernalia of crises that can at will be created—or not created, by presidential actions.''[46]

In the end, President Johnson opted for Cleveland's approach. On 23 July Undersecretary of State George Ball notified the assistant secretaries of state that the president was ''anxious to present the decision which might be made in the next few days in a low-key manner in order to avoid an abrupt challenge to the communists, and undue concern and excitement in the congress and in domestic public opinion.''[47] Shortly thereafter, General Wheeler cabled Westmoreland to inform him that McNamara's recommendations had been approved but that he should not be ''surprised or disappointed if the public announcement does not set forth the full details of the program but instead reflects an incremental approach.''[48]

[46] Memo, Harlan Cleveland for Secretary Rusk, 22 Jul 65, sub: Vietnam, FAIM/IR. Cleveland's memo summarizes Bundy's recommendations.

[47] Memo, George Ball for the Assistant Secretaries of State, 23 Jul 65, sub: Actions Agreed on at Special Meeting of the Assistant Secretaries, FAIM/IR.

[48] Msg, Wheeler JCS 2800-65 to Westmoreland, 28 Jul 65, Westmoreland Papers, CMH.

Whatever the Johnson administration's intentions, there was little chance that the war would stay out of the headlines for long. The Saigon correspondents may have accepted Zorthian's guidelines for the press, but they remained ill disposed toward giving officialdom the benefit of a doubt where the war itself was concerned. In addition, a confrontation during McNamara's trip between Arthur Sylvester and a number of important newsmen had already reduced official credibility to a new low.

Sylvester had accompanied McNamara to Saigon to coordinate relations with the press while the secretary was in South Vietnam. In hopes that an informal meeting between the assistant secretary and newsmen might promote better understanding, Zorthian one evening invited Sylvester, Edward White of the Associated Press, Morley Safer and Murray Fromson of CBS News, Keyes Beech of the *Chicago Daily News*, Jack Langguth of the *New York Times*, and several other correspondents to his villa for what he called a "bull session."[49]

The meeting went poorly from the start, with the exchange between Sylvester and the newsmen becoming increasingly bitter and personal as time passed. When Sylvester said he failed to understand "how you guys can write what you do while American boys are dying out there," one of the reporters suggested that he was attempting to be deliberately provocative. "I don't even have to talk to you people," Sylvester responded. "I know how to deal with you through your editors and publishers back in the States." Although the newsmen switched the subject to practical matters—the need for better communications and transportation for the news media—Sylvester continued in the same vein. "Do you guys want to be spoon fed? Why don't you get out and cover the war?" That remark prompted a question from Langguth about the credibility of official spokesmen, to which Sylvester replied, "Look, if you think any American official is going to tell you the truth you're stupid." In time of war, he added, the news media had the obligation to become the "handmaiden" of government.

Sylvester later said his remark about handmaidens had been a joke, but by then Safer and Fromson had stalked out of the house, indignantly slamming the door, and several other correspondents were preparing to follow. In all, White later told his bureau chief, Malcolm Browne, it had been "a long, disagreeable night." After the reporters left, Zorthian asked Sylvester why he had allowed a confrontation to develop. Sylvester's only response was that "They needed it. It was good for them."

Soon after that encounter, disagreement flared between officials and newsmen over the first American air attack on a North Vietnamese surface-to-air missile site. Seeking to decrease the impact of what some might consider an escalation, the Defense Department directed the Military Assistance Command to withhold news of the event. Then, contrary to the usual practice of releasing information

[49] This account of the meeting is drawn from Morley Safer, "Television Covers the War," in U.S. Congress, Senate, Committee on Foreign Relations, *News Policies in Vietnam, Hearing, 17 and 31 Aug 66*, 89th Cong., 2d sess., p. 90. Barry Zorthian corroborated Safer's account of the event in an interview with the author on 10 February 1984 but noted that Sylvester had not been drinking.

about the air war in Saigon, it made an announcement of its own in Washington.[50]

The Saigon correspondents reacted angrily. During a bitter session with MACV spokesmen, the newsmen berated the command's refusal to allow participants in the raid to brief them, charging that the entire system of voluntary cooperation would break down unless officials followed their own rules. Information officers responded that they had received no prior notice that the announcement would be made in Washington. In an attempt to placate the newsmen, they asked the deputy commander of the 2d Air Division, Maj. Gen. Gilbert S. Meyers, to brief the press. Although Meyers met some of the reporters' objections by revealing aspects of the raid unmentioned in the Washington announcement, a number of newspapers in the United States carried stories the next day criticizing MACV's refusal to discuss the raid and speculating that domestic political considerations had entered into the decision to make the announcement from Washington.[51]

Information officers at the U.S. mission in Saigon were almost as chagrined as the correspondents, especially since the State and Defense Departments had recently reaffirmed the practice of making Saigon the main point of release for news of the air war. They protested that while they were unaware of all the factors entering into the decision, they would be remiss in their duty if they failed to warn that "grave repercussions" could be expected to follow any attempt to set aside normal announcement procedures.[52]

Civilian Casualties: Incident at Cam Ne

The outcry over the raid on the missile site had hardly begun to subside when a new and graver problem arose involving the treatment of South Vietnamese civilians by U.S. marines. Since their arrival in South Vietnam in March, the marines had patrolled in generally unpopulated areas north and west of the air base at Da Nang, leaving responsibility for a densely populated region to the south in the hands of the South Vietnamese. After the Viet Cong launched a damaging raid against the base from the South Vietnamese sector in early July, the commander of the III Marine Amphibious Force, Lt. Gen. Lewis Walt, decided the marines should patrol that area as well. Concerned that the Americans would be unable to tell friend from foe in an area dominated by the Communists for generations, South Vietnamese officials at first refused to agree but in the end ceded responsibility for a zone stretching six kilometers southward from the perim-

[50] U.S. Department of Defense News Release, 27 Jul 65, sub: 27 July Missile Site Raid, DDI Protective Reaction file; Msg, Saigon 302 to State, 28 Jul 65, FAIM/IR.

[51] Msg, Saigon 301 to State, 28 Jul 65, FAIM/IR; Msg, Saigon 302 to State, 28 Jul 65; Jack Langguth, "U.S. Silences Aides in Saigon on Missile Site Raid," *New York Times*, 29 Jul 65.

[52] Msg, Saigon 301 to State, 28 Jul 65.

eter of the base. The decision gave the marines their first extended contact with a hostile civilian population.[53]

Marine units took control of the new area on 12 July. Immediately they encountered stiff resistance from a number of hamlets and villages that contained enemy fortifications. Since an aggressive response seemed the best solution, Marine commanders in early August decided to subdue one of the enemy's main strongpoints, the village of Cam Ne. Planning a conventional assault on a fortified town, they told their men to "overcome and destroy" every "hedgerow, trench line, bunker, spider-trap, [and] hut" from which the enemy fired.[54]

The operation began on 3 August, with the marines arriving on the outskirts of Cam Ne, where—according to the officers in charge—they received occasional fire from an estimated one hundred Viet Cong hiding in and around the village. They returned the fire with rockets and M79 grenade launchers, setting off a number of secondary explosions among mines and booby traps ringing the village. When they entered Cam Ne, they found most of some four hundred huts surrounded by trenches, concealed firing positions, and connecting tunnels.[55]

Although enemy fire slackened once the marines were inside the village, sniping continued throughout the day. During the course of the fight many huts went up in flames. Others, in the words of the battalion commander, Lt. Col. Verle E. Ludwig,

> were burned or damaged incidentally, as a result of flame thrower action or demolitions . . . which were used to neutralize bunkers, trenches, and firing positions actually in use by the VC. My people in the town tell me that before they blew caves and tunnels in the houses, they made sure that all civilians were out. As far as we know, only three civilians were wounded and a child, a boy, approximately ten years old, was killed while in a hut occupied by a VC who was exchanging fire with marines.

As the marines prepared to leave Cam Ne that afternoon, Ludwig continued, enemy fire became so heavy that he had to call in artillery and mortar support to enable his men to withdraw. "The VC were able to pop right back up out of the ground and fire . . . ," Ludwig said. "That gives some indication of the extent of the fortifications in the town."[56]

One of the first television newsmen to be stationed permanently in South Vietnam, Morley Safer, had accompanied the operation with his cameraman, Ha Tue Can. Safer presented a different version of events. When he cabled CBS News in New York to inform his producers that a film report would be arriving within

[53] Jack Shulimson and Major Charles M. Johnson, USMC, *U.S. Marines in Vietnam: The Landing and the Buildup, 1965* (hereafter cited as *The Landing and the Buildup, 1965*) (Washington, D.C.: History and Museums Division, USMC, 1978), pp. 50–65.

[54] Ibid.; Msg, CG FMFPAC to Commandant Marine Corps, 7 Aug 65, History and Museums Division, HQ, U.S. Marine Corps, 7A22065, Cam Ne file (hereafter cited as HQMC Cam Ne file). For the order, see Memo, Counsel for Commandant, U.S. Marine Corps, 3 Sep 65, sub: Cam Ne, HQMC Cam Ne file.

[55] Memo, HQ Marine Corps for ASD (PA), 9 Aug 65, sub: Mr. Morley Safer's Report of Marine Attack on the Village of Cam Ne, HQMC Cam Ne file.

[56] Ibid.

a few days, he said that an officer at the scene had told him the marines at Cam Ne had orders "to burn the village to the ground if [they] . . . received even one round of enemy fire." After the enemy fired an automatic weapon from an unidentified direction, the marines responded with rockets, grenades, and machine guns. Despite the pleas of elderly villagers, they used cigarette lighters and flamethrowers to destroy 150 dwellings. He concluded,

> I witnessed the foregoing and heard that another marine unit on the opposite side of the village wounded three women and killed one child. . . . Two marines were wounded by their own fire. Marine sources deny this. Prior to the burning, townspeople urged to abandon their shelters in English. [Not understanding,] . . . they remained in their positions. This reporter offered services of South Vietnamese cameraman to give desired instructions in native tongue. Marines had no official interpreters, only three Vietnamese who spoke no English. Defense Department says all our troops constantly reminded of need to protect civilians. Marines have lost men helping civilians in Danang area.

Safer's dispatch so disturbed officials at CBS that they instructed Harry Reasoner to introduce that evening's newscast with a reading of the cable.[57]

A flurry of news stories on civilian casualties followed Safer's story. The next night, the ABC Evening News ran a film clip of marines leveling an unidentified village and accidentally killing civilians. Although he noted that General Walt deplored the deaths and had ordered precautions to prevent similar incidents in the future, commentator Bob Young nevertheless stated that some marines were "getting tired of being told when to shoot and when not to." On WABC Radio in New York, Edward P. Morgan told his listeners that increased civilian casualties seemed a by-product of the growing American commitment to South Vietnam. "One marine hurling a grenade yelled, 'I got me two VC,' found he got two children. At the risk of death, the Viet Cong force . . . civilians to mingle with them. The marines and paratroopers there have not had special training in handling such situations. The French in Indochina experienced this situation and soon faced the wrath and hatred of the civilian population."[58]

The report prompted Bankson to telephone Colonel Legare in Saigon for the command's view of the situation. Had marine operations south of Da Nang produced a string of civilian casualties, he wanted to know, and was it necessary for U.S. troops to go into villages? After checking with the III Marine Amphibious Force, Legare responded that the marines never burned houses and villages unless they doubled as fortifications. The hut ignited with a cigarette lighter had concealed the concrete entrance to a tunnel. Since the Viet Cong frequently used villages as heavily fortified hideaways, it was important for U.S. troops to enter and search them.[59]

Although CBS carried Legare's version two nights after the first broadcast,

[57] CBS Evening News, 3 Aug 65, Radio-TV-Defense Dialog. Garrick Utley of NBC News was the first television correspondent to be stationed permanently in Saigon. See "TV's First War," *Newsweek*, 30 Aug 65, p. 32.

[58] WABC Radio News, 4 Aug 65, Radio-TV-Defense Dialog.

[59] Msg, MACOI to NMCC, 5 Aug 65, HQMC Cam Ne file.

Safer's film of the operation at Cam Ne appeared on the same program and dominated the evening's news. The film, which showed a marine, his rifle hung casually at his waist, lighting a hut with a cigarette lighter, appeared to dispute the official contention that the marines had faced heavy opposition in the village and that most of the huts had been destroyed in the exchange of fire. According to Safer,

> It first appeared that the marines had been sniped at before and that a few houses were made to pay. Shortly after, one officer told me he had orders to go in and level the string of hamlets that surround Cam Ne village. And all around the common paddy fields [camera focuses on a roof being lit by a flamethrower] a ring of fire. One hundred and fifty homes were leveled in retaliation for a burst of gunfire. In Vietnam like everywhere else in Asia, property, a home, is everything. A man lives with his family on ancestral land. His parents are buried nearby. These spirits are part of his holdings. . . . Today's operation shows the frustration of Vietnam in miniature. There is little doubt that American fire power can win a military victory here. But to a Vietnamese peasant whose home means a lifetime of backbreaking labor, it will take more than presidential promises to convince him that we are on his side.[60]

The next morning newspapers across the United States carried an Associated Press photograph of a marine igniting a hut with a cigarette lighter. The caption read, "Marines are under orders to burn any village from which sniper fire is received." That night CBS ran another film report by Safer on the accidental killing of a South Vietnamese youth by a Marine patrol. As the film showed villagers mourning the victim, Safer commented, "These are the people to whom the war is a curse. Intimidation and atrocity by the VC, and now, to them, equal brutality by the government and its allies."[61]

Information officers in Saigon responded that there was no evidence that the Marine command had ever issued an order to burn villages. They also released a directive from General Westmoreland to all U.S. Army and Marine Corps combat units in South Vietnam stipulating that American fighting men were to use "the utmost discretion, judgment, and restraint . . . in the application of . . . fire power."[62]

The broadcast of another film by Safer followed shortly, this one showing Safer interviewing marines who had participated in the action at Cam Ne. "You're up against a lot of women, children and old men," Safer said to one marine; "how do you feel about it, corporal?" The soldier responded:

> Well, this is what makes it hairy, being against these women and children, . . . but you treat everyone like an enemy until he's proven innocent. That's the only way you can do it. . . .
>
> *Q*. Yesterday, we were in that village of Cam Nanh [*sic*], we burned all the houses, I guess. Do you think that was necessary to fulfill the mission?

[60] Marine Corps, Transcript of CBS Evening News Broadcast of 5 Aug 65, HQMC Cam Ne file.

[61] The AP photo is mentioned in MFR, 9 Aug 65, sub: Leonard F. Chapman Conversation With Editor of the *Washington Post*, HQMC Cam Ne file. Safer's report is in Radio-TV-Defense Dialog, 6 Aug 65.

[62] Msg, Saigon 384 to State, 5 Aug 65, HQMC Cam Ne file. See also Jack Langguth, "Marines Defend Burning of Village," *New York Times*, 6 Aug 65.

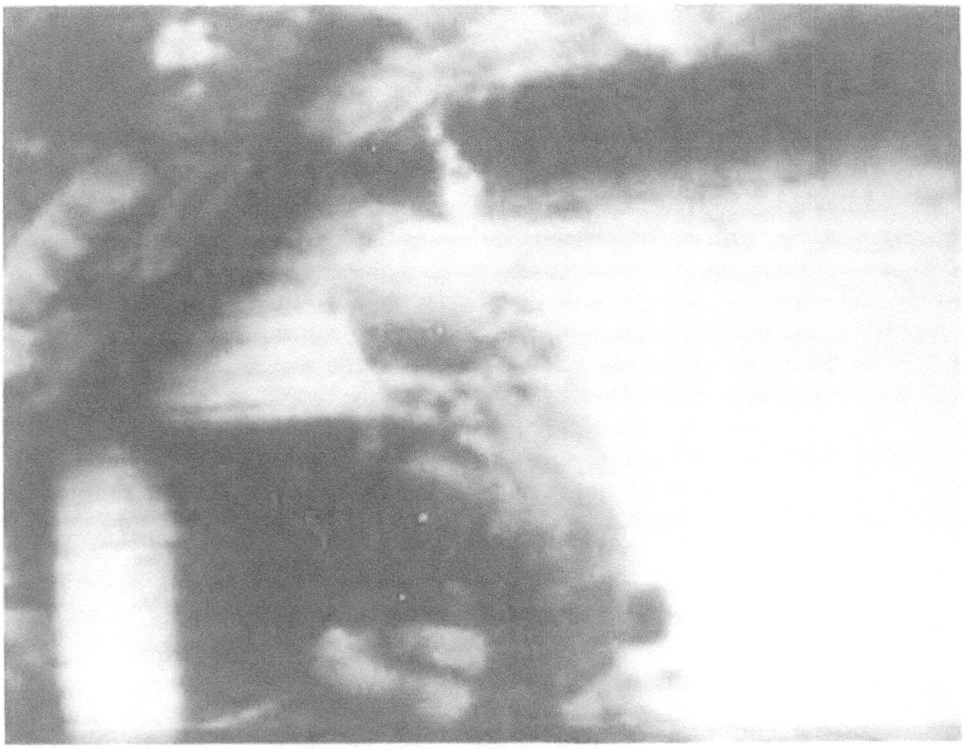

Marine Ignites Hut at Cam Ne

A. Yes I do. . . . We are the only company that went in there that hasn't had people killed . . . and I feel we . . . done a good job right there. And then we're going to have to show these people over a period of time that we're done playing with them. . . . These other companies moved through [on 12 July] and left that stuff stand and they got people killed. . . . We went in and we done our job and destroyed the villages and we took four casualties. So I think we proved our point.[63]

"Do you have any private doubts . . ." Safer asked next, "any private regrets about some of these people that you are leaving homeless?" One marine said there seemed no way around the problem because "everybody's caught in the middle and nobody knows what to do about it." A second marine, who had responded to Safer's earlier questions, was more blunt. "You can't have a feeling of remorse for these people. I mean, like I say, they are an enemy until proven innocent. They are an enemy. . . . I feel no remorse. I don't imagine anybody else does. You can't do your job and feel pity for these people." Safer continued the questioning. "After the marine patrol's been through there and someone's been sniped at or wounded or killed," he asked, "do you go in with revenge

[63] The interviews are contained in U.S. Marine Corps, Transcript of the CBS Evening News Broadcast of 5 Aug 65. See also Radio-TV-Defense Dialog, 7 Aug 65.

in your hearts into those villages?" A marine replied that he did: "I mean, I don't like to see a fellow marine shot, wounded, or even as much as scratched over here in this country."[64] Interviewed by MACV investigators later in the week, all of the men Safer had filmed excused themselves by claiming that they had been lured into their indiscreet statements by misleading questions.[65]

General Walt responded to Safer's reports by banning the reporter from the I Corps Tactical Zone, but Colonel Legare requested that he rescind the order, arguing that the Military Assistance Command had responsibility for disciplining reporters. Walt complied, but Safer was soon in further trouble. On 11 August he violated MACV's new ground rules by revealing that U.S. airborne troops were on the move to Pleiku and might relieve a besieged Special Forces camp at Duc Co. That disclosure led to a warning from the MACV Office of Information to the Saigon correspondents that any reporter breaking the ground rules in the future would face disaccreditation.[66]

Tired of Safer's continual probing, Arthur Sylvester contacted the president of CBS News, Fred Friendly, to have the reporter recalled. "Canadian military friends of mine who know Mr. Safer personally . . . ," he wrote, "tell me he has long been known . . . as a man with a strong anti-military bias. They say the record shows that he shafted the Canadian defense establishment in the sense that he did not present a balanced account of controversial situations. That would be my complaint about his reports, picture and verbal, on Cam Ne." Perhaps as a Canadian, Safer had

> no interest in our efforts in Vietnam and no realization that the Vietnam conflict is not World War II or Korea, but a new kind of political, economic, military action. But since this is a new kind of war, balance is a prerequisite in any presentation of actions out there, particularly since we are acting not only militarily but also politically before the world. I think that an American reporter and an American photographer, rather than the Vietnamese photographer Mr. Safer used, would be more sensitive to those considerations.

The premature revelation of the movement of American paratroopers, Sylvester concluded, amply demonstrated Safer's ill will toward the military.[67]

Although Sylvester included with his letter a detailed rebuttal of Safer's version of the events at Cam Ne, Friendly refused to recall the reporter. While the marines contended that they had merely burned huts and bunkers actually in use by the Viet Cong, Friendly noted in his response, Safer's film showed clearly that a number of huts had been set afire with cigarette lighters and flamethrowers without any indication of enemy resistance. Even if it was correct that the Communists had attacked in strength as the marines withdrew, that was hardly proof

[64] Ibid.

[65] Shulimson and Johnson, *The Landing and the Buildup, 1965*, pp. 50-65.

[66] Msg, Saigon 462 to State, 12 Aug 65, FAIM/IR; Msg, Defense 7945 to COMUSMACV, 11 Aug 65, DDI PA OPNS ORGNS file; Msg, Westmoreland MAC 4123 to Sylvester, 14 Aug 65, Westmoreland Papers, CMH.

[67] Ltr, Sylvester to Friendly, 12 Aug 65, HQMC Cam Ne file. See also Msg, Saigon 462 to State, 12 Aug 65.

of tunnel fortifications. The enemy might have come upon the marines from the surrounding area without using any tunnels at all. As for the incident at Pleiku, Safer had made every effort to comply with MACV's regulations. After trying unsuccessfully to get through to the MACV Office of Information in Saigon, the reporter had obtained clearance for his transmission from the most senior officer available in the field. Any doubt that might have remained was dispelled by officers at the scene, who told Safer that the Viet Cong already knew airborne troops were arriving in the area.[68]

Sylvester's letter, Friendly concluded, was "a matter of pure and simple character assassination." The suggestion that an American might be more sensitive to the situation than a Canadian was tantamount to saying that an American would be "more 'sympathetic'" to the official line. "The essence of our dispute is quite simple. You don't want anything you consider damaging to our morale or our world-wide image reported. We don't want to violate purely *military* security with reports which could endanger the life of a single soldier but, by the same token, we must insist upon our right to report what is actually happening despite the political consequences. . . . In the long term, this, too, will help enhance our nation's position in the eyes of the world."

Sylvester's attempt to have Safer recalled had hardly ended before another controversy involving the marines arose. On 14 August an Associated Press photographer gave the U.S. mission in Saigon a number of photographs of South Vietnamese soldiers torturing enemy captives while U.S. marines looked on. When General Westmoreland saw the pictures he contacted General Walt. Alluding to "the unfortunate press and TV coverage of actions of your command," he directed the general to "do everything humanly possible to disassociate our presence with any indiscriminate use of force, brutality, or violations of the spirit of the Geneva Conventions on the handling of prisoners of war." Although the photographs gave no indication that the marines had participated in the torture, Westmoreland continued,

> their presence . . . could implicate them since there is no evidence to suggest that they attempted to moderate the actions by the Vietnamese. I admit that this is a difficult problem since we have no command authority over Vietnamese troops . . . , we must try to moderate their treatment of prisoners so that it conforms to the spirit of the Geneva Conventions, which the GVN has agreed to in principle. In any case, we should attempt to avoid photographs being taken of these incidents of torture and most certainly . . . try to keep Americans out of the picture.[69]

Under pressure from the Joint Chiefs of Staff, which reflected a general concern in Washington that news stories of atrocities might hamper public acceptance of larger U.S. troop commitments, Westmoreland directed the MACV staff to develop a new set of guidelines to govern military relations with noncombatants. Although formal regulations could never substitute for the common sense of a

[68] This section is based on Ltr, Friendly to Sylvester, 16 Aug 65, HQMC Cam Ne file.
[69] Ltr, Westmoreland to Walt, 14 Aug 65, Westmoreland History, bk. 17, tab 23, CMH.

good officer, Westmoreland reasoned that detailed rules might contribute to restraint.[70]

Published on 7 September, the MACV directive specifically prohibited indiscriminate destruction of populated zones. Commanders were to select landing sites and artillery targets only after giving due regard to the lives and property of noncombatants. Forward air controllers and helicopter pilots were likewise to inform themselves of areas that were politically sensitive or off limits to military action. Whenever security allowed, units operating in the field were to use loudspeakers and leaflet drops to warn nearby villagers of impending air and ground assaults. Qualified South Vietnamese officers were also to accompany large operations both to help identify the enemy and to ensure the close coordination of American units with South Vietnamese officials and troops. Where possible, South Vietnamese units were to fight alongside Americans down to battalion and company level to assist in searching dwellings and to indicate to the peasants that the government of South Vietnam endorsed the operation. In all cases, U.S. forces were to demonstrate their "concern for the safety of non-combatants, their compassion for the injured, their willingness to aid and assist the sick, the hungry, and the dispossessed."[71]

By the time Westmoreland's directive appeared, the marines, at Walt's behest, were already putting many of its provisions into practice. On 9 August, for example, while the controversy over Safer's report on Cam Ne was still raging, another Marine unit operating near the village took enemy fire. Losing two men killed and twenty-one wounded in the incident, the marines decided to secure the area once and for all. Morley Safer accompanied the operation that resulted, but filed a dispatch with his bureau in Saigon that contrasted sharply with his earlier reports.[72]

Safer's colleague in Saigon, Jack Laurence, transmitted the substance of the reporter's findings to New York.

> A postscript to the incident at Cam Ne is being written by the U.S. Marines. Morley Safer, who revealed the burning of a Vietnamese village by marines two weeks ago and touched off a major controversy over U.S. military policy, has just sent in a follow up report. He went back to the same village today . . . and watched marines rebuilding Cam Ne. They completed the mission of searching for VC hideouts, but this time, as they did, the villagers were given full warning by leaflets dropped from helicopters. Shelters were built for the homeless, and this time everybody was happy. He quoted a high-ranking marine officer as saying, "All of that bad publicity generated by the action at Cam Ne has done more good than harm."[73]

General Westmoreland would hardly have agreed. He was visiting his family in Hawaii when Safer's Cam Ne story appeared. After viewing a number of tele-

[70] Msg, Westmoreland MAC 4328 to Wheeler, 28 Aug 65, Westmoreland Papers, CMH. JCS concerns are mentioned in Msg, Wheeler JCS 3196 to Westmoreland, 26 Aug 65, Westmoreland Papers, CMH.

[71] MACV Directive 525-3, 7 Sep 65, sub: Combat Operations: Minimizing Non-Combatant Battle Casualties, DDI Rules of Engagement file.

[72] MFR, USMC, 9 Aug 65, sub: Cam Ne, HQMC Cam Ne file.

[73] Radio-TV-Defense Dialog, 18 Aug 65.

vision reports on the war, he cabled his deputy in Saigon, Lt. Gen. John L. Throckmorton, to comment that he was

> beginning to appreciate the many comments we have received in Saigon on the distorted and unfavorable publicity coming out of Vietnam. Last evening, on the ABC News broadcast . . . , there was . . . [an] interview . . . by a couple of young reporters who stuck the microphone in the faces of three young, surly marines . . . and asked them a few leading questions. The first marine alleged that he was not happy . . . and wanted to go home; the second . . . could not see why additional troops were coming over because those that were there were not being allowed to exercise their capabilities; the third . . . stated that there were too many restrictions . . . , to wit: that they could not fire unless fired upon and were not allowed to load their weapons until ordered to do so. On balance, I consider the performance misleading. . . . It suggests to me that the [MACV chief of information] . . . is not exercising [the] controls available to him in that the press is apparently allowed to free-wheel as they please.

Westmoreland wanted General Walt and Colonel Legare to look into the possibility of placing television reporters under constraint but urged discretion: "I do not want a cause celebre made of this for the simple reason that if it comes to the attention of the press that the command is 'investigating,' there would be unproductive repercussions."[74]

Recognizing that any attempt to discriminate between television and print journalists would cause trouble, the MACV Office of Information passed the issue to the Office of the Assistant Secretary of Defense for Public Affairs. The Defense Department responded six days later with a major policy statement for all major military commands throughout the world. There was to be no distinction between print journalists and television reporters, that guidance specified: information was to be "equally available on the same basis to all media and all media representatives."[75]

Censorship Reconsidered, August 1965

General Westmoreland apparently had no intention of proposing formal censorship, but talk of the subject nevertheless increased in Washington following Safer's violation of the MACV ground rules. It became so prevalent that Sylvester felt constrained to have a plan for censorship on file just in case Secretary McNamara asked for one. He assigned the drafting to the commander of the U.S. Army Reserve's field press censorship detachment, Col. Ervin F. Kushner, and asked Bankson to do another study of whether censorship in South Vietnam would work.[76]

[74] Msg, Westmoreland to Throckmorton, 6 Aug 65, Westmoreland Papers, CMH.

[75] Msg, Defense 8096 to Secretary of the Army et al., 12 Aug 65, DDI Releasing Authority in RVN file.

[76] Memo, Bankson for Chief of Army Information, 12 Aug 65, sub: Censorship, DDI Censorship file; Intervs, author with Bankson, 28 Aug 75, and with Charles W. Hinkle, Chief of the Defense Department's Office of Freedom of Information and, at the time, Chief of the department's Office of Security Review, 11 Apr 78, both in CMH files.

That Sylvester was again considering restraints on the press disturbed information officers at the State Department. One of them, Joseph Lumen, wrote a memorandum to Bankson pointing out that formal censorship posed the possibility of international repercussions. Should a non-American newsman antagonize the South Vietnamese government, Lumen said, South Vietnamese officials might assert their prerogatives as censors either to destroy the newsman's copy or to expel the reporter from the country. That, in turn, could lead to representations from the newsman's home government, complicating President Johnson's effort to seek international support for the war.[77]

The MACV Office of Information also objected. Colonel Legare pointed out that the news media would give the United States full credit and full blame for whatever happened in South Vietnam no matter what controls the Military Assistance Command imposed. To stifle critical comment, censorship would have to go beyond strictly military information into the political sphere. That step might lead to restraints on editorials and news analyses, which would violate the First Amendment to the Constitution and deny the traditional right of the American public to be informed. A powerful coalition would come into being, Legare said, uniting Congress, the public, and the news media in opposition to the president and the war.[78]

Bankson filed his report during the last week of August. He repeated all the objections to censorship first raised at the Honolulu Conference in March and added many of the arguments provided by the State Department and the Military Assistance Command. By so doing, he reaffirmed Sylvester's conviction that further restraints upon the press were unthinktable.[79]

Colonel Kushner's censorship plan was less thorough. Rushed to completion, it took little account of the U.S. government's inability to control all means of communication out of South Vietnam or to prevent correspondents from filing dispatches from points outside the country. If only to be prepared to respond to queries from the Joint Chiefs of Staff and the secretary of defense, Sylvester decided he needed a more credible plan and assigned the task of preparing one to the director of the Defense Department's Office of Security Review, Charles W. Hinkle.[80]

Hinkle had difficulty devising an effective plan. He could discern no way, for example, to get around the two problems Colonel Kushner had failed to solve. In addition, since there were no facilities for developing television film in South Vietnam, the Military Assistance Command would have to ship network news footage to the Philippines or Hawaii for processing before review. That would

[77] Memo, Joseph Lumen, State Department Office of Public Affairs, for OSD (PA), 24 Aug 65, sub: Hazards of Field Press Censorship, DDI Censorship file.

[78] Msg, MACV 29892, MACOI to OSD (PA), 25 Aug 65, DDI Censorship file.

[79] Intervs, author with Bankson, 28 Aug 75, and with Hinkle, 11 Apr 78.

[80] Ibid. Col Ervin F. Kushner, Study of Field Press Censorship, 28 Aug 65; Memo, ASD (PA) for Assistant Secretary of State for Public Affairs, 16 Nov 65. Both in DDI Censorship file.

put television news at a marked disadvantage in relation to the print media and cause a major outcry.[81]

Work on the study dragged into December with Hinkle unwilling to sign his name to a program he knew would never work. When the commander in chief, Pacific, inquired about the status of the plan, Sylvester returned Hinkle to his regular duties in the Office of Security Review and assigned the project to the newly designated Special Assistant for Southeast Asia, Col. Winant Sidle, who assumed the position when Bankson left to become chief of the MACV Office of Information.[82]

Convinced that censorship would be counterproductive, Sidle purposely drew up a plan so ponderous that it could never become a serious alternative to the voluntary guidelines already in effect. The State Department would have to negotiate the program with the South Vietnamese, Sidle noted. All the nations concerned—South Vietnam, the United States, South Korea, Thailand, Australia, the Philippines, and New Zealand—would have to inaugurate the system simultaneously. The South Vietnamese would require a huge organization to check material for publication within their own country; the Military Assistance Command would need an even larger, multilingual establishment to review dispatches destined for the United States and elsewhere. The necessity to screen all television reports would require elaborate film laboratories and viewing facilities, and the thousands of letters and packages mailed daily from South Vietnam would have to be opened and inspected. In all, Sidle implied, the system would require the services of hundreds of military personnel and civilians.[83]

Transmitting an information copy of the plan to the State Department in August 1966, a year after he had directed its preparation, Sylvester noted that the document was designed only to cover contingencies and that the Defense Department had no intention of instituting censorship.[84] The Assistant Secretary of State for Public Affairs, Dixon Donnelly, replied that he too considered censorship unwise. "In the highly likely event that the existence of this plan should become known," he added, testifying to the political sensitivity of the subject, "we shall, of course, refer inquiries to the Department of Defense."[85]

With that, all consideration of field press censorship in South Vietnam ended. As the American buildup proceeded and American forces continued on the offensive, the press corps in Saigon would report the war as it saw fit, under only the lightest official scrutiny.

[81] Interv, author with Hinkle, 11 Apr 78.
[82] Ibid.
[83] Interv, author with Maj Gen Winant Sidle, 6 May 73, CMH files. Sidle's plan may have been preceded by a draft originally authored by Bankson. See Ltr, Rodger Bankson to the author, 26 Oct 79, CMH files. ASD (PA), Draft Plan for Field Press Censorship [probably Jul 65], DDI Censorship file.
[84] Ltr, Sylvester to Dixon Donnelly, 6 Aug 66, DDI PA OPNS ORGNS file.
[85] Ltr, Dixon Donnelly to Sylvester, 12 Aug 66, DDI PA OPNS ORGNS file.

9

Problems With the Press

As the war had grown in South Vietnam, so had the corps of correspondents in Saigon. From some 40 at the beginning of 1964, it had reached 282 in January 1966. By August of that year 419 newsmen were accredited to the Military Assistance Command.[1]

Of the 282 reporters present in South Vietnam at the beginning of 1966 only 110 were Americans. Sixty-seven were South Vietnamese, 26 Japanese, 24 British, 13 Korean, 11 French, and 7 German. The remainder were from countries as diverse as Ceylon, India, Canada, Italy, Australia, Ireland, Thailand, Denmark, the Republic of China, and New Zealand. The correspondents were considerably older than might have been expected. Of the Americans present, 72 were more than thirty-one years old, and of them 60 were over the age of thirty-six. The same was true of the 143 non-Americans. One hundred thirteen were over the age of thirty-one and of them 66 were thirty-six years old or older.[2]

According Rodger Bankson, no more than one-third of the correspondents serving in South Vietnam during 1965 and 1966 were true working reporters. The rest were support personnel—secretaries, managers, interpreters, television sound technicians, television cameramen. A few were reporters' wives who had gained accreditation in order to use the post exchange in Saigon when their husbands were away on assignment. Others were hangers-on—stringers who represented small magazines and newspapers but rarely went into the field, attended briefings, or wrote stories.[3]

Most of the rest, however, were hardworking professionals: Beverly Deepe,

[1] With the fall of Saigon in 1975 and the loss of the records of the MACV Office of Information, it became difficult to determine the number of correspondents present in South Vietnam at any one time. The number for the pre-1964 period and for August 1966 are from U.S. Congress, Senate, Hearings Before the Committee on Foreign Relations, *News Policies in Vietnam*, 89th Cong., 2d sess., pp. 66f. The figure for January 1966 is from JUSPAO, Breakdown of News Correspondents as of 18 Jan 66, Papers of Barry Zorthian, copies in CMH files.

[2] JUSPAO, Breakdown of News Correspondents as of 18 Jan 66.

[3] Draft Memo, Rodger Bankson for Frank Olcott, 1 Jun 69, CMH files.

who became a freelance correspondent when the *New York Herald-Tribune* went out of business in 1966; Wendell Merick, who had arrived for a ten-day visit in 1964 but stayed on, first as a stringer for ABC News and the *London Daily Express*, later as a much respected correspondent for *U.S. News & World Report;* and Garrick Utley, the first television correspondent to be stationed full-time in Saigon, who arrived in early 1965 to become chief of the NBC News Bureau. Others included Jack Foisie of the *Los Angeles Times;* Ron Nessen of NBC News; Keyes Beech of the *Chicago Daily News;* Dan Rather of CBS; Charles Mohr of the *New York Times;* Richard Critchfield of the *Washington Star;* Francois Pelou of Agence France Presse; Frank McCulloch, bureau chief for *Time;* and Ward Just of the *Washington Post*. Although David Halberstam had departed Saigon by 1966 to report from Paris, many of the correspondents who had been prominent in earlier years were still present: Neil Sheehan, who left UPI in 1964 to join the *New York Times;* Pulitzer Prize winner Malcolm Browne, reporting for NBC News; *Newsweek* stringer Francois Sully; and Peter Arnett and Horst Faas of the Associated Press.[4]

As they grappled with the increasingly difficult, fluctuating situation in South Vietnam, those reporters lost much of the camaraderie that had characterized the press corps in Saigon during the early years of the war. Competing strenuously for every scrap of information and under pressure from home offices to produce, they became a constant source of irritation to the U.S. mission in Saigon. Yet in the absence of a practical censorship program, there was little officials could do to impose a solution. They dealt with problems piecemeal, addressing each as it occurred.[5]

Nuisance Stories

Sometimes even a favorable press could be a liability, as happened during August 1965, when the U.S. marines launched the largest American ground operation of the war to that time. Code-named STARLITE, the attack trapped a major portion of the *1st Viet Cong Regiment* on a peninsula near Chu Lai, an important Marine base a hundred kilometers southeast of Da Nang. Although the marines encountered heavy resistance from an enemy who hid in caves and tunnels until bypassed and then attacked from behind, by the end of the two-day operation they had accounted for some six hundred enemy dead. Odors rising from the battlefield long afterward indicated that many more had perished, sealed in their bunkers by demolitions, air strikes, and artillery fire.[6]

The MACV Office of Information escorted correspondents to the battlefield

[4] "Covering Vietnam: Crud Fret & Jeers," *Time*, 10 Jun 66, p. 54; "Femininity at the Front," *Time*, 28 Oct 66, p. 73.

[5] "Covering Vietman: Crud, Fret & Jeers," *Time*.

[6] Msg, DA 729354 to CG USCONARC et al., 23 Aug 65, CMH files; Msg, MACV 31635 to CINCPAC, 9 Sep 65, Gas file, CMH.

and allowed officers who had participated in the attack to brief the press. Although fifty-five marines had been killed—the largest American toll for any operation to that date—the news media treated the event as a major victory. A small number of dissenters such as James Reston of the *New York Times* pointed out that Americans could win every battle, but the South Vietnamese would have to win the war. Most commentators nevertheless agreed with Mark Watson of the *Baltimore Sun,* who called STARLITE "a true textbook example of an air-sea-land attack under ideal conditions." The *Kansas City Star* noted that American military professionalism was beginning to tell in South Vietnam, and the *Detroit News* remarked that the victory would drive home to the Communists the fact that there were "no easy pickings in Southeast Asia."[7]

Although news stories of that sort were always welcome to the U.S. command in Saigon, the heavy press coverage and the fact that the marines had fought and won without the assistance of South Vietnamese units stirred professional jealousies among some members of the South Vietnamese Joint General Staff. Remembering the mauling their troops had received in May at the hands of the *1st Viet Cong Regiment,* a few of those generals began to question the large American body count and to disparage the marines' handling of the operation.[8]

Both Ambassador Henry Cabot Lodge, who had succeeded Taylor during late July, and General Westmoreland recognized that a problem existed. Because he realized that the complaints stemmed from South Vietnamese sensitivity to any suggestion that American troops would take over the fighting, Lodge emphasized to the State Department the importance of portraying all future military operations as joint U.S.–South Vietnamese endeavors. "General Westmoreland plans to do all possible," he said, "to include in the early stages of any large operation at least some Vietnamese elements to whom a large share of the credit can be attributed. We also intend to pay particular attention to this aspect in our [press] briefings and public output . . . and would hope that Washington and other interested parties would be able to do the same."

Following Lodge's line of reasoning, Zorthian in early September prevailed upon the South Vietnamese to create a national press center and to begin weekly briefings for the Saigon correspondents, but neither effort produced any increase in the news media's coverage of South Vietnamese operations. On the day the center opened, the briefing began fifteen minutes late and consisted entirely of a lieutenant's reading from an uninformative mimeographed summary that was then distributed to correspondents. Restless and uncomfortable because a power failure in the building had cut off all fans, the newsmen had to submit their questions in a block, wait until they were translated into Vietnamese, and then wait again while the answers were rendered into English. Before a month had passed,

[7] James Reston, "Chu Lai: The Politicians and the Marines," *New York Times,* 22 Aug 65; [AP], "More Marines Land in Vietnam Build-up," *Baltimore Sun,* 24 Aug 65; "Corner Viet Cong Force," *Kansas City Star,* 19 Aug 65; [AP], "Cong Losses May Spur Peace Bid," *Detroit News,* 22 Aug 65.

[8] This section is based on Msg, Saigon 595 to State, 24 Aug 65, FAIM/IR.

the reporters had begun sending their Vietnamese assistants to the sessions while they concentrated as much as ever upon the American portion of the war.[9]

The nature of the conflict in South Vietnam compounded the problems of both the press and the Military Assistance Command. Characterized by sudden flare-ups and widely scattered action, the war taxed newsmen's ability to interpret events on a day-to-day basis. Under instructions to report combat, scrambling for colorful leads and headlines to gain an edge on the competition, correspondents concentrated on what they knew, not only emphasizing the American role in the war but also ascribing to engagements involving companies, platoons, and even squads the importance of encounters that had involved the divisions and regiments of World War II and Korea.[10]

The stories that resulted were a constant nuisance to the Military Assistance Command because senior officials in Washington paid excessive attention to what the press said and questioned the U.S. mission on every report that threatened to upset either the American public or the Congress. Labeled "rockets" by information officers, the messages containing those queries forced the command to investigate and justify events and decisions that in earlier wars would have been taken for granted.

On one occasion in mid-September, for example, the Associated Press reported that the men of the 1st Brigade, 101st Airborne Division, were going into battle with "gaping holes" in their boots and wearing tennis shoes because regulation footwear was in short supply. After a query from the Office of the Assistant Secretary of Defense for Public Affairs, Colonel Legare reported that the whole affair had been blown out of proportion. Although there had indeed been twelve instances of boots wearing out during a recent airborne operation and follow-up inspections had uncovered another fifty-seven cases where footwear was in doubtful condition, there had never been a critical shortage. Unaware of how quickly leather would deteriorate in a jungle environment, the men had merely left their spare boots in their base camp. The problem would not recur, Legare said, because the Military Assistance Command had taken the precaution of airlifting five hundred pairs of boots to the unit.[11]

On the same day the boot story appeared, the CBS Evening News prompted another query from Washington by broadcasting a filmed report on South Vietnamese civilians scavenging in a U.S. Marine garbage dump. During the sequence, the camera played upon peasants sifting through piles of rubble while the narrator, Jack Laurence, observed that even entering a dump was dangerous because live ammunition had become mixed with the refuse. As if on cue, a grenade went off, wounding a South Vietnamese boy.[12]

[9] Msg, Saigon 825 to State, 9 Sep 65, FAIM/IR; Msg, Saigon 1908 to State, 27 Nov 65, Press Policy file, CMH.

[10] Memo, Bankson for Sylvester, 22 Nov 65, sub: Backgrounders for Newsmen, DDI News from Vietnam file.

[11] [AP Dispatch], 14 Sep 65; ASD PA Response to AP Query [Sep 65]. Both in DDI Equipment file.

[12] This section is based on Msg, Wheeler JCS 3423-65 to Westmoreland, 16 Sep 65, Westmoreland Papers, CMH. The cable mistakenly attributes the story to Peter Lawrence.

Garrick Utley of NBC News *(lower right) interviews troops in the field.*

In previous wars the local command would have handled the problem in a routine manner. In South Vietnam, with Washington agencies observing the smallest details through the eye of a television news camera, it became a matter of significance at the national level. Having South Vietnamese scavenging in American dumps was sordid enough to those unacquainted with the grim realities of war, General Wheeler cabled Westmoreland, without adding an ammunition problem to it. The mixing of munitions with garbage had to stop, because it was a hazard to Americans as well as South Vietnamese and because it wasted valuable resources. Westmoreland was to inform the Joint Chiefs of Staff on the actions he was taking to remedy the situation, so that the chiefs could pass that word to the civilian leadership.

Westmoreland replied that he was well aware of the problem and had published guidance a month before the incident to ensure proper garbage disposal. He would amplify those instructions by once more directing commanders at all levels to keep ammunition under strict control and to screen garbage daily for explosives.[13]

[13] Msg, Westmoreland MAC 4647 to Wheeler, 18 Sep 65, Westmoreland Papers, CMH.

The Tear Gas Issue, September 1965

If the news media's penchant for eye-catching headlines was a source of concern to the Military Assistance Command, there were nevertheless occasions when the tendency worked to the command's advantage. An example occurred in the first week of September, during a Marine operation near Qui Nhon, a provincial capital some three hundred kilometers southeast of Da Nang. The marines encountered Viet Cong entrenched in bunkers and tunnels along with women and children. The battalion commander, Lt. Col. Leon M. Utter, uninformed of MACV's ban on riot control agents, employed tear gas. The move drove the Viet Cong and the civilians—some four hundred people—into the open, eliminating any need to fire into the caves and saving many lives. Recalling Secretary McNamara's earlier promise that U.S. forces would never again use riot control agents in South Vietnam, an Associated Press correspondent covering the operation asked Utter whether the authority to use gas had come from the Military Assistance Command. Utter replied that he had made the decision himself.[14]

Word of the incident and of Utter's response reached Saigon quickly. While the command started an investigation into why the marines had failed to follow instructions, information officers held a special briefing to put what had happened into the best possible light. Stressing the humanitarian motives of the battalion commander, they pointed out that whether riot control agents were proscribed or not, their use was militarily and morally preferable to flamethrowers and grenades, especially when women and children were involved. They asked the newsmen to refrain from publicizing the incident in order to protect Utter's military career.

When the story nevertheless appeared the next day, officials in Washington braced for an onslaught, believing that "the trickiest aspect of . . . [the] affair is not the use of tear gas *per se* but the implication that U.S. subordinate commanders do not know the terms under which weapons in their arsenal can be used." The expected uproar never came. Instead, United Press International noted blandly that Utter had apparently been unaware of MACV's instructions; Reuters spoke of the incident's humanitarian aspects; and the Associated Press commented that "there are military men here and other persons familiar with the Vietnamese war who believe the use of tear gas is the best method of dealing with . . . situations . . . such as [the one at] Qui Nhon."[15]

With the news media none too alarmed over the incident, General Westmoreland sensed an opportunity. He asked Admiral Sharp for authority to use riot control agents to clear caves, tunnels, and underground shelters. The tactic would have reduced American casualties at Cam Ne and during Operation STARLITE, he said; it was also preferable to using high explosives and flamethrowers where civilians were involved. Ambassador Lodge and Admiral

[14] This section is based on Msg, MACV to NMCC, 7 Sep 65, DDI Gas file. See also Shulimson and Johnson *The Landing and the Buildup, 1965*, pp. 90–91.

[15] MFR, 8 Sep 65, sub: Public Affairs Policy Committee for Vietnam Meeting of 7 Sept, ISA 092 VN, 70A3717, box 44, WNRC. The AP, UPI, and Reuters dispatches are all in the DDI Gas file.

Sharp both endorsed the proposal. Lodge told the State Department that "we must not be dissuaded from doing something . . . essentially constructive because of a few tendentious writers . . . out merely to make a sensation." Sharp even suggested that the news media's low-keyed reaction indicated receptivity to the idea that riot control agents were more humane than the usual weapons of war. Perhaps the military services could channel that impulse into outright support for the tactic.[16]

As if to confirm Sharp's assessment, the *New York Times* on 11 September published an editorial advocating the use of tear gas in South Vietnam. Although war was never humane, the newspaper avowed, the employment of riot munitions was "obviously more humane than any other effective type of action." If the United States abandoned that tactic, it would "condemn to death or injury many more Americans and Vietnamese than the absolute necessities of war demand."[17]

Although the U.S. delegation to the United Nations objected that the employment of gas in South Vietnam would provide grist for Communist propagandists, the *Times* editorial and the lack of critical comment in the American news media convinced President Johnson that he could safely make tear and nausea gases available to the Military Assistance Command on a case-by-case basis. McNamara notified Westmoreland of the decision on 23 September, authorizing him to employ the agents during an operation scheduled to begin two days later. "This has been a most difficult and complicated hassle," General Wheeler told Sharp and Westmoreland shortly thereafter. "Nevertheless, . . . I am satisfied that we are on the way to achieving a satisfactory policy which will untie . . . [Westmoreland's] hands and permit him to use riot control agents when he believes it necessary."[18]

Both the State and Defense Departments prepared carefully for the event. The State Department instructed its embassies to conduct briefings along the lines of the 11 September *New York Times* editorial in advance of the operation for any foreign government that appeared likely to make an official protest. McNamara meanwhile presented Westmoreland with detailed guidance on handling the Saigon correspondents. At a briefing to precede the attack, official spokesmen were to explain the reasons for using gas, avoid all debate on the subject, and stress the lack of risk to both Americans and South Vietnamese. If the question of Colonel Utter's conduct arose, the spokesmen were to make the point that the officer had exercised his own initiative in as humane a manner as possible.[19]

[16] For Lodge's comment, see Msg, Saigon 820 to State, 9 Sep 65, FAIM/IR. See also Msg, Westmoreland MACJ00 31635 to Sharp, 9 Sep 65, and Msg, CINCPAC to JCS, 10 Sep 65, both in DDI Gas file.

[17] The *New York Times* article is reprinted in Cir 567, State to All Diplomatic Posts, 7 Oct 65, DDI Gas file.

[18] Msg, USUN 760 to State, 15 Sep 65, FAIM/IR; Msg, Wheeler JCS 3528 to Sharp and Westmoreland, 22 Sep 65, Miles Policy/Strategy files, CMH; Msg, Defense 2425 to Westmoreland, 23 Sep 65, DDI Gas file; Msg, Wheeler JCS 3548-65 to Westmoreland, 23 Sep 65, Miles Policy/Strategy files, CMH. See also MACV History, 1965, p. 443, CMH files.

[19] Msg, State 823 to Saigon, 22 Sep 65, FAIM/IR; Msg, Defense 2425 to Westmoreland, 23 Sep 65, DDI Gas file.

Although the Military Assistance Command canceled the 25 September operation because of a breach of security, Westmoreland rescheduled the first use of gas for 8 October, when contingents of the 173d Airborne Brigade were to investigate an enemy tunnel complex near the Iron Triangle, a Communist base area thirty kilometers northwest of Saigon. McNamara's earlier instructions all applied, together with a new requirement that the command notify the State Department as soon as gas was used.[20]

Except in the Communist press, news stories that followed the event bore little resemblance to the angry commentaries that had appeared after Peter Arnett's revelations in March. In the Soviet Union, *Izvestia* condemned the incident as an American attempt to slaughter women and children. The New China News Agency in Peking charged that Pentagon procurement of poison gas and chemicals had been increasing steadily. Italian Communist Party newspapers ran headlines alleging "Monstrous War Crimes." Yet throughout most of the rest of the world the reaction was subdued. In West Germany the press concentrated on a major U.S. offensive in South Vietnam, mentioning tear gas only in passing, well down in the story. In England most newspapers carried word of the event on inside pages under quiet headings. Even the Communist *London Daily Worker* was subdued, conceding in a story generally critical of the United States that the gases involved had no lasting effect.[21]

Newspapers in the United States also said little, generally confining their criticism to the public relations effort accompanying the event. John Maffre of the *Washington Post* noted that while the Military Assistance Command obviously intended to clear enemy positions "with as little bloodshed as possible," it had "rigidly schooled" the soldiers of the 173d Airborne to speak of "tear gas" rather than "gas." In the same way, he said, the operation in question had occurred "not only to flush Viet Cong . . . but also to assuage world opinion, . . . with as much detailed planning in public relations as normally goes into a major operational assault."[22]

The press thus restored in a figurative sense what it had earlier taken away. On the day after the favorable news reports appeared, General Westmoreland cabled the Joint Chiefs of Staff for leave to employ tear and nausea gases at his own discretion. When the Joint Chiefs on 3 November granted the request, he moved systematically to reinstate the tactic, at first delegating authority to the three major American field commands in South Vietnam but within a month authorizing unrestricted use of riot control agents whenever local commanders saw fit.[23]

[20] Msg, Saigon 1175 to State, 5 Oct 65, and Msg, State 964 to Saigon, 6 Oct 65, both in FAIM/IR.

[21] Memo, Leonard H. Marks, Director, USIA, for McNamara, 12 Aug 65, sub: Daily Reaction Report, DDI Gas file; Msg, Rome 910 to State, 12 Oct 65, and Msg, London 1571 to State, 9 Oct 65, both in FAIM/IR.

[22] For samples of news reporting, see Peter Kumpa, "Drive Opens in Jungle Near Saigon," *Baltimore Sun*, 9 Oct 65; "U.S. Uses Tear Gas in Viet Offensive," *New York Herald-Tribune*, 9 Oct 65; John Maffre, "U.S. Publicizes Tear Gas Attack in Vietnam," *Washington Post*, 9 Oct 65.

[23] Msg, Westmoreland MAC 5056 to Sharp, 10 Oct 65, and Msg, Wheeler JCX 4207-65 to

Information Policy Tightens, August–September 1965

Although press coverage thus proved beneficial where the use of riot control agents was concerned, officials in Washington had few illusions that the news media's agreeable mood would last. Convinced for some time that press and public support for the war would erode once casualties increased, President Johnson had for months sought to thwart those segments of the press such as the *New York Times* that continued to criticize the war. His preoccupation overlapped MACV's own concern for military security, contributing from August on to a tightening of information released to the press.[24]

The first restrictions concentrated on protecting information of value to the enemy. When the MACV Office of Information announced in mid-August that it would no longer release figures on the number of aircraft attacking North Vietnam and the tonnage of the bombs dropped, the reason it gave was straightforward. The information would help the enemy to learn American bombing techniques and assist him in adjusting his defenses.[25]

A second set of restrictions appeared later in the month, shortly after South Vietnamese Army units conducted a multibattalion sweep 120 kilometers southwest of Saigon in the Mekong Delta. Information officers characterized South Vietnamese casualties as moderate but then revealed off the record that fifty-nine men had been killed and ninety-nine wounded. Reporters suspected that a successful ambush had occurred and began to speculate on whether the losses had been concentrated in a single battalion or spread among several (the former circumstance would have indicated to them that casualties had been heavy rather than moderate). Hoping to reduce damaging speculation of that sort in the future, the Military Assistance Command several days later announced an end to the practice of revealing specific American and South Vietnamese casualty figures.[26]

Restrictions aimed at avoiding aid to the enemy led, perhaps inevitably, to pressure for others that were less justifiable on grounds of military security. News stories on 10 September, which revealed an American air strike in North Vietnam a scant seventeen miles from the Chinese border and well within a previously established thirty-mile buffer zone, provided the occasion. They set off a discussion between the commander in chief, Pacific, and the Joint Chiefs on whether the Military Assistance Command was required to incur self-inflicted wounds by publicizing its own mistakes. When "rockets" began arriving from Washington, Westmoreland explained that an information officer had passed erroneous Air Force map coordinates to the press and that in the future the command would report air strikes only in relation to distance from Hanoi, but General Wheeler remained dissatisfied. He saw no reason to report errors to the press,

Westmoreland, 3 Nov 65, both in Miles Policy/Strategy files, CMH; MACV History, 1965, pp. 443f.

[24] MFR, 26 Aug 65, sub: Points Discussed in the President's Luncheon With the Secretary and Senior State Department Officers, 19 Aug 65, FAIM/IR.

[25] Msg, Saigon 484 to State, 14 Aug 65; Msg, Saigon 489 to State, 15 Aug 65; and Msg, Saigon 511 to State, 17 Aug 65, all in FAIM/IR.

[26] Msg, Saigon 670 to State, 28 Aug 65, and Msg, Saigon 685 to State, 30 Aug 65, both in FAIM/IR.

if only because the specific instructions pilots received defining the limits they were to observe might be of significance to the enemy.[27]

Admiral Sharp disagreed. "We can get away with concealing mistakes from the press some of the time but by no means all of the time," he told Wheeler; ". . . a lack of credibility could cause problems far more serious than result from the revelation of occasional mistakes." Having made the point, Sharp nevertheless suggested a compromise—the Military Assistance Command might from time to time omit from its announcements "some of these incidents which we prefer not to have known." General Westmoreland concurred. "Since the press has no way of finding out about strikes in North Vietnam until we announce them," he told Sharp, "an error in target can be protected until I feel it is to our advantage to notify the press."[28]

A reference later in the month by UPI to "secret radio detection equipment aboard U.S. surveillance planes" produced more urgent messages and further discussion of restrictions. "What troubles me most," General Wheeler told Westmoreland, "is the degree of close and intimate contact between military personnel . . . and the press, and the repeated indications that our people in uniform feel free to talk with these members of the press about military matters which are, or should be, classified." To Wheeler, the policy of maximum candor had "had its day." He recommended a tighter control of information and a more formal manner in dealing with the press.[29]

Westmoreland urged caution. He assured Wheeler that the presence of radio-direction-finding aircraft in Southeast Asia had been common knowledge since the French Indochina War and pointed out that since maximum candor had been reaffirmed as official policy in April, it would have to remain in effect until Washington agencies issued other instructions. He agreed nevertheless that newsmen should be kept at greater distance and noted that in addition to instructing unit information officers to maintain "a friendly but dignified" relationship with the press, he had also issued guidance emphasizing the importance of withholding classified material from anyone lacking a security clearance. The Military Assistance Command and the U.S. mission had also changed the format of their evening news conference from an informal gathering to the kind of formal briefing usually given to military staff officers. As for his own background briefings, he had moved them from a casual setting to the command conference room.[30]

The Saigon correspondents for the most part accepted the new restrictions. When questions arose at the UPI Editors Conference in Washington during October about whether the Military Assistance Command was impeding the flow of information to the press, Phil Newsom, a correspondent recently returned from

[27] Msg, Wheeler JCS 3377-65 to Sharp, 11 Sep 65, and Msg, Westmoreland MACV 4558 to Sharp, 11 Sep 65, both in Miles Policy/Strategy files, CMH.

[28] Msg, Sharp to Wheeler, 14 Sep 65, and Msg, Westmoreland MAC 4620 to Sharp, 15 Sep 65, both in Miles Policy/Strategy files, CMH.

[29] Msg, Wheeler JCS 3479 to Westmoreland, 18 Sep 65, Miles Policy/Strategy files, CMH.

[30] Msg, Westmoreland MAC 4690 to Wheeler, 20 Sep 65, Miles Policy/Strategy files, CMH.

South Vietnam, reported that newsmen were receiving "the straight stuff." Although he complained that MACV's briefings lacked detail, sometimes occurring twelve to twenty-four hours after events in the field, Newsom said that most of the changes had been for the better. Officers no longer appeared to believe that the press was out to sabotage military and diplomatic plans, and newsmen had little reason now to suspect that military spokesmen were withholding details by design. As evidence, Newsom compared MACV's attempt early in the year to conceal the use of tear gas with the candid handling of the incident at Qui Nhon.[31]

Official Washington was less sanguine about how the press was performing. Arthur Sylvester declared in a cable to Legare that on-the-spot battlefield reports seemed to imply that the sole objective of American operations was to kill Viet Cong. In the absence of a large body count, he said, newsmen tended to conclude that an operation had failed. He wanted MACV's briefers to begin stressing that military operations had many objectives, from disrupting enemy communications to freeing South Vietnam's peasantry from Communist domination.[32]

Most reporters did indeed believe that numbers were "the name of the game," Legare responded, but since newsmen who attended the briefings were not necessarily the ones who covered the battles and wrote stories, the approach Sylvester advised would have little if any effect. Legare might have added that officials in Washington also emphasized the body count—both because few other measures of progress existed and because they believed with the commander in chief, Pacific, that "figures reflecting Viet Cong casualties are of great significance in estimating Viet Cong capabilities." At a 21 October background briefing for the press, for example, Secretary of Defense McNamara drew upon enemy casualty rates to demonstrate that the introduction of U.S. troops into South Vietnam had resulted in marked progress. From six to seven hundred enemy were being killed every week, McNamara told newsmen—an increase of 75 percent over the previous year.[33]

As in all earlier wars involving the United States, American correspondents in South Vietnam concentrated upon what their readers wanted most—stories on the activities of U.S. troops, particularly in combat—but as the end of 1965 approached, President Johnson became increasingly concerned lest that kind of reporting disturb the American public. He believed that besides attracting criticism from abroad, it would blur the image he wanted most to convey of the United States helping the people of an endangered and depressed nation to create a viable society. Anxious to see more publicity for programs that constructed roads, built schools, dug wells, and distributed rice, Johnson instructed the U.S. mission in Saigon to organize a special staff at the Joint U.S. Public Affairs Office

[31] UPI Reporter, 14 Oct 65, CMH files.

[32] Msg, Sylvester Defense 4109 to MACV IO, 14 Oct 65, DDI News from Vietnam file.

[33] Msg, MAC 36846 to OSD, 18 Oct 65, DDI News from Vietnam file; Msg, CINCPAC to COMUSMACV, 30 Oct 65, DDI Body Count file; Msg, DA 737529 to All Military Commands, 22 Oct 65, sub: Secretary's 21 October Backgrounder, Public Affairs Messages for 1965–1966, CMH.

to pinpoint newsworthy nonmilitary items, arrange visits by the press to observe them, and provide specially written news stories for distribution throughout the world by the U.S. Information Service. In addition, he wanted province advisers to keep files describing nonmilitary projects within their jurisdictions for visiting journalists to consult. The U.S. mission was meanwhile to release statistics showing that American assistance to South Vietnam flowed downward to the level of the peasants in the villages and hamlets.[34]

Although most of the programs Johnson wanted were already in effect, Zorthian moved to do more to publicize civil affairs activities. He would expand JUSPAO's Media Division, he told the State Department, to enable the embassy to increase the number of handouts to newsmen. The Public Affairs Office would also begin preparing fact sheets and photo stories for use by reporters traveling into the field and would update and enlarge its file on continuing nonmilitary programs. The ambassador and other American dignitaries in South Vietnam were already attending civic events, Zorthian added. In the future, the mission would advise visiting delegations from the United States to do the same and to pay as much attention to civic action projects as they usually did to strictly military activities.[35]

Although a slackening of combat for a time drew the Saigon correspondents in the direction the president wanted, old habits in Washington were difficult to overcome. At the end of October, officials within the Johnson administration announced that the United States would shortly achieve a solid beachhead extending the length of the northern coast of South Vietnam. That statement prompted Ambassador Lodge to complain to the State Department that while American military commanders had been scrupulous in avoiding any exaggeration of the U.S. military contribution to the war, officials in Washington were sounding "much too shrill a self-laudatory American note." Lodge requested that the State and Defense Departments issue firm guidance to all their employees emphasizing that Americans were not taking over the war and that statements downgrading the efforts and suffering of the South Vietnamese only played into Communist hands.[36]

The Battles of Plei Me and the Ia Drang Valley, October–November 1965

On 4 November State and Defense issued the instructions Lodge sought, but the move had little effect upon the way the press reported. For in the month to follow, American troops fought a number of sharp engagements that once more

[34] MFR, sub: Public Affairs Policy Committee Meeting of 20 Sep 65, ISA 092 VN, 70A3717, box 44, WNRC; Msg, State 951 to Saigon, 5 Oct 65, FAIM/IR.
[35] Msg, Saigon 1256 to State, 11 Oct 65, FAIM/IR.
[36] Msg, Saigon 1501 to State, 30 Oct 65, DDI News from Vietnam file.

drew attention away from South Vietnamese contributions and toward the U.S. role in the war.[37]

The pace of the war quickened on the evening of 19 October, when Communist troops launched a major attack against Plei Me, an American-advised South Vietnamese Special Forces camp located in the Central Highlands forty kilometers south of Pleiku City and thirty kilometers east of the Cambodian border. Aware that enemy infiltration had increased in the highlands in previous months but expecting the Communists to operate mainly in the heavily populated coastal region, the American and South Vietnamese commands at first interpreted the attack as an enemy training exercise designed to give inexperienced troops time under fire. Yet as the siege grew in intensity, the commands revised that estimate, concluding that two enemy regiments were involved and that the Communists intended both to overrun the camp and to ambush any relief force that appeared.[38]

The South Vietnamese knew that a trap awaited, but supported by units of the U.S. 1st Cavalry Division (Airmobile), they dispatched an armored column to Plei Me from Pleiku City on the morning of 23 October. The force struck the North Vietnamese ambush some eight kilometers from the besieged camp, beating the enemy off after a fierce engagement. Badly shaken but bolstered by nearby American artillery, the South Vietnamese proceeded to the camp and broke the siege on the twenty-fifth.[39]

Newsmen in Saigon followed the action closely, relying at first upon information from the Military Assistance Command but later going out to the camp themselves. The story they told gave little credit to the South Vietnamese. Charles Mohr of the *New York Times,* after interviewing the American advisers at Plei Me, wrote of the enemy's prowess in battle while belittling the camp's Montagnard defenders. One enemy soldier, he wrote, emerged from a hole clutching a single grenade, charged two platoons of Montagnards, and routed them both. Mohr quoted an American adviser who compared a single enemy captive to Plei Me's entire garrison. "We ought to put this guy on the north wall and throw out these government troops," that adviser had said. "He would probably hold it alone." Mohr's story and others like it infuriated those South Vietnamese who could read English. The editor of a Saigon newspaper told Keyes Beech that, according to an account of the battle he had read in *Time* magazine, "twelve American Special Forces men held off six thousand communists. The fact that there were four hundred Vietnamese troops who also took part was passed over lightly."[40]

[37] Memo, McNamara for the Secretaries of the Military Departments and the Joint Chiefs of Staff, 3 Nov 65, sub: The Role of U.S. Forces in Vietnam, ISA 092 VN, 70A3717, box 44, WNRC.

[38] Combat After Action Report, U.S. 1st Cavalry Division (Airmobile), Pleiku Campaign, 4 Mar 66, p. 10 (hereafter cited as CAAR, Pleiku Campaign); Lt. Gen. Harry W. O. Kinnard, "A Victory in the Ia Drang: The Triumph of a Concept," *Army Magazine* 17 (September 1967): 72.

[39] CAAR, Pleiku Campaign.

[40] Charles Mohr, "The Siege of Plei Me: Americans Marvel at Tough Foe," *New York Times,* 28 Oct 65; Keyes Beech, "Vietnamese Want Proper Credit," *Washington Post,* 25 Nov 65.

Relief Force Evacuates 1st Cavalry Troops *killed in ambush at Landing Zone Albany.*

Convinced that the attack at Plei Me was the prelude to a Communist attempt to seize the Central Highlands, General Westmoreland on 27 October directed the 1st Cavalry Division to find and destroy the enemy regiments that had threatened the camp. The month-long campaign that followed—code-named SILVER BAYONET—further diverted reporters from the South Vietnamese portion of the war. For although the operation started slowly, with air cavalry units establishing widely scattered patrol bases and sweeping the area west of Plei Me for any sign of the enemy, it soon developed into a series of sharp encounters.[41]

The first of those engagements came on the morning of 1 November. An American force sighted a number of enemy soldiers attempting to hide in foxholes along the edge of the Ia Tae, a stream flowing to the south of a rugged group of hills known as the Chu Pong Massif. The force landed to investigate and surprised a large Communist hospital. The enemy battalion defending the installation regrouped and counterattacked, but U.S. reinforcements arrived in time to regain the offensive. Before leaving the area, the Americans seized more than $40,000

[41] Kinnard, "A Victory in the Ia Drang," p. 72; Msg, Westmoreland MAC 5358 to Sharp, 27 Oct 65, Westmoreland Papers, CMH.

worth of medicine and discovered a map of the region detailing the enemy's bases and trails.[42]

Guided by that map, the air cavalry narrowed its search to areas heavily traveled by the enemy, springing an ambush on 4 November that destroyed the better part of a North Vietnamese weapons company. By 6 November the *33d North Vietnamese Regiment* was a shambles, having lost over a three-week period 890 killed, 500 wounded, and 100 missing out of an original complement of 2,000 men.

The Communists struck back on the afternoon of 14 November, when one of their reinforced regiments—four battalions—surprised a battalion of the air cavalry near the Chu Pong Massif at a landing zone code-named X-RAY. The Americans spent a harrowing day and night fending off the attack, calling air strikes and artillery to within one hundred meters of their position. They counterattacked the next day, shortly after reinforcements arrived, driving the enemy from the field and winning a major victory.

With the enemy apparently in retreat, two fresh American battalions took to the field on the morning of 16 November. While moving to a new position, Landing Zone ALBANY, to make room for a B–52 strike in the vicinity, one of those units stumbled into a hastily prepared enemy ambush. The melee that followed raged the entire afternoon, with friend and foe in such close combat that air strikes and artillery were useless. By midafternoon the opposing forces had separated enough for fire support to have some effect, but enemy attacks continued until dark.

The pressure eased during the night while the enemy policed the battlefield, rifling the pockets of the dead and executing the American wounded. Aware of what was going on, small groups of Americans made continued forays beyond their perimeter, in their anger sometimes committing atrocities of their own. By morning the enemy was gone, having inflicted heavy losses of 151 killed, 121 wounded, and 4 missing upon the U.S. battalion, destroying it as a fighting unit. The air cavalry would later claim to have killed 400 of the enemy in the engagement; since the Communist force had consisted of only one already bloodied North Vietnamese battalion, that estimate seems exaggerated.[43]

Word of the costly operation was slow to move up the U.S. chain of command. Although General Westmoreland visited the 1st Cavalry command post at Pleiku on the morning after the ambush, he received little more information than had *New York Times* reporter Neil Sheehan the night before, when the 1st Cavalry Division's headquarters told newsmen that two U.S. battalions in the Ia Drang valley had repulsed a determined North Vietnamese assault, killing thirteen of the enemy while incurring light casualties. Westmoreland began to suspect something was wrong when he met some of the survivors from Landing Zone ALBANY

[42] This section is based on CAAR, Pleiku Campaign, p. 11, and Kinnard, "A Victory in the Ia Drang," pp. 78f.

[43] George L. MacGarrigle, Pleiku Campaign—Operation LONG REACH, CMH MS [1972], CMH files.

at a field hospital in Qui Nhon that afternoon. Yet only that evening were his suspicions fully aroused. On his return to Saigon he learned from Legare that correspondents, basing their conclusions upon reports from newsmen who had witnessed the battle, were writing stories critical of the air cavalry's conduct of the operation.[44]

An assistant division commander of the 1st Cavalry Division, Brig. Gen. Richard G. Knowles, would later tell newsmen that possible reinforcements had been standing by as the fighting had raged but had never been used because no one had called for them. "As of this morning [18 November]," he said, "through checks with my forward command post, I was told we had suffered one killed and forty-eight wounded. . . . I thought the situation was in hand. . . . I was delighted." General Westmoreland himself would later write that "nobody, to include the brigade commander, had any knowledge of what actually happened." Those contentions to the contrary, brigade headquarters must have known an extremely serious engagement was in progress, even if the details were vague. The brigade commander reinforced the battalion at ALBANY twice, airlifting a company to the scene from Pleiku at 1825 on the evening of the attack, as soon as enemy firing died down enough to allow helicopters to land, and at 2200 ordering a second company to make a risky, three-kilometer nighttime march into the area.[45]

If the brigade commander consciously withheld information about the engagement, he may have been waiting until the relief forces could set the situation right and change the complexion of what had happened. Since the concept of a highly mobile helicopter-borne division was for the first time being tested in combat, he may have reasoned that the air cavalry could ill afford an embarrassment. In addition, there was considerable concern on the part of some American commanders that a demoralizing defeat might occur the first time U.S. units went into combat on a large scale, as had happened during World War II at the Kasserine Pass and during the first commitment in Korea.

In the event, information about the battle was impossible to repress and difficult to control. Although General Westmoreland ferried a large contingent of the Saigon correspondents to Pleiku to give the press an accurate picture of what had happened, the news stories that followed in the United States were, from his point of view, unfortunate in the extreme. The *Washington Post* and the *Washington Star* printed headlines implying that the air cavalry had suffered a defeat. *U.S. News & World Report* said that the enemy had forced the air cavalry to fall back. The *New York Times* published an article by Neil Sheehan asserting that although MACV information officers had described U.S. casualties as moderate, observers at the scene classed them as very serious because an entire company at the center of the ambushed battalion had suffered near annihilation. Vivid eyewitness

[44] Westmoreland Historical File, vol. 2, tab D, CMH files; Neil Sheehan, "G.I.'s Under Fire Again in Valley in South Vietnam," *New York Times,* 18 Nov 65.

[45] [UPI], "G.I. Plight Unknown, Reserves Stood By," *New York Times,* 19 Nov 65; Westmoreland Historical File, vol. 2, tab D; CAAR, Pleiku Campaign, pp. 93f.

accounts of the battle also appeared, many of them noting that Americans as well as the enemy had committed atrocities.[46]

In all, press coverage of the event was relatively accurate in detail, but General Westmoreland was incensed. At a background briefing in Saigon on 20 November in which he singled out the *Post* and *Star* articles for special comment, he told newsmen that he was sympathetic to the press but had no intention of allowing it to jeopardize the U.S. effort in South Vietnam. He accused the newsmen of informing the enemy of American mistakes and vulnerabilities, of discrediting the United States before its allies, and of lowering the morale of both the troops in the field and their families at home. In fact, Westmoreland said, the operation in the Ia Drang valley had been an "unprecedented victory." Far from withdrawing, "when the dust . . . settled, the American troops were present to clear the battlefield. . . . The enemy had fled the scene. American casualties were heavier than in any previous engagement but small by comparison with [those of] the enemy."[47]

Credibility Declines, November–December 1965

The Pleiku campaign had indeed thwarted Communist ambitions in the highlands temporarily, but Westmoreland's apparent attempt to put the best face possible on the one portion of the operation that had not gone well, together with MACV's statement that U.S. casualties had been moderate, led newsmen to doubt the official interpretation. By the end of the week, correspondents in both South Vietnam and the United States were questioning whether an "unprecedented victory" had occurred and suggesting that the Military Assistance Command was attempting to whitewash its losses. American policy during the Ia Drang campaign had been to "hunt and destroy the enemy" rather than merely to occupy ground, *New York Times* commentator James Reston wrote, and American commanders had found the enemy more than willing to "stand and fight and take seven casualties for every one of ours." Victory thus depended upon how long both sides would be willing and able to incur such losses. "The 'find, fix, and fight' strategy may decimate four divisions of enemy troops and break the will of the enemy, and then again it may lead to four more divisions from the north and as many multiples of four as the North Vietnamese wish to commit."[48]

On the same day Reston's comment appeared, the *Times* published an article on official credibility by Charles Mohr. Claiming that "a steady stream of misin-

[46] Westmoreland Historical File, vol. 2, tab D; "Now It's a Major War in Vietnam," *U.S. News & World Report*, 29 Nov 65, p. 37; Neil Sheehan, "Battalion of G.I.'s Battered in Trap; Casualties High," *New York Times*, 19 Nov 65.

[47] Westmoreland Historical File, vol. 2, tab D; Msg, MACV 41188 to OSD PA, 20 Nov 65, and Msg, Saigon 1820 to State, 20 Nov 65, both in FAIM/IR.

[48] James Reston, "Washington, The Casualty Controversy," *New York Times*, 26 Nov 65.

formation about the war in Vietnam is reaching the American public,'' Mohr cataloged many of MACV's most recent transgressions against the press, giving prominent position to the command's reporting of the body count. As an example, he cited an incident that he said had occurred early in the Pleiku campaign, when a battalion commander had filed an estimate of 160 enemy dead for a two-day operation, only to see the number grow to 869 by the time it reached the press. Obviously, Mohr conceded, many more of the enemy had perished at Plei Me than had been counted, but the pressure for large kills had become so great that soldiers of the 1st Cavalry Division were beginning to joke about ''wild-eyed guesses'' whenever requests for body counts reached them from higher headquarters.[49]

In an article widely paraphrased in newspapers throughout the United States, Australian correspondent Denis Warner, a veteran of every conflict in Southeast Asia from World Var II to Vietnam, charged that the MACV Office of Information was ''engaged in the business of turning defeats into victories.'' Giving no credence to MACV's contention that the enemy habitually retrieved weapons before attempting to recover the dead and wounded, Warner alleged that the discrepancy between the large number of enemy casualties claimed by the Military Assistance Command and the small number of enemy weapons actually captured during most American operations—1,264 versus 317 during the week ending 3 November—indicated that someone was lying. ''Military difficulties and reverses are acceptable to most nations,'' Warner said. ''What no one will accept indefinitely and especially in a war of this sort, is the persistent attempt to win by pretense what has not been won on the ground.''[50]

Newsweek meanwhile commented on MACV's policies for announcing American casualties. The command usually described U.S. losses as light or moderate, its editors observed, but the practice was becoming more and more difficult to justify as the lists of American dead lengthened. Because the Military Assistance Command always described losses in relation to the total military force involved in an operation, the very terms *light, moderate,* and *heavy* could mean whatever the command wanted. The enemy might thus annihilate an entire platoon, but if the battalion to which that unit belonged had lost only 1 or 2 percent of its strength, casualties would be announced as light.[51]

The furor over the air cavalry's engagement could hardly have come at a more inopportune moment for President Johnson. Throughout November and December 1965 his administration's credibility had been under attack. During November, at the height of the battle of the Ia Drang valley, secret administration testimony before Congress had leaked to the press indicating that the U.S. intervention in the Dominican Republic earlier in the year, although publicly justified as an attempt to rescue Americans stranded in a foreign revolution, had in fact been an attempt to prevent a Communist take-over. The revelation caused a stir

[49] Charles Mohr, ''War and Misinformation,'' *New York Times*, 26 Nov 65.
[50] Denis Warner, ''Army's Word Suspect in Vietnam,'' *Denver Post*, 7 Dec 65.
[51] ''Moderation in All,'' *Newsweek*, 6 Dec 65, p. 42.

in the news media. Syndicated columnist Marquis Childs joined others in questioning administration pronouncements on a number of important subjects. "Office holders from the time of Aristides the Just have done their best to put themselves in a good light," Childs said, "but when government fails to make its account believable with enough of the truth there is bound to be trouble."[52]

As if to bear out Childs' suspicions, the 30 November issue of *Look* magazine carried an article by CBS News commentator Eric Sevareid reporting a conversation the reporter had held with U.S. Ambassador to the United Nations Adlai Stevenson just two days before Stevenson's sudden death. The ambassador had revealed that, despite public protestations to the contrary, the United States had twice during 1964 rejected North Vietnamese offers to discuss peace.[53]

The State Department played down the report, insisting that "on the basis of the total evidence available to us we did not believe . . . North Vietnam was prepared for serious peace talks." The press, for its part, conceded that negotiations at the time would probably have been counter to U.S. interests; nevertheless, most commentators objected with James Reston that "From beginning to end in the Vietnamese war there has been a serious and widespread lack of trust in the government's statements about how well the war was going, what role our men were playing and how well the South Vietnamese government was doing." The administration's first problem, Reston said, was not how to talk to the North Vietnamese, "but how to talk candidly to the American people."[54]

Taking up that theme, newspapers throughout December criticized the administration's public statements, scoring what *Washington Post* reporter Murrey Marder called "President Johnson's obsession with secrecy." During the same period, every time the United States took steps in secret either to protect U.S. troops in battle or to stem the flow of North Vietnamese into South Vietnam, the press found out and published the story, in effect casting doubt on the administration's continuing assertions that it sought only peace in Southeast Asia.[55]

During Operation SILVER BAYONET, for example, General Westmoreland, with Ambassador Lodge's concurrence, requested permission from State and Defense for American and South Vietnamese troops to enter Cambodia to destroy enemy sanctuaries. The inability of Cambodia's ruler, Prince Norodom Sihanouk, to control the areas in question, Westmoreland argued, gave the Communists secure bases from which they could attack American troops operating across the border in South Vietnam. State denied the request. It would permit American troops to fire into Cambodia only if fired upon from across the border and to enter the country only when that step was essential for self-defense. Since U.S. relations

[52] Marquis Childs, "Government News Lacks Credibility," *Washington Post*, 17 Nov 65.

[53] Eric Sevareid, "The Final Troubled Hours of Adlai Stevenson," *Look*, 30 Nov 65, p. 81.

[54] Msg, State Circular 912 to All Diplomatic Posts, 15 Nov 65, FAIM/IR; James Reston, "Washington: Candor Compels Me To Tell You," *New York Times*, 17 Nov 65.

[55] Murrey Marder, "Greater Skepticism Greets Administration Declarations," *Washington Post*, 5 Dec 65. See also R. H. Shackford, "Officialdom Seen Deepening the 'Crisis in Confidence,' " *Washington Daily News*, 29 Dec 65.

with Cambodia were already strained and since Sihanouk had recently denied in public that the North Vietnamese were using his territory, the State Department cautioned the U.S. mission in Saigon to handle all public announcements of border crossings with care. Official spokesmen were to avoid making a statement to the press for as long as possible and then were to stress that the action had been a matter of strict self-defense. Whenever plausible, they were also to note that it was often difficult to determine the exact location of the border.[56]

The U.S. mission in Saigon never had an opportunity to put the State Department's instructions into practice. On 17 December, before a border crossing had even occurred, Seymour Topping of the *New York Times* and Keyes Beech of the *Chicago Daily News* published stories revealing that the Johnson administration had authorized American troops to enter Cambodia in self-defense. Inclined to say as little as possible about the new rules, the State Department and the Military Assistance Command at first responded to questions with "no comment." Yet as speculation in the press increased, they recognized that some statement was necessary. On 21 December, therefore, while the command held a backgrounder attributable to "authoritative sources," the State Department issued a terse announcement affirming that "It continues to be the policy of the United States . . . to respect the sovereignty, the independence, and the territorial integrity of Cambodia and not to widen the war. . . . As to news reports concerning instructions issued to United States forces . . . American military commanders throughout the world have authority to take those actions essential to the inherent right of self-defense to protect their forces."[57]

Neither MACV's backgrounder nor the State Department's announcement did anything to dampen speculation that the United States was about to widen the war. Reporters in Washington asked sarcastically how the exercise of self-defense squared with the U.S. government's desire to respect the sovereignty, independence, and territorial integrity of Cambodia. In Saigon, *New York Times* reporter R. W. Apple, Jr., although willing to accept MACV's explanation that U.S. troops would cross the border if a failure to do so would jeopardize their lives, nevertheless commented that the United States was obviously "edging closer to . . . confrontation . . . throughout Southeast Asia."[58]

A number of embarrassing news stories appearing at the time bolstered Apple's conclusion. On 13 December, using leaked information, correspondent Joseph Fried of the *New York Daily News* disclosed that American aircraft had begun defoliation operations in Laos along the Ho Chi Minh Trail. Two days later the reporter revealed that a recent air strike against a large North Vietnamese power plant,

[56] Fact Sheet, DCSOPS, 18 Jan 66, sub: Rules of Engagement, Southeast Asia, CMH files; Msg, Westmoreland to Sharp, 9 Dec 65, Miles Policy/Strategy files, CMH; Msg, State 1634 to Saigon, 11 Dec 65, and Msg, State 1399 to Saigon, 20 Nov 65, both in FAIM/IR.

[57] Msg, State 1697 to Saigon, 17 Dec 65, FAIM/IR; Msg, State 1739 to Saigon, 21 Dec 65, DDI Cambodia file; R. W. Apple, Jr., "U.S. To Let Forces Go Into Cambodia in Self-Defense," *New York Times*, 21 Dec 65.

[58] Msg, State 1739 to Saigon, 21 Dec 65, FAIM/IR; Apple, "U.S. To Let Forces Go Into Cambodia in Self-Defense."

far from being the normal mission information officers had announced, had actually been planned as a reprisal for the enemy's destruction on 4 December of the Metropole Hotel, a U.S. enlisted billet in Saigon. Finally, on 18 December UPI correspondent Ray Herndon, relying on more leaks, divulged that the United States was conducting a series of B–52 strikes against Communist infiltration routes in Laos.[59]

The two Fried articles were embarrassing, revealing as they did that the United States was conducting defoliation operations in a supposedly neutral country at a time when American diplomats had just gone on record to disapprove of similar raids in Yemen, Algeria, and Israel. Much more damaging to U.S. interests, however, was Herndon's revelation of B–52 strikes, for the U.S. Ambassador to Laos, William H. Sullivan, had neglected to inform the Laotian government.[60]

Information officers in Saigon inadvertently compounded Sullivan's problems by responding to questions from the press on the subject with a vigorous "no comment," in effect confirming that the raids had occurred. Incensed, Sullivan instructed his air attache to tell the Laotians that the strike had not taken place on their territory at all, but along the border in South Vietnam. Then he cabled the State Department to complain, as he had in the past, that the Johnson administration was entirely too concerned about its public relations and that information officers should be dissembling as much as possible whenever sensitive operations in Laos were concerned. "I need not . . . reiterate," he said, "how difficult it is for us to be helpful in the Ho Chi Minh Trail area when we are constantly having to compete with those who seem to consider their press relations more important than the operations at stake."[61]

Convinced that any departure from a policy of no comment on operations in Laos would undermine newsmen's confidence in official statements, the U.S. mission in Saigon refused to compromise. As for the leaks, there seemed little anyone could do. Although the problem was indeed troubling, Ambassador Lodge told the State Department,

> in an operation as big as this, it is virtually impossible to carry out an effective anti-leak policy. . . . leaking to the press is one of the prerogatives of the president and of his duly appointed representatives, whoever takes it upon himself to leak, therefore, is, in effect, usurping a presidential prerogative and taking the conduct of foreign relations into his own hands. About the most practical thing I can think of is for me to discuss this at length with U.S. agency chiefs because, in the last analysis, they are the ones on whom execution of the policy depends.[62]

Because they realized that much of the problem stemmed from unrealistic policies governing what could be said to the press about operations in Laos, both

[59] Msg, State 391 to Vientiane, 15 Dec 65, FAIM/IR; Msg, Saigon 2257 to State, 16 Dec 65, DDI Operations file; and Msg, State 1709 to Saigon, 18 Dec 65, FAIM/IR.

[60] Msg, State 1602 to Saigon, 9 Dec 65, FAIM/IR; Msg, Vientiane 665 to State, 20 Dec 65, Miles Policy/Strategy files, CMH.

[61] Msg, Saigon 2220 to State, 21 Dec 65, FAIM/IR; Msg, Vientiane 665 to State, 20 Dec 65; Msg, Vientiane 580 to State, 29 Nov 65, ISA 380.01, 70A5127, box 11, WNRC.

[62] Msg, Saigon 2220 to State, 21 Dec 65; Msg, Saigon 2183 to State, 18 Dec 65, FAIM/IR.

Rodger Bankson and William P. Bundy suggested that the time had come for a change of policy. In a cable to Ambassador Sullivan, Bundy in particular noted that as the war in South Vietnam grew and attention focused on North Vietnam's infiltration through Laos, pressure would rise from both Congress and the public not only for a greater U.S. role in the area but also for a policy of acknowledging that the United States had no intention of giving North Vietnam unopposed entry into the South. The press was already aware of what was going on in Laos, he said. "Our public position of no comment has . . . worn very thin and [is] likely to be even tougher to stick to in the coming months."[63]

Sullivan disagreed, observing that the Soviet ambassador "has made it clear to both Souvanna and me that his government is willing to continue to wink at everything it knows provided they [*sic*] are not officially acknowledged. This, it seems to me, is a considerable gain to be had for a small measure of continence on our part." The gain would multiply if the Soviet Union were to seek to enhance its influence over Hanoi in order to persuade North Vietnam to "step back from its active, high-risk, Chicom-type policy to a more subtle line. [We] must not trade the substance of what we have going for us for the shadow of easier public relations."[64]

Unwilling to tell outright lies yet constrained by Souvanna Phouma's desire to preserve an appearance of neutrality, the Johnson administration decided to skirt the issue for the time being. When the next B–52 strike set out for Laos on 11 January, a target was selected that straddled the Laotian–South Vietnamese border. In that way, should the press inquire, information officers would be able to affirm in all honesty that the strike had taken place in South Vietnam.[65]

Bombing Halt in North Vietnam, December 1965

Although willing to hedge on the matter of B–52 strikes in Laos, the Johnson administration recognized that good public relations were essential to any task it undertook in South Vietnam. As Senator Robert F. Kennedy told Secretary of Defense McNamara on 9 December, a disturbing political mood appeared to have developed in the United States, the result of lengthening casualty lists and the American public's lack of understanding of the alternatives available in Southeast Asia. If the administration intended to enlarge the war, Kennedy said, it would first have to build up support at home and abroad by making a bold move for peace.[66]

[63] Memo, Bankson for Phil G. Goulding, Deputy Assistant Secretary of Defense for Public Affairs, 14 Dec 65, sub: State's Proposed Joint Reply, DDI Operations file; Msg, State 420 to Vientiane, 24 Dec 65, FAIM/IR.

[64] Msg, Vientiane 687 to State, 27 Dec 65, FAIM/IR.

[65] Msg, State 451 to Vientiane, 7 Jan 66; Msg, Vientiane 731 to State, 8 Jan 66; Msg, State 455 to Vientiane, 10 Jan 66; and Msg, Vientiane 638 to State, 11 Jan 66, all in FAIM/IR.

[66] Memo, Robert McNamara, 9 Dec 65, sub: Telephone Conversations With Members of Congress, FAIM/IR.

When he polled important members of Congress on what that maneuver should be, McNamara found that Senators John Pastore, Mike Monroney, Warren Magnuson, and Sam Ervin, along with Kennedy and others, all favored a halt in the bombing of North Vietnam. As Ervin observed, the United States should give the Communists a chance to stop their aggression—but bomb them out of existence as soon as their ill will became apparent to the world.[67]

That the *New York Times* was also calling for a bombing halt added weight to the congressmen's suggestion. In a series of editorials and running comments on the war throughout December, the paper contended that there were alternatives to escalation as yet "unexhausted . . . in the eyes of many sincere and patriotic Americans, quite aside from any lingering sense of uneasiness they may have because of recent revelations that the administration rejected peace feelers put out by Hanoi a year ago." While a bombing halt might fail to produce substantial concessions from North Vietnam, there was nevertheless a possibility that it might "lead to the saving of untold American and Vietnamese lives."[68]

Although the administration doubted that a bombing halt would lead to a major breakthrough, it had been considering a pause since July, when McNamara had proposed the step as a test of North Vietnam's intentions. Secretary Rusk had opposed the idea at the time in the belief that so important a card should be played only when chances were greater that Hanoi would respond; yet by early December the idea had taken new life. The military forecast for 1966 indicated an accelerated deployment of American fighting men to South Vietnam, foreshadowing still larger increases in 1967 and requiring a $10 billion supplement to the fiscal 1966 budget. The magnitude of those increases, William Bundy observed in a cable to Saigon, would "hit Congress and the U.S. public hard and could trigger prolonged and difficult debate" on whether the United States had exhausted all the avenues to a peaceful solution of the war. "Unfortunately," Bundy continued, "such episodes as Sevareid article have significantly weakened our peace-seeking posture, which we regarded as extremely strong in early fall." Without some "major diplomatic initiative—for which we see only a pause as truly effective and sufficient—noise level could reach point that would seriously damage our basic posture of firmness and determination."[69]

To test the American public's receptivity to a bombing halt as well as the North Vietnamese reaction, President Johnson and Secretary Rusk made a number of public statements during December in which they hinted broadly that a pause could be arranged under certain circumstances. On 10 December Johnson told the biennial convention of the AFL-CIO that he would explore all prospects for peace before taking the "other hard steps" he had in mind. When reporters contacted the State Department's Office of Public Affairs for an explanation, information officers referred them to a recent speech by Rusk in which the secretary

[67] Ibid.
[68] "Escalation Goes On," *New York Times*, 3 Dec 65.
[69] Draft Msg, State to Saigon, 11 Dec 65, FAIM/IR; *Pentagon Papers*, 4: 32f.

had refused to rule out a bombing halt, provided the North Vietnamese gave some indication that they were interested in negotiating. The Communists responded with a thinly veiled message of their own. While the American news media speculated on the possibility of a halt, in mid-December a spate of articles and commentaries appeared in the North Vietnamese press labeling all American peace overtures tricks and asserting that the United States was less interested in peace than in finding an excuse to widen the war.[70]

Although both the State and Defense Departments finally recommended a bombing pause, President Johnson reserved judgment on the matter while officials within the administration continued to discuss pros and cons. The opponents of any further attempt to start negotiations included the former Deputy U.S. Ambassador in Saigon, U. Alexis Johnson. In a 24 December memorandum, Ambassador Johnson told Rusk that North Vietnam would reject any offer to negotiate that was American in origin, if only to save face before the other Communist nations of the world. The enemy would interpret the move as a response to political pressures from within the United States and would conclude that "we are more anxious than they to find a compromise solution. This would tend further to persuade them of the validity of their view that if they can hold on long enough, . . . we will gradually be forced to whittle back our position."[71]

Assistant Secretary of State for International Organizations Joseph Sisco, on the other hand, argued that the case for a pause was strong. Besides helping to justify further escalation, he told Secretary Rusk, "it would be viewed as a move from strength not weakness; it would neutralize the innumerable peacemakers—amateurs and professionals, domestic and international; . . . it would help place the onus on the communists for failure to start talks; it would give the Soviets a handle to play a more active role with Hanoi. In short, such a proposal would meet with universal acceptance and help keep us where we belong—at the helm of the peacemakers."[72]

President Johnson decided on 27 December to extend a brief bombing halt that had begun on the twenty-fourth as part of a thirty-hour Christmas truce with the Viet Cong. The State Department informed the U.S. mission in Saigon of the decision that night, adding that the president was keeping track of developments and had yet to determine how long the pause would last.[73]

With Ambassador Lodge's backing, General Westmoreland questioned the decision. He told Admiral Sharp that although American air strikes had succeeded in forcing the Communists to travel at night, the enemy was still moving troops

[70] Msg, Saigon 2113 to State, 12 Dec 65; Msg, State 1644 to Saigon, 13 Dec 65; and Msg, Saigon 2127 to State, 13 Dec 65, all in FAIM/IR. Lyndon Johnson, "Why We Are in Vietnam," Speech to AFL-CIO Convention, 10 Dec 65, Department of State *Bulletin*, 27 December 1965, p. 1024; John D. Pomfret, "Hard Steps Await Results," *New York Times*, 10 Dec 65.

[71] Memo, U. Alexis Johnson for Rusk, 24 Dec 65, sub: Diplomatic Initiatives With Respect to Vietnam, FAIM/IR.

[72] Memo, Sisco for Rusk, 24 Dec 65, sub: Some Year End Thoughts on Vietnam: Need for Diplomatic Initiative, FAIM/IR.

[73] Memo, Rusk for the President [late 1966], sub: Christmas Truce, FAIM/IR.

A Destroyed Bridge in Laos, 1965

into South Vietnam at a rate unmatched by the United States. Air attacks against the enemy's entire line of communications, from the Chinese border in the north to the point where the Ho Chi Minh Trail entered South Vietnam, were thus essential. Far from reducing the tempo of the air war, the Johnson administration needed to step it up.[74]

Secretary of State Rusk cabled the U.S. mission in Saigon to explain the situation. A yet to be released Harris poll, he said, would show that 73 percent of the American people, "including 64 percent of Goldwater voters," would favor a renewed effort for a cease-fire and that 59 percent would favor a bombing pause, including 48 percent of Goldwater voters. The same poll showed that 61 percent would favor increasing the bombing if a cease-fire or pause failed to elicit the interest of the other side. Those figures, Rusk said, illustrated the "need to prepare our people for major sacrifices" by placing the blame for the war where it belonged, upon the North Vietnamese. "The simple fact is that we must sustain support for what has to be done in months ahead. . . . Compared to this overriding requirement, the destruction of the limited targets which would otherwise be struck during this period is a secondary matter."[75]

[74] Msg, MACV 45265 to CINCPAC, 27 Dec 65, FAIM/IR.
[75] For a retransmission of Rusk's message to the U.S. mission, see Msg, McNamara Defense 5041

American policy makers were also aware that bad weather over North Vietnam during January would preclude most air strikes, making a bombing halt at that time a minimal risk. Since attacks previously programmed for North Vietnam could be diverted to targets in Laos, strikes against infiltration routes could also continue. In fact, by the end of the bombing halt in January 1966 American sorties against the Ho Chi Minh Trail in Laos were running at 8,000 for the month, compared to 3,023 for December.[76]

The Saigon correspondents surmised that a bombing halt was in effect shortly after attacks on North Vietnam failed to resume at the end of the Christmas truce. When they pressed MACV spokesmen for an explanation, they received only avowals that the command would announce the resumption of bombing when it occurred. Unsure themselves of what was going on, information officers later cautioned the newsmen that there had been lulls in the fighting before, followed by sharp outbreaks. Although channels were always open for diplomatic contacts, they knew of no unusual diplomatic activity taking place in connection with the pause.[77]

That some sort of initiative was in progress was nevertheless shortly apparent to the entire world. As President Johnson wrote personal letters to many heads of state underlining his desire for peace and stressing the sincerity of U.S. motives, five prominent American emissaries began highly visible diplomatic missions around the world in obvious search for an opening that would bring North Vietnam to the negotiating table. Vice President Hubert Humphrey visited the Republic of China, Japan, the Philippines, and Korea; Ambassador W. Averell Harriman traveled to Eastern Europe; U.S. Ambassador to the United Nations Arthur Goldberg commuted between New York, Paris, Rome, and London; Assistant Secretary of State for African Affairs G. Mennen Williams conferred with African leaders; and an expert on Latin America, Thomas Mann, visited a number of South American capitals. In all, the United States contacted some 115 countries, taking pains at each step to reemphasize the American desire for negotiations and North Vietnam's unwillingness to stop its aggression.[78]

The Johnson administration repeated that point in a continuing series of speeches and public statements throughout the halt. On 30 December, for example, before the public announcement that a bombing pause was in effect, Secretary Rusk told Canadian television viewers that he knew of no past attempt by the North Vietnamese to make peace. Vice President Humphrey said almost the same thing when he returned from the Far East a few days later. G. Mennen Williams, at the personal request of President Johnson, told newsmen upon

to Wheeler, Sharp, Westmoreland et al., 29 Dec 65, Miles Policy/Strategy files, CMH. McNamara addressed a similar message to Westmoreland. See Msg, McNamara Defense 5038 to Westmoreland, 28 Dec 65, Miles Policy/Strategy files, CMH.

[76] Msg, McNamara Defense 5038 to Westmoreland, 28 Dec 65. The sortie statistics are in U.S. Congress, Senate, Committee on Armed Services, *Hearings on S. 1263: Fiscal Year 1974 Authorization for Military Procurement*, 93d Cong., 1st sess., p. 427.

[77] Msg, Saigon 2279 to State, 26 Dec 65, and Msg, Saigon 2291 to State, 27 Dec 65, both in FAIM/IR.

[78] Lyndon B. Johnson, *The Vantage Point* (New York: Holt, Rinehart and Winston, 1971), p. 238.

President Johnson Briefs Congressional Leaders on the Bombing Halt

returning from Africa that the United States had been restraining its military strength in order to leave a hand free for diplomacy and to test whether a bombing halt would bring peace as some critics of the war had suggested. President Johnson himself took the stage on 12 January. During his annual State of the Union message he requested a $5.8 billion supplemental appropriation to run the war, then added, "We have carried our quest for peace to many nations and peoples because we share this planet with others. . . . We have found understanding and support. And we know they wait with us tonight for some response that could lead to peace."[79]

As expected, no response of any significance was forthcoming. Although an encouraging contact with North Vietnam developed through the U.S. embassy in Rangoon, Burma, on 4 January the North Vietnamese Ministry of Foreign Affairs rejected any peace overture from the United States. The Communist world's attitude toward the entire halt was, perhaps, typified by Soviet Ambassador to the

[79] Lyndon B. Johnson, "State of the Union Address, 12 Jan 66," Department of State *Bulletin*, 31 January 1966, p. 150. See also "Rusk Discusses Vietnam on Canadian T.V.," 30 Dec 65, Department of State *Bulletin*, 17 January 1966, p. 87; "Vice President Humphrey Returns From Far East Mission," Department of State *Bulletin*, 24 January 1966, p. 115; Memo, G. Mennen Williams for Secretary Rusk, 8 Jan 66, sub: Meeting With the President, FAIM/IR.

United States Anatoly Dobrynin, who told the Hungarian Charge d' Affaires in Washington, Janos Radvanyi, that the dispatch of peace envoys around the world "imparted a rather theatrical flavor" to President Johnson's peace initiative.[80]

As the bombing halt continued and diplomats maneuvered, the American news media followed events closely, chronicling the comings and goings of the various ambassadors and watching the war in South Vietnam for any sign that the enemy might be reducing the level of hostilities. Some of the stories that resulted were potentially embarrassing to the president. Reporters immediately discovered, for example, that the United States was escalating the air war in Laos and began speculating that the event was merely the prelude to larger operations in that country if American moves to achieve peace went unheeded. Information officers responded to questions on the subject with their usual no comment, in effect killing the story by refusing to nourish it. Meanwhile the *New York Times* and other papers greeted the halt optimistically, expressing the hope that the Johnson administration would pursue the peace initiative to a successful conclusion.[81]

If the news media were well disposed toward the halt, Ambassador Lodge and other members of the U.S. mission in Saigon continued to have reservations. The Military Assistance Command had recorded 1,133 overt acts of enemy aggression between 26 December and 1 January, Lodge told President Johnson on 5 January, the highest total for one week since the United States had entered the war. If the halt went on much longer, the resumption of bombing would coincide with the Vietnamese New Year, Tet. Such timing would leave the United States open to the charge that it observed the Christian spirit of Christmas but disregarded the Vietnamese people's most important holiday. "Of course," Lodge concluded, "I have no first hand contact with whatever results the bombing pause may be achieving outside of this area and do not know how widespread the appeal of the pause is as regards American public opinion. . . . But in this area, the pause has not only done us no good, it has definitely caused losses."[82]

General Wheeler also objected. The Communists had succeeded in countering the air campaign in Laos by dispersing troop concentrations under the trees and breaking supply depots into small, easily camouflaged caches, he told Secretary McNamara. The situation was beginning to deteriorate as the North Vietnamese repaired their communications and improved their air defenses. They were probably also increasing the flow of men and equipment into South Vietnam. Whatever the impact of the peace offensive upon American and world public

[80] Msg, State 950 to Warsaw, 29 Dec 65, FAIM/IR; *United States–Vietnam Relations, 1945–1967: A Study Prepared by the Department of Defense*, 12 vols. (Washington, D.C., 1971), vol. 6, sec. C, pt. 1, p. 1b; Janos Radvanyi, *Delusion and Reality: Gambits, Hoaxes, and Diplomatic One-Upmanship in Vietnam* (South Bend, Ind.: Gateway Editions, Ltd., 1978), p. 110.

[81] Msg, State 1139 to Bangkok, 6 Jan 66, FAIM/IR; Msg, Defense 1392 to CINCPAC, 11 Jan 66, DDI Laos file; Msg, Westmoreland MACV 592 to Sylvester, 22 Jan 66, Westmoreland Papers, CMH; "A Bombing Pause," *New York Times*, 27 Dec 65; Memo, Leonard Meeker for Secretary Rusk, 22 Jan 66, sub: Length of the Pause, FAIM/IR.

[82] Msg, Saigon 2399 to State, 5 Jan 66, FAIM/IR.

opinion, continuation of the stand-down would shortly place U.S. forces under a "serious and progressively increasing military disadvantage."[83]

Despite evidence that the enemy had begun to move large convoys southward in broad daylight, policy makers in Washington doubted the gravity of Wheeler's and Lodge's complaints. The State Department's Bureau of Intelligence and Research issued a study which contended that the level of enemy activity in South Vietnam was little changed since Christmas and actually lower than during the three to four weeks prior to the Christmas truce—evidence that American fighting men were in little immediate danger. The point was also made by at least a few officials that just as the Communists were resupplying their troops, the United States had augmented its own forces during the same period.[84]

President Johnson also appeared willing to tolerate risks for the sake of a stronger public commitment to the war. Meeting with G. Mennen Williams shortly after the ambassador's return from Africa, he spoke with feeling of the dilemmas the halt had caused for him: of the military on one side insisting that he lift the suspension and of prominent senators and others wanting it continued. Nevertheless, he dismissed the problem, telling Williams that although the 1,500 Americans already lost in the war represented a "terrible responsibility," as many lives were lost monthly in the United States in traffic accidents.[85]

The State Department put the matter more delicately in its response to Lodge. The United States was reaping "considerable dividends from this whole effort in terms of both present and future support," it said. That fact, together with a "strongly affirmative response from forty or fifty nations and widely encountered pleas that we give adequate time for a response from Hanoi," made resumption impossible for the time being.[86]

Although the United States was unwilling to resume bombing for the moment, there was little doubt that sooner or later attacks on North Vietnam would continue. By 24 January the Johnson administration was preparing actively for the event. While the president wrote letters to world leaders explaining North Vietnam's recalcitrance and his own lack of options, administration officials continued making speeches and public statements that placed the onus for any resumption squarely upon the Communists. On 25 January the Defense Department also released a detailed intelligence report to the news media documenting North Vietnam's use of the halt to increase the flow of men and materiel into South Vietnam.[87]

The approach apparently worked, for when the bombing finally resumed on 30 January, few Americans were dismayed, and most of the news media expressed only regret that the initiative had failed. To the Johnson administration's satis-

[83] JCSM 16-66, Wheeler for the SECDEF, 8 Jan 66, attached to Ltr, McNamara to Rusk, 19 Jan 66, FAIM/IR.

[84] Msg, MACV 1580 to CINCPAC, 17 Jan 66, and Memo, Leonard Meeker, State Department Legal Adviser, for Rusk, 20 Jan 66, sub: Length of Pause, both in FAIM/IR.

[85] Memo, G. Mennen Williams for Rusk, 8 Jan 66, sub: Meeting With the President.

[86] Msg, State 1907 to Saigon, 6 Jan 66, FAIM/IR.

[87] The infiltration report was carried in the *Washington Post*, 25 Jan 66.

faction, a Harris poll released that week indicated that 61 percent of the public appeared ready to accept all-out bombing if the Communists failed to negotiate and that 60 percent advocated deploying up to 500,000 American troops to South Vietnam if it would shorten the war.[88]

The effects of the bombing halt upon the security of American fighting men in South Vietnam were more difficult to assess. As Westmoreland, Wheeler, and Lodge had all alleged, the Communists had used the stand-down to repair their positions and to resupply their troops. Yet although the enemy had rebuilt at least two important rail lines in North Vietnam and had restored a number of bridges and roads, it was almost impossible to determine how much of that activity would have gone on anyway. The weather over North Vietnam was bad throughout the halt and for several months thereafter; as late as April most air strikes scheduled for North Vietnam were being diverted to Laos. Those American aircraft that did venture into North Vietnam met more resistance: 1 out of every 3.5 missions encountered ground fire after the halt as opposed to 1 out of 12 before. Yet overall losses of aircraft remained a constant 4 percent of all sorties flown.[89]

Of perhaps greater importance, in the eyes of the Johnson administration, was the fact that the president's policies on the war appeared to have at least the fragile support of a substantial portion of the American public. Despite the continual complaining of the press and incessant leaking at all levels, the administration seemed to have reached a point where it could begin gearing for the larger war it contemplated.

[88] Msg, State 699 to Santiago, Chile, 3 Feb 66, FAIM/IR; Harris, *Anguish of Change*, p. 59.

[89] Msg, Saigon 3302 to State, 11 Mar 66, DDI Operations file; Msg, 2d Air Division to PACAF, 21 Feb 66, cited in PACAF Command Center Chronological Log, 20 Jan 66 to 7 Mar 66, no. 7873, A. L. Simpson Research Center Archives, Maxwell Air Force Base.

10

Gearing for a Larger War

Although public support for the war seemed to have increased, the administration remained concerned that an angry outcry might develop in the United States when President Johnson moved to enlarge the American commitment in South Vietnam. For that reason, officials in both the United States and South Vietnam prepared throughout the bombing halt and for many months thereafter for any public relations problems that might arise. Their objective was to limit adverse coverage of the war to preserve a public consensus for strong action in Southeast Asia.

Public Opinion, January–February 1966

By the end of the halt, American public opinion appeared to be running in favor of the president's Vietnam policies. Private polls commissioned by the White House and released to the press during February 1966 showed that 63 percent of those interviewed approved of President Johnson's handling of the war and that 66 percent thought he had done everything he could to make peace. Harris polls taken about the same time said much the same thing, and the Gallup poll was also encouraging. To the question, "In view of developments since we entered the fighting in Vietnam, do you think the United States made a mistake sending troops to fight in Vietnam?" Gallup reported that 59 percent of those interviewed responded, "No."[1]

Although the public seemed willing to support the war, there were still indications that many Americans were ambivalent. According to Louis Harris,

[1] John D. Pomfret, "President Finds Backing on War," *New York Times*, 18 Feb 66; Harris, *Anguish of Change*, p. 58; Hazel Erskine, "The Polls: Is War a Mistake," *Public Opinion Quarterly* 34 (Spring 1970): 134.

73 percent of those answering one of his questionnaires in early 1966 voiced deep concern about America's involvement in South Vietnam, and another 73 percent said they favored a cease-fire.[2]

Little noticed at the time, those figures corresponded to the results of a poll conducted during late February and early March 1966 by a group of social scientists working at Stanford University. In the belief that commercial polling organizations had failed to probe public opinion of the war deeply enough, the group joined with the National Opinion Research Center of the University of Chicago to interview a carefully selected cross-section of the American public. It found that while a majority of Americans approved of the president's handling of the conflict, that same majority also favored deescalation. Eighty percent were willing to bargain with the Viet Cong, 70 percent favored free elections in South Vietnam even if the Viet Cong should win, and 52 percent were prepared to accept a coalition government that included Communists. Asked to choose between continuing the existing situation, fighting a full-scale war that might involve hundreds of thousands of casualties, and withdrawing, 49 percent preferred the status quo, 23 percent full-scale war, and 19 percent withdrawal. When those composing the 49 percent were asked to choose between withdrawal and full-scale war, 60 percent chose to fight. Yet a majority of all respondents rejected increasing taxes or cutting social programs to finance the effort.[3]

The survey's findings received varying interpretations when they appeared, each commentator judging according to his own preconceived ideas. Tom Wicker of the *New York Times* wrote that the 60 percent favoring full-scale war represented "overwhelming" sentiment when that choice was in fact a fallback position for 60 percent of the 49 percent who preferred the status quo. Philip Potter of the *Baltimore Sun* and syndicated columnists Rowland Evans and Robert Novak also concluded that the poll had uncovered strong support for the war and attributed those findings that disagreed with their point of view to rigged and leading questions. The poll's authors were obviously of an antiwar persuasion, Potter suggested, because all lived in California's San Francisco Bay area, "a hotbed of anti-war demonstrations."[4]

In fact, as both the poll's authors and a 17 March *New York Times* editorial noted, far from revealing either a prowar or an antiwar consensus, the Stanford survey showed that the American people favored moderation and reason where South Vietnam was concerned. "There is little support in the country . . . for the extreme alternatives of withdrawal or all-out war," the *Times* said. "But there is substan-

[2] Harris, *Anguish of Change*, p. 59.

[3] For the survey's findings, see Sheldon Appleton, ed., *United States Foreign Policy: An Introduction With Cases* (Boston: Little, Brown & Co., 1968), p. 336. See also Nelson W. Polesby, "Political Science and the Press: Notes on Coverage of a Public Opinion Survey on the Vietnam War," *Western Political Quarterly* 22 (March 1969):46 (hereafter cited as "Political Science and the Press").

[4] Tom Wicker, "Confusion in Vietnam," *New York Times*, 15 Mar 66; Philip Potter, "Poll Shows Strong Vietnam War Support," *Baltimore Sun*, 15 Mar 66; Rowland Evans and Robert Novak, "Rigging the Polls," *Washington Post*, 7 Apr 66.

tial support for a policy of holding military operations at the present level while taking new initiatives to seek peace.''[5]

Although in dispute, the poll's findings approximated the Johnson administration's own interpretation of American public opinion. Unwilling to appear more bellicose than necessary, the president sought to stress that Communist actions were responsible for American escalations. If the administration gave an appearance of moderation, so the reasoning went, it would alienate fewer voters.

The Department of Defense took the lead in attempting to soften the impact of the war. To reduce speculation in the press on the size of American troop commitments, it issued an order on 29 December 1965 forbidding the release of advance information on the movement of American military units to South Vietnam. Justified as an attempt to deny valuable intelligence to the enemy, the rule had the effect of presenting critics with a fait accompli each time more American troops entered the war. In the same way, since vehement discussions seemed certain to erupt in Congress and the press once the war escalated, presidential adviser McGeorge Bundy asked that the Military Assistance Command avoid intensifying the debate by choosing unprovocative titles for all future military operations. General Wheeler relayed that request to Westmoreland, citing an operation in Binh Dinh Province code-named MASHER as an example of what Bundy wanted to avoid. Wheeler instructed the command to select the sort of titles that would deprive ''even the most biased person'' of a theme for a public speech. Westmoreland complied but not without a touch of irony. He changed MASHER to WHITE WING and from then on made the naming of operations an important function of his staff.[6]

In late January 1966 President Johnson even attempted to draw Westmoreland into the politics of the war. When the AP Managing Editors Association invited the general to speak at an April luncheon in the United States, Johnson himself suggested that he attend. The association provided ''an unusually good forum,'' the president said, to explain the values at stake in Southeast Asia and to assist in bringing domestic public opinion into line with administration policy.[7] In the belief that it was improper for a commanding general to make public appearances while his men were fighting, Westmoreland declined the invitation. He had already turned down a similar request from the Overseas Press Club, he told the president. In addition, the end of the rainy season in April usually meant a return to heavy fighting in South Vietnam, making any trip by him to the United States highly inadvisable.[8]

[5] ''The Vietnam Debate,'' *New York Times*, 17 Mar 66; Polesby, ''Political Science and the Press,'' p. 54.

[6] Msg, Dept of the Army 745350 to All Army Commands, 29 Dec 65, and Msg, Defense 9509 to MACV, 29 Dec 65, both in CMH files. For Wheeler's comment, see Msg, Wheeler JCS 460-66 to Westmoreland, 1 Feb 66, Miles Policy/Strategy files, CMH. Interv, George MacGarrigle with Maj Gen John C. F. Tillson III, MACV J-3, 28 May 76, CMH files.

[7] Msg, Wheeler JCS 457-66 to Sharp, Westmoreland, 31 Jan 66, Miles Policy/Strategy files, CMH.

[8] Msg, Westmoreland MACV 885 to Wheeler, 1 Feb 66, Miles Policy/Strategy files, CMH.

The MACV Information Center at Da Nang

Refinements to the Information Program, 1965–1966

The prospect of an escalating war was also on the mind of Arthur Sylvester, who as early as October 1965 had informed Secretary McNamara that the arrangements for handling the press in South Vietnam needed adjustment. Fashioned at a time when only forty correspondents had covered the activities of a few thousand American advisers, the system, according to Sylvester, had failed to cope with the nearly three hundred newsmen resident in South Vietnam and had no chance to accommodate those who would arrive when the pace of the war increased. The MACV Office of Information clearly required a more skilled staff if it was to continue to handle the press effectively.[9]

High on the list of problems Sylvester wanted solved was the lack of communications facilities for newsmen and information officers in South Vietnam. Since the local telephone system was primitive and U.S. military circuits overloaded, a telephone call to almost any point more than fifteen miles outside Saigon took hours to complete. Reporters in the field thus had difficulty relaying news to their

[9] Msg, Saigon 1899 to State, 26 Nov 65, FAIM/IR. This section is based on Memo, Sylvester for McNamara, 26 Oct 65, sub: Information Program, DDI Censorship file.

bureaus in Saigon, and information officers in Saigon had no quick way to check on rumors originating in the field.[10]

Air transportation for newsmen was also meager, consisting of a single helicopter based in Saigon. When a reporter wanted to cover an event outside of the city, he had little choice but to hitch a ride on the first available conveyance going in his direction. Once outside the city, correspondents made their own arrangements with American or South Vietnamese pilots. Connections were so poor, Peter Kalischer of CBS News complained, that leaving the capital for the field usually demanded "a combination of . . . gall, contacts, and steady nerves." The lack of regular transportation thus remained a constant source of irritation.[11]

With McNamara's approval Sylvester took a number of steps during November 1965 to remedy the situation. To improve communications, he authorized the construction by April 1966 of a special public affairs teletype system linking all the major military bases in South Vietnam with MACV headquarters in Saigon. Then, to increase the prestige of the MACV Office of Information, he approved raising the position occupied by its chief of information from colonel to brigadier general and notified Westmoreland that Bankson would be available for the job after the first of the year. The change had little practical meaning since Bankson remained a colonel, but it allowed for the future assignment of a general officer to the position.[12]

Solving the transportation problem proved more difficult. Although McNamara authorized a special news courier flight once each day from Saigon to Nha Trang, Qui Nhon, Da Nang, and Pleiku and back, the Military Assistance Command objected that a shortage of cargo aircraft made the arrangement impractical. Only on 24 January 1966 were the flights able to begin, and they were replaced within a month by a more flexible system in which C–130 aircraft flew the circuit four times a day with fifteen seats reserved for newsmen. To ride the planes, a reporter had merely to notify the MACV Office of Information a day in advance.[13]

While these improvements were taking shape, the Military Assistance Command was continuing work begun by the U.S. marines earlier in 1965 to assist correspondents covering events in the field. Before 1965 there had been little need for special facilities for the press outside of Saigon because most of the news had occurred within easy reach of reporters stationed in the capital. With the expansion of the air war and the arrival of the marines at Da Nang, however, the attention of the press had shifted northward, prompting a number of correspondents

[10] Charles Mohr, "This War—and How We Cover It," *Dateline* 10 (April 1966): 20.

[11] Albert R. Kroeger, "Television's Men at War," *Television Magazine*, Jul 65, p. 38.

[12] Memo, Cyrus Vance, Deputy Secretary of Defense, for Sylvester, 12 Nov 65, and Msg, CINCPAC to JCS, 22 Dec 65, both in DDI Censorship file. Msg, Sylvester Defense 4510 to Westmoreland, 22 Nov 65, Westmoreland Papers, CMH; Interv, Maj Robert H. Van Horn, CO, 3d MHD, with Col Robert J. Coakley, USARV IO during 1966, VNIT-27, in CMH files (hereafter cited as Coakley Interv).

[13] Msg, JCS 7342 to CINCPAC, 27 Nov 65, and Msg, MACV 44150 to JCS, 17 Dec 65, both in Information Advisory files, 69A702, box 13, WNRC. Msg, MACV 1994 to 315th Air Commando Squadron, 20 Jan 66, and Msg, MACV 4797 to CINCPAC, 15 Feb 66, both in DDI Correspondence with MACV file.

to take up station in Da Nang. To accommodate the influx while ensuring that newsmen formed a good opinion of the U.S. effort in the I Corps Tactical Zone, the Military Assistance Command during May 1965 had established an information center. Located at Da Nang in a former French brothel and staffed largely by U.S. Marine Corps personnel, the center provided the press with overnight lodging, briefings, and tips on upcoming stories.[14]

MACV Chief of Staff Maj. Gen. William B. Rosson knew that poor communications between Da Nang and Saigon would cause long delays before news from the I Corps Tactical Zone could reach Saigon for release to the press. He therefore designated the information center at Da Nang the sole releasing authority, under the policy guidance of the MACV Office of Information, for all news of American actions in the region. Besides providing correspondents with accommodations, the press center thus became a semi-independent extension of the MACV Office of Information, holding daily briefings of its own and releasing a daily communique.[15]

Toward the end of 1965 and early in 1966, as U.S. involvement in the war continued to grow, the Military Assistance Command followed the precedent it had set at Da Nang by approving the establishment of three more press centers: at Nha Trang, the U.S. Army headquarters for the II Corps Tactical Zone; at Long Binh, the III Corps Tactical Zone headquarters; and at Can Tho in the Mekong Delta. U.S. Army information officers administered all three. In the end, only the facility at Nha Trang survived for any length of time, the others being too near Saigon to be popular with correspondents. The Army later set up other centers at Qui Nhon, An Khe, and Pleiku but never allowed any of them to be as independent as the press center at Da Nang, a fact that weighed heavily upon Army information officers.[16]

Although major American military units serving in South Vietnam supplemented MACV's efforts by making their own provisions for visiting newsmen, the quality of those accommodations varied from unit to unit. A reporter spending the night in the compound of a well-established American advisory group might find a comfortable room, good food, a bar, and a movie; but a reporter moving with units in the field rarely received more than a sleeping bag, a poncho, and the right to share the rations of the troops for a small fee. On those occasions the newsman's lot was exactly the same as that of the soldiers he was covering. As *New York Times* correspondent Charles Mohr recalled, "My own

[14] Memo, Combat Information Bureau, Da Nang, for ACS G-3, III MAF, 2 Jun 65, sub: Command Diary Submissions, III MAF Command Diary for 1 to 31 May 65, Office of Marine Corps History, Navy Yard; Ltr, Rodger Bankson to the author, 26 Oct 79, CMH files.

[15] Memo, Brig Gen Winant Sidle for Chief of Staff, MACV, 14 Oct 67, sub: Danang Press Center, Sidle file, 73A0243, box 19, WNRC; Msg, MACV to CG, III MAF, 18 May 65, Incl to Memo, Combat Information Bureau for ACS G-3, III MAF, 2 Jun 65, sub: Command Diary Submissions, III MAF Command Diary for 1 to 31 May 65, Office of Marine Corps History.

[16] Coakley Interv; Transcript, USARV Information Officer Conference, Mar 67, Comments of the USARV IO, Col Joseph R. Meacham, 13 Mar 67, USARV Daily Staff Journal Source files, 70A748, box 28, WNRC.

worst nights were spent in Plei Me Special Forces Camp, where rats kept running over our chests . . . and in a flooded sugar cane field in Hau Nghia Province, where Jack Foisie of the *Los Angeles Times* bitterly contested the single, tiny, hip-sized patch of dry ground I had found.''[17]

Although the MACV Office of Information professed absolute impartiality in its handling of newsmen and impressed that requirement on all of its personnel, some newsmen received better treatment than others. In the field, *Washington Post* correspondent Peter Braestrup observed, reporters for the major print organizations—*Time*, the *New York Times*, the *Washington Post*, to name a few—obtained considerable help and hospitality from division commanders and information officers. Television and wire-service reporters, on the other hand, received much less. Angling for the main chance and working under short deadlines, they tended to be more competitive and abrasive than reporters for large newspapers and news magazines, who often wrote interpretive articles and could spend time on research.[18]

Observing that information officers soon learned whom to trust, Rodger Bankson spoke candidly of his own relationships with newsmen during the time he served as MACV chief of information during 1966 and 1967. ''You can,'' he said,

> take Wendell (Bud) Merick of *U.S. News & World Report* and schedule him for an interview round of every important staff member before his year-end trip to the U.S. for his debriefing and appearances on major news shows. You can vouch for him and he can get the earthiest and most honest word from staff chiefs, and you don't have to worry about any breach of security. Beverly Deepe was another good news person. She had the smartest Vietnamese news assistant in Saigon. She also was smart. John Maffre, Ward Just, Keyes Beech, Jim Lucas, Bob Tuckman—the list could go on.[19]

Bankson admitted that he felt little kindness for several reporters, especially Peter Arnett, Morley Safer, and CBS News correspondent Don Webster, each of whom had, in one manner or another, caused embarrassment to the Military Assistance Command. Such reporters, it is clear, received what was due them in the way of services and little more.[20]

While increasing official assistance to newsmen, the Military Assistance Command also changed the way it accredited reporters to cover the war. Before 1965 the process had been a mere formality. Since all of the correspondents arriving in South Vietnam had prior accreditation by the Department of Defense to cover the Pentagon, the command left most of the processing to local South Vietnamese officials, who required newsmen to obtain clearance from the presidential palace and the Directorate of Psychological War. The resulting accreditation

[17] Coakley Interv; Mohr, ''This War—and How We Cover It,'' p. 20.
[18] Ltr, Peter Braestrup to Dr David Trask, Chief Historian, U.S. Army Center of Military History, 19 Nov 84, CMH files.
[19] Ltr, Bankson to the author, 26 Oct 79.
[20] Ltr, Bankson to Frank Olcott, 1 Jun 79, CMH files.

was supposedly good for only one month, but as Scripps-Howard reporter Jim Lucas noted, enforcement of the rule was so lax that few correspondents ever bothered renewing more than once.[21]

With nearly three hundred correspondents present in South Vietnam at the end of 1965, more than half of them non-Americans, the Defense Department decided to liberalize and decentralize its process for accrediting reporters to cover the war. At the end of November, therefore, the Military Assistance Command announced that beginning in December, newsmen with valid Defense Department accreditations had merely to show their credentials to receive a card giving them access to military transportation and facilities. All others were to present a letter from their editors stating that they represented a bona fide news-gathering organization which would take responsibility for their conduct. Freelance correspondents, who sold their stories and photographs independently to any newspaper or wire service willing to pay for them, were required to produce a letter from one of their clients affirming that agency's willingness to purchase their work. Information officers in Saigon were unhappy with the new arrangements since they allowed the most marginal of newsmen access to MACV facilities and briefings, but there was little anyone could do. The old system, with its background checks and sometimes classified files, carried the potential of becoming an embarrassment should a newsman denied access to the war for any reason other than security decide to sue.[22]

Along with the change in accreditation procedures came a series of attempts to tighten information policy. In the past, reporters had been able to roam the JUSPAO headquarters building in Saigon at will. On 27 November, maintaining that newsmen were disrupting the conduct of business, Zorthian restricted access to the facility and put visiting correspondents under escort. When the press complained that the restriction was a subtle form of censorship, officials responded that similar rules were in effect at most U.S. embassies throughout the world and that the Joint U.S. Public Affairs Office was involved in a number of highly classified projects. "After all," one of them told John Maffre of the *Washington Post*, "we do give out accreditation cards rather freely around here."[23]

A more serious attempt to restrict the press occurred in mid-December, when General Wheeler complained that the daily MACV communique for the press contained too much information of value to the enemy. In the belief that most of the news in question played little part in day-to-day coverage of the war, the Office of the Assistant Secretary of Defense for Public Affairs went along with Wheeler. It told the Military Assistance Command to begin gradually cutting the release down to 350 words and to start summarizing military engagements with-

[21] Jack Foisie, "News Gap in Vietnam," *Nation*, 14 Jul 62, p. 13; Jim Lucas, *Dateline Vietnam* (New York: Award House, 1966), pp. 11f.

[22] Ltr, Bankson to the author, 26 Oct 79; Beverly Deepe, "Revised Accreditation Plan Outlined for Viet by MACV," *Overseas Press Club Bulletin* 20 (18 December 1965):3.

[23] John Maffre, "Saigon Censors Restore 'Pigeon' Post Popularity," *Washington Post*, 27 Nov 65. See also Msg, Saigon 1914 to State, 27 Nov 65, FAIM/IR.

Billy Graham Holds a Press Conference at JUSPAO Headquarters

out revealing details that would help the enemy assess trends, double-check losses, or estimate the success or failure of his operations.[24]

Although agreeing that the enemy might find some value in the communique, neither the Military Assistance Command nor the Pacific Command thought the danger great enough to warrant restrictions. The Military Assistance Command warned that the press would notice even a gradual reduction in the size of the news release and that another assault on official credibility would result. If the Defense Department decided nevertheless to institute the policy, announcement of the change ought to originate in Washington so that reporters would recognize that the restriction was national policy rather than the product of an arbitrary decision in the field.[25]

The Pacific Command was even more emphatic. With the large number of correspondents present in South Vietnam, it told the Defense Department, a flood of stories on the war was bound to appear in the press. If the flow of officially released news stopped, reporters would turn to rumor and hearsay for informa-

[24] Memo, Wheeler CSM 1044-65 for ASD PA, 17 Dec 65, DDI News from Vietnam file; Msg, Defense 9088 to CINCPAC, 21 Dec 65, DDI Releasing Authority file.
[25] Msg, MACV 45225 to OASD PA, 27 Dec 65, DDI Releasing Authority file.

tion. All in all, the American people and the world at large were receiving an accurate picture of the war. "If we are to retain the support of the public, it is essential that the people receive and believe they are receiving the facts of the situation."[26]

The arguments had the desired effect. Unwilling to alienate the American public or to cause another uproar in the press, the Defense Department backed away from Wheeler's suggestion. For its part, the MACV Office of Information kept the format of the daily communique but began to scrutinize carefully the news it released to ensure that nothing of value to the enemy slipped through.[27]

Restricting Still Photography and Television News, 1965

Although the Military Assistance Command differed with the Defense Department on whether to reduce the amount of news released to the press, the two agreed on another issue: the question of combat photography. Neither wanted pictures appearing in the news media that would disturb or alienate the American public. From 1964 onward they grappled with the problem, attempting various expedients that never quite seemed to work.

At first, during 1964 and early 1965, official attention had centered on news photographers' penchant for depicting military action to the virtual exclusion of nonmilitary programs. In July 1964 a special ten-man team from the Department of Defense had arrived in South Vietnam with orders to fill that gap by photographing little-noticed areas of the war for later release to the press. Nothing had come of the effort at the time because the U.S. mission's public affairs program was still evolving and incapable of absorbing the new function. The Military Assistance Command resurrected the idea in early fall 1965, adding a motion picture unit to the team to balance the combat-oriented footage broadcast by the major U.S. television networks with stories on subjects they had little interest in covering. Although most professional news-gathering organizations preferred the work of their own correspondents, considering the pictures and films generated by the command little more than propaganda, the program still found willing patrons among the many regional newspapers and television stations that could not afford to send correspondents to South Vietnam. In August 1966 Arthur Sylvester reported to Congress that during the previous eight months the Military Assistance Command had released to the press 641 still photographs and 157 television news films.[28]

In the meantime, the command and the Department of Defense remained concerned that guidelines had yet to be developed for news photography. Sooner

[26] Msg, CINCPAC to OSD PA, 28 Dec 65, DDI Releasing Authority file.
[27] Interv, author with Barry Zorthian, 4 Jan 79, CMH files.
[28] Msg, Saigon 1032 to State, 24 Sep 65, FAIM/IR; U.S. Congress, Senate, Hearings Before the Committee on Foreign Relations, *News Policies in Vietnam*, 89th Cong., 2d sess., p. 68.

or later, Bankson told Sylvester, a picture of a dead or wounded American soldier appearing in a newspaper in the United States would shock the man's family with the first word of his injury or death. Both good taste and troop morale dictated that the Defense Department issue guidance to keep that from happening.[29]

During the Korean War, when censorship had been in effect, problems of that sort had been easy to remedy. The censors had merely withheld all pictures of the wounded for thirty days or until the next of kin had been notified. Pictures showing large numbers of wounded and dead had also been banned unless some chance existed that they might, as the policy guidance of the day stipulated, "inspire patriotism or determination or otherwise contribute to the war effort." In the absence of censorship in South Vietnam, similar procedures were impossible.[30]

Information officers were more concerned about television than still photography, but a number of problems stood in their way. Convinced that sound-on-film pictures of dying Americans could have a strongly adverse emotional impact on families with husbands and sons serving in the war, they nevertheless found that to restrict the television networks in some manner without doing the same to the rest of the press would have serious consequences for official credibility. Even a system of voluntary guidelines similar to the ones already in effect for reporting the ground and air wars seemed unworkable because individual news cameramen had little control, in the heat and confusion of combat, over the shots they took. Their job was to cover the war and to leave selection of the footage to be telecast to producers in New York. Since there were no facilities for developing motion picture film in South Vietnam, they also had no opportunity to review their work before it left the country.[31]

In the end, the Department of Defense and the Military Assistance Command adopted different approaches for military cameramen and civilian newsmen. For the military, the Defense Department issued policy guidance on 17 December 1965, prohibiting the release of pictures of recognizable dead or wounded until the next of kin had been notified. Pictures of disfigured wounded, of amputees, or of men in severe shock were also to be withheld unless the permission of the individual involved had been obtained first.[32]

Dealing with civilian cameramen took longer, with the Military Assistance Command and the Defense Department drafting and then rejecting formal guidelines on the subject before finally adopting an informal approach. On 24 April 1966, Zorthian and Bankson, who had replaced Legare as MACV Chief of Information on 5 February, met with representatives of the three television networks in Saigon to warn that if complaints about film footage of the dead and wounded

[29] Memo, Bankson for Sylvester, 28 Sep 65, sub: Proposed Ground Rule Covering Release of Casualty Information, DDI Casualties file.

[30] Dept of Defense, Office of Public Information, Security Review Branch, Public Information Security Guidance no. 3, 14 Aug 50, DDI Casualties file.

[31] Msg, MACV 14147 to CINCPAC, 24 Apr 66, DDI Casualties file.

[32] Msg, Defense 8911 to MACV et al., 17 Dec 65, DDI Casualties file.

arose, commanders in the field would undoubtedly deny cameramen the right to accompany troops into combat. Representatives of the Defense Department meanwhile met with executives from NBC, CBS, ABC, UPI Newsfilm, Metro-Goldwyn-Mayer, and the Mutual Broadcasting Company to emphasize the need for discrimination when selecting film footage for broadcast.[33]

The news media went along, either because of the threat implied by Zorthian and Bankson, or, more likely, because they feared that gruesome pictures broadcast into homes at the dinner hour would prompt viewers to switch stations. The result was that the American public, although treated nightly to scenes of combat and men in battle—the subjects news producers considered necessary to attract and keep viewers—rarely, if ever, before 1968 and the Tet offensive saw the war in all its bloody detail. Even then the scenes of combat that appeared paled in comparison with the choreographed violence of such popular television dramas as ''Gunsmoke.''[34] As commentator Michael Arlen observed, what appeared nightly in American living rooms during the war was a ''generally distanced overview of a disjointed conflict . . . composed mainly of scenes of helicopters landing, tall grasses blowing in the helicopter wind, American soldiers fanning out across a hillside on foot, rifles at the ready, with now and then (on the soundtrack) a far-off ping or two, and now and then (as the visual grand finale) a column of dark, billowing smoke a half mile away, invariably described as a burning Viet Cong ammo dump.''[35] From August 1965 to August 1970, only 76 out of more than 2,300 television news reports originating in Vietnam depicted heavy fighting—soldiers in combat, incoming artillery, dead and wounded on the ground.[36]

Further Refinements to the Information Program, May–July 1966

Upon arrival in Saigon, Rodger Bankson encountered a number of administrative problems that threatened the low profile the Johnson administration sought for the war. Bankson had few complaints about the quality of the officers and enlisted men working for him but found staff members arriving and departing at such poorly spaced intervals that, at the very time when the press was clamoring for more and better information, office efficiency was diminishing. In addition, officers were arriving without notice and sometimes without the skills required for the positions they were to fill, having been matched to a specific line requirement by their services without thought to their experience or time

[33] Msg, MACV 14147 to CINCPAC, 24 Apr 66; Msg, Defense 1722 to CINCPAC, COMUSMACV, 11 May 66, DDI Casualties file.

[34] Robert MacNeil, *The People Machine* (New York: Harper & Row, 1968), pp. 65f; Edward Jay Epstein, *News From Nowhere* (New York: Random House, 1973), p. 178; Lawrence Lichty, "Comments on the Influence of Television on Public Opinion,'' in Peter Braestrup, ed., *Vietnam as History* (Washington, D.C.: Woodrow Wilson International Center for Scholars, 1984), p. 158.

[35] Michael Arlen, ''The Falklands, Vietnam, and Our Collective Memory,'' *New Yorker*, 16 Aug 82.

[36] Lawrence W. Lichty, ''Comments on the Influence of Television on Public Opinion,'' in Braestrup, ed., *Vietnam as History*, p. 158.

"The Five O'Clock Follies"

in grade. As a result, a lieutenant colonel assigned by the Air Force to be chief of a branch within the Office of Information might find his Army or Navy assistant senior to him in date of rank, a situation almost certain to rankle both officers.[37]

Bankson remedied the rotation problem by adopting the unpopular but necessary expedient of curtailing some tours of duty while extending others. As for staffing, he informed the Army, Navy, Air Force, and Marine Corps that despite organization tables assigning to each service responsibility for definite positions within the information office, only qualifications and rank would determine who occupied what position. Perennially concerned about the image the military conveyed to the American public and unwilling to surrender by default positions of influence within the information program, the services began giving greater thought to assignments.

Turning to the nightly MACV briefings, Bankson discovered that the Saigon correspondents disputed many of the assertions made by official spokesmen, putting the officers in charge under considerable strain. He also found that the reporters' acceptance of the briefings—newsmen called them "The Five O'Clock Follies"—was in direct proportion to the professionalism of the briefers and that

[37] This section is based on Memo, Rodger Bankson for ASD PA, 13 May 66, sub: Personnel Problems, DDI Correspondence with MACV IO (36a) file.

an officer who had never served in combat in South Vietnam rarely possessed the insight necessary for satisfactory performance.[38]

Since the briefings were the single most important source of day-to-day news for the Saigon correspondents—none of whom could be everywhere in South Vietnam at once—Bankson moved quickly to remedy what was wrong. To relieve the tension that sometimes developed, he increased the number of MACV spokesmen from 1 to 4, creating two alternating teams of 2 briefers each. One officer on a team specialized in the ground war, and the other dealt only with the war in the air. Later Bankson added a third briefer to each team to cover actions involving the U.S. Navy. He discovered with time that switching subjects and personalities at set points during a session served almost invariably to cut off debate, much to the benefit of the officers presiding.[39]

With the assistance of the U.S. Army, Vietnam, and the Seventh Air Force, Bankson concentrated on finding briefers who had both public relations credentials and combat experience in Vietnam. He discovered that as the new men took charge the credibility of MACV's information operations improved steadily, if only because correspondents tended to place more trust in the word of line officers than they did in men who had served only in staff positions. His changes were so well received by the press that on one occasion the Saigon correspondents jeered the author of a story criticizing MACV's information effort because he continued to use the command's facilities.[40]

After reorganizing MACV's information services, Bankson, at Westmoreland's request, reviewed the MACV guidelines for the press. On 31 October 1966, he issued revised standards that confirmed those rules already in effect but added others that appeared necessary to protect troops in the field. Reporters were no longer to reveal the amounts of ordnance and fuel on hand in combat units, or the activities, locations, and methods of operation of intelligence teams. Deliberate violations of the guidelines, Bankson stressed, would result in the cancellation or suspension of the offender's credentials.[41]

While aware of the need to preserve military security, Bankson also recognized that obvious information was being withheld from the press to the detriment of official credibility. He raised the issue during April 1966, when he notified the Office of the Commander in Chief, Pacific, which had jurisdiction over the release of information on the air war in North Vietnam, that Wendell Merick was about to file a story on the Shrike, an air-to-ground missile used to suppress enemy radar. He asked permission to acknowledge the missile's existence and to reveal its uses without going into details that might be of value to the enemy.[42]

[38] Memo, Bankson for Chief of Army Information, 18 Apr 68, sub: Briefings, 73A0243, box 19, WNRC.

[39] Ibid.; Memo, Lt Col D. C. Biondi, USA, for D. Z. Henkin, OASD PA, 29 Jun 67, DDI Correspondence with MACV IO (36) file; Ltr, Bankson to the author, 26 Oct 79.

[40] Ltr, Bankson to Sylvester, 27 Jun 66, DDI Correspondence with MACV (36a) file; Memo, Bankson for the Chief of Army Information, 18 Apr 68, sub: Briefings.

[41] Westmoreland Historical File, bk. 7, tab D; Memo, MACV for Correspondents, 31 Oct 66, sub: Rules Governing Release of Military Information. Both in CMH files.

[42] Msg, MACV 14402 to CINCPAC, 26 Apr 66, DDI SHRIKE and SAMS file.

He followed that request on 6 May with a cable to the Defense Department suggesting a review of all the policies that restricted the release of information on subjects such as napalm, defoliation, and newly developed weapons. The character of the war had changed drastically over the previous year, he said, bringing devices such as cluster bombs and flechette ammunition into the hands of the troops and before the eyes of newsmen visiting field units. In the past American officials had refused to confirm the existence of those weapons in order to deny the Communists a pretext for charging that the United States was employing terror tactics; yet the lack of any outcry against the continuing use of tear and nausea gases in South Vietnam demonstrated that the propaganda potential of such subjects was insignificant. On the other hand, an affirmation that the United States was indeed employing sophisticated weaponry would reassure the families of the 255,000 men serving in South Vietnam and reduce speculation by the press.[43]

In neither case was Bankson successful. The Commander in Chief, Pacific, Admiral Sharp, refused to allow any announcement on the Shrike missile because he believed the information might be of value to the enemy. For the same reason, the policies restricting the release of information on napalm, defoliation, and cluster bombs also remained in force.[44]

On 16 July, Bankson took up the subject again, observing in a message to Sharp that captured enemy documents proved the Communists were fully aware of the existence and capabilities of cluster bombs and that continued silence on the subject would only harm official credibility. Cluster bombs had been dropped recently by accident on both an American unit and a South Vietnamese village, causing a large number of casualties. Yet official spokesmen had been able to refer in news releases only to "fragmentation weapons." As a result, stories were beginning to appear in American newspapers alluding to "some exotic new weapons system."[45] Once again Sharp refused. He was unwilling to release information that might be of even the slightest value to the enemy and remained concerned that what he considered emotional topics might fuel antiwar sentiment in the United States.[46]

Clandestine operations in Laos were a particular problem for information officers. As the U.S. Navy succeeded in thwarting Communist attempts to supply enemy troops in South Vietnam by sea, enemy supply routes through Laos became increasingly important American targets. Yet although B-52 bombers had begun attacking the trail in early 1966 and special ground reconnaissance patrols—code-named SHINING BRASS—made regular forays into the region, U.S. mission spokesmen, out of deference to Souvanna Phouma, continued to tell newsmen only that U.S. aircraft were conducting reconnaissance missions in Laos with authority to fire if fired upon. They never mentioned ground patrols.[47]

[43] Msg, MACV 15628 to CINCPAC, SECDEF, 6 May 66, CMH files.

[44] Msg, CINCPAC to OSD PA, 7 May 66, and Memo, Col Winant Sidle, Special Assistant for Southeast Asia, for Sylvester, 12 May 66, both in DDI SHRIKE and SAMS file.

[45] Msg, MACV 42527 to CINCPAC, 16 Jul 66, DDI CBU/BLU file.

[46] Msg, CINCPAC to MACV, 9 Aug 66, DDI CBU/BLU file.

[47] Msg, Vientiane 1037 to State, 31 May 66, 69A702, box 4, WNRC; Msg, Vientiane 721 to State, 6 Jan 66, and Msg, State 514 to Vientiane, 1 Feb 66, both in FAIM/IR.

The Johnson administration was prepared to accept occasional news stories on the war in Laos, reasoning that they served to assure the American public that the United States was doing all it could to protect its troops. The policy of never volunteering information on that portion of the war nevertheless led almost inexorably to embarrassment, as in May 1966, when the family of an airman killed in Laos went to the newspapers with the story. Lacking alternatives, the Defense Department admitted at that time that some eleven airmen had died in Laos over the previous two years, but, in an attempt to keep the lowest possible profile, neglected to add that another eleven had been killed prior to 1964. The omission disturbed Arthur Sylvester, particularly after R. W. Apple of the *New York Times* and Jack Foisie of the *Los Angeles Times* wrote in July about what they called "the growing clandestine war in Laos." Since official spokesmen could hardly deny allegations that were beginning to arise in the press that the United States was concealing casualties, Sylvester appealed to the secretary of defense for a change in policy. Although revising the way casualties were counted was bound to cause criticism in the press, especially since six more airmen had died since the announcement in May, he found the alternative even more unattractive. "Our credibility is under attack, and the severe blow which could be inflicted by the revelation that we are hiding casualties could be a telling one in the November elections."[48]

McNamara authorized the change. On 3 August he notified the Military Assistance Command that beginning the next day, the lead sentence of the weekly statistical summary of casualties in South Vietnam would be changed from "casualties incurred by U.S. military personnel in Vietnam" to "casualties incurred by U.S. military personnel in connection with the conflict in Vietnam." Although newsmen would inevitably link the new terminology with the addition of casualties from Laos, there was to be no acknowledgment that anything was different. Official spokesmen were to continue the policy of refusing to admit that the United States was operating in Laos.[49]

Although it agreed to the change, the State Department specified that casualties among the 125 American military personnel assigned to the military attache's office in Vientiane were to be excluded. Concerned that the naming of any casualties from the group might reveal that the attaches were really advisers to the Laotian armed forces, State justified the policy on grounds that those individuals were in no way involved with the war in South Vietnam. The Defense Department had little choice but to go along, commenting only that "the burden is upon State at all times to verify or support the fact that such casualties are not in any way connected with the conflict in Vietnam."[50]

[48] Memo, Arthur Sylvester for the SECDEF, 19 Jul 66, sub: Credibility and the Release of Casualty Statistics From Laos, DDI Casualties file; Msg, State 514 to Vientiane, 1 Feb 66.

[49] Msg, Defense 8294 to CINCPAC, COMUSMACV, 3 Aug 66, DDI Laos Policy file; Msg, Defense 8329 to CINCPAC, 4 Aug 66, DDI Casualties file.

[50] Memo, Deputy Assistant Secretary of State for Public Affairs for Deputy Assistant Secretary of Defense for Public Affairs, sub: Public Affairs Handling of U.S. Service Casualties in Laos, attached to Memo, Cdr Williams, SA/SEA, for Sylvester, 3 Nov 66, sub: Laos Casualties, DDI Laos Policy file.

Interservice Rivalry and the Practice of Leaking, January–August 1966

Less easy to remedy was the continual leaking of sensitive, classified information to the press, a practice that reached epidemic proportions during the first half of 1966. When the Johnson administration approved B–52 strikes on the Mu Gia Pass in North Vietnam, the funnel through which the enemy moved most of the supplies for the war in the South, officials in Saigon and Washington took pains to preserve military security and to coordinate press releases because the target straddled North Vietnam's border with Laos. Yet on the day before the strike occurred, *Life* magazine's Saigon bureau notified its home office to be prepared for the event. In the same way, shortly before U.S. marines launched a secret operation into the Rung Sat, an enemy base area that threatened the main shipping channel to Saigon, reporters got word and attempted to go along. The press also learned that the United States was building a base in the Mekong Delta at Dong Tam, information that led to damaging speculation on whether American troops would shortly begin operations in the region. When MACV intelligence analysts discovered enemy troops massing north of the Demilitarized Zone, Joseph Fried had the story before the command was prepared to release it. This story on B–52 raids in Laos was perhaps the worst of the leaks, so potentially harmful that for a time U.S. officials feared Souvanna Phouma would demand an end to all U.S. air operations in his country.[51]

Some of those revelations may have been the result of chance slips of the tongue, others of good detective work by skillful reporters, but most occurred because someone in a position to know purposely released the information. Before 1966, leaks of that sort had often been expressions of disapproval of official policy or attempts to influence a course of action by bringing public pressure to bear upon decision makers. As the U.S. Army, Air Force, Navy, and Marine Corps began full-scale operations in South Vietnam, another factor entered in: parochial service pride. Constrained by rules restricting the release of information on the more sensitive aspects of the war, individual members of the various services came to believe that the sacrifices of their comrades were going unnoticed by the American public. They compensated by leaking word of what was actually going on.[52]

The various service information offices in South Vietnam contributed at times to the problem. With the MACV Office of Information the sole release point for news of the war, those offices lacked any outlet for publicizing the units and services they represented. Although they submitted a stream of press releases to

[51] Memo, COMUSMACV for VMAC, 11 May 66, CMH files; MFR, 17 Aug 66, sub: MACV Commanders' Conference, 24 Jul 66, Westmoreland Historical File, Aug 66, CMH; Msg, Defense 6993 to CINCPAC, 18 Jul 66, for Sharp from Sylvester, DDI Laos Policy file.

[52] Memo, COMUSMACV for VMAC, 1 May 66. Multiple leaks also occurred from sources other than the military. Correspondents boasted of their connections at the embassy and among employees of the Central Intelligence Agency and the Agency for International Development. See also Memo, Sidle for the author, sub: Public Relations, 7 Nov 84, CMH files.

the command on all aspects of their service's activities, they found that the Office of Information held back communiques that it considered too blatantly self-serving. Concerned mainly, as a result, with troop information—the publishing of unit newspapers and the coordination of publicity campaigns directed at the individual U.S. serviceman—those offices sometimes engaged in a subterranean struggle for public attention that fostered disregard for regulations. During a March 1967 meeting of Army information officers, for example, Chief of Information, U.S. Army, Vietnam, Col. Joseph R. Meacham, spent considerable time describing the MACV information center served by the Marine Corps at Da Nang. Meacham claimed that the center was little more than a public affairs operation for the marines and asked rhetorically, "How did they get all this? Quite frankly, they got it by General Walt giving a gung-ho colonel the mission of getting the job done, giving him a license to steal and promising that he'd keep the Inspector General and the auditors off his back until the function was running smoothly and sorta' legal like." Meacham said that he had tried to work within the system, but, having failed, had himself begun circumventing regulations in order to compete.[53]

Much of the rivalry between service information offices involved petty, harmless complaining. Meacham, for example, used a content analysis of the daily MACV communique to argue that the Army, with 53 percent of the forces in South Vietnam, had received only 23 percent of the space in the release while the much smaller Navy and Air Force contingents received 25 and 35 percent respectively. At times, however, the elbowing was serious. After an important series of air strikes against the North Vietnamese transportation system, Air Force information officers, without waiting for the official MACV news release, issued a communique awarding sole credit to their own service even though U.S. Navy aircraft had been equal participants. Since the incident might affect the morale of naval air units operating in North Vietnam and incurring casualties as heavy as those of the Air Force, the Military Assistance Command reproved the officers involved, pointing out that one service had no right to aggrandize itself at the expense of another.[54]

Competition for the public eye nevertheless continued. On 29 July 1966, for instance, an Air Force search-and-rescue helicopter recovered a U.S. Navy pilot, Lt. (jg.) Dieter Dengler, who had been shot down and captured in Laos on 1 February but had escaped five months later during transport to a prisoner-of-war camp in North Vietnam. Problems began shortly after the rescue team deposited Dengler at a field hospital near Da Nang. Congressman Robert Taft of Ohio and two reporters, Richard Kirkpatrick of the *Cincinnati Enquirer* and Rod Williams of WSAI Radio in Cincinnati, happened to be in the receiving room. Unaware that the pilot had been shot down over Laos or that any revelation of his story might threaten

[53] Transcript, USARV IO Monday Staff Meeting, 13 Mar 67, USARV Daily Staff Journal files, 70A748, box 28, WNRC.

[54] MFR, sub: Conference of USARV Information Officers, 3 to 5 May 67, USARV Daily Staff Journal files, 70A748, box 28, WNRC; Interv, author with Cdr Joseph Lorfano, 1 Jun 73, CMH files; Msg, Defense 1664 to CINCPAC, 11 May 66, DDI Press Flaps file.

U.S. relations with Souvanna Phouma, Kirkpatrick wrote an article on Taft's visit to the hospital that included a reference to recovery of an unidentified American pilot in North Vietnam near the Laotian border.[55]

When contacted by the Military Assistance Command, Kirkpatrick and Williams readily agreed to withhold further information until the command issued a communique, but Joseph Fried got word of what had happened and telephoned the MACV Office of Information for details. The reporter had obviously talked to a knowledgeable insider. He knew the name of the pilot, how long the officer had been in captivity, when the recovery had occurred, and the sequence of events leading up to the rescue. When the command refused to verify the leak, Fried filed a story anyway, using what he considered nonsensitive facts. Although the Defense Department prevailed upon his employer, the *New York Daily News*, to withhold publication, by that time the need for some sort of official statement had become obvious. The Director of Information, Seventh Air Force, Col. William McGinty, attempted to fill the vacuum by drafting a lengthy communique for MACV's approval and release. On instructions from the Defense Department, however, the command held its announcement to the barest minimum, affirming only that a Navy pilot had been recovered after a period of Communist captivity and that his family was being notified. McGinty protested that the statement failed to give the Air Force proper credit for the rescue, but the command refused to make any change. There was to be one authorized announcement, Bankson replied, and it had been made.

That afternoon information officers in Saigon watched the story develop as it came from New York on the UPI ticker. At first, the filing appeared to be a reworked version of Fried's original, but as the afternoon progressed additional details began to appear, including a revelation that a U.S. Air Force "Jolly Green Giant" helicopter had made the rescue. With that, the Saigon correspondents began complaining loudly that the Military Assistance Command was favoring a few reporters to the detriment of the many. Bankson could do little to quiet the furor but did finally release the name of the pilot and the fact that the officer was undergoing debriefing.

Convinced that the Office of Information for the Seventh Air Force was to blame for the leak, Bankson sought out McGinty. Although the colonel denied any wrongdoing, he admitted that he had released prerecorded tapes on the subject to a few newsmen and justified his action by stating that the tapes had only amplified details of the rescue itself, a matter of public record. Bankson reminded him that the release of details of the rescue might reveal the procedures pilots used to evade capture, making future recoveries more difficult. Once higher headquarters had decided against an announcement, he added, officers in the field had no right to change the policy.

Although he believed that McGinty had shown bad judgment, Bankson was

[55] Msg, CDR NAVSUPCOM, Da Nang, to CNO, 20 Jul 66, 206-02(1966), III B (3) Discipline, 69A702, WNRC. This section is based on MFR, Bankson, 9 Aug 66, sub: Release of Information About Navy Pilot, DDI Correspondence with MACOI (36) file.

willing to admit that extenuating circumstances existed, if only because the airmen who had made the rescue had undoubtedly been proud of their achievement and had probably talked freely. Then, only a few hours after the meeting with McGinty, Bankson learned that a Seventh Air Force information officer had been caught passing a copy of McGinty's tapes to Fried. Armed with evidence that leaking was continuing, he sought General Westmoreland's concurrence and then presented the facts to McGinty's superior, General William W. Momyer, Commander, Seventh Air Force. Momyer promptly arranged for McGinty's reassignment to a post unrelated to public affairs in the United States. While never able to prove the contention, Bankson and other officers concerned with the information program suspected that McGinty had acted on instructions, and that the Air Force Chief of Information, Maj. Gen. E. B. LeBailly, had sent him to Saigon to open the air war to public view despite State and Defense Department rules to the contrary.[56]

After McGinty's departure the Joint Chiefs of Staff attempted to cut back further on the information the Military Assistance Command was releasing to the press. Citing the many leaks that had occurred during previous months, they pointed out that continued boasting about American technological and tactical successes might goad Russia and China into providing the enemy with even more sophisticated weaponry. In addition, highly dramatized news stories on potentially controversial weapons such as cluster bombs might increase pressure on the Johnson administration to restrict operations in South Vietnam. For those reasons, all information on U.S. air bases in Laos and Thailand was to be classified, along with the use of new or significantly improved munitions and weapons systems; changes in the capabilities of U.S. and allied troops that altered significantly the relative effectiveness of friendly and enemy forces; and the circumstances surrounding the capture, imprisonment, escape, evasion, or recovery of U.S. and free world personnel.[57]

The rules added almost nothing to regulations already in effect at the Military Assistance Command and apparently had little impact. The practice of leaking continued, giving rise to innumerable investigations that almost always ended without result. Except for the episode involving McGinty and another involving Defense Department analyst Daniel Ellsberg, who in 1971 released a top secret manuscript history of policy making during the early years of the war, few if any of the offenders were ever caught. The problem of interservice rivalry also remained. At one point during 1966 the MACV Office of Information even stationed an official observer at the press center in Da Nang to keep Marine information officers there from trying to make the war seem an exclusively U.S. Marine Corps affair. Although the Johnson administration continued to seek the lowest possible profile for the war, it was obvious that the war itself activated far too many parochial concerns for the policy to succeed.[58]

[56] Coakley Interv. Intervs, author with Rodger Bankson, 30 Aug 79, and with Cdr Lorfano, 1 Jun 73, both in CMH files.

[57] Msg, JCS 8489 to CSA et al., 5 Aug 66, DDI Security of Information file.

[58] Memo, Sidle for Chief of Staff, MACV, 14 Oct 67, sub: Danang Press Center.

11

First Gusts of the Whirlwind
The Buddhist Crisis of 1966

Although the Johnson administration streamlined its handling of the press to ensure that as few controversies as possible accompanied the American buildup, opposition to the war in the United States was too strong and the situation in South Vietnam too unwieldy for the policy to have much chance of success. It began to break down, indeed, shortly after the January bombing halt. Just as the president was beginning to increase the pressure on North Vietnam, the Chairman of the Senate Foreign Relations Committee, J. William Fulbright of Arkansas, during congressional hearings questioned the legality of using the Gulf of Tonkin Resolution to justify further escalations. Citing the Foreign Relations Committee's jurisdiction over a foreign aid appropriation for South Vietnam, he also announced that he would shortly hold public hearings on the president's Vietnam policies.[1]

The Fulbright Inquiry, February 1966

The Johnson administration took immediate steps to dilute the impact of Fulbright's investigation. On the grounds that classified information was involved, McNamara and Wheeler refused to testify in public session, forcing the committee to save face by responding that it would hold no closed hearings. Then President Johnson announced that he and his top military and civilian advisers would meet with South Vietnam's leaders in Honolulu to discuss the war's social and economic dimensions, a move interpreted by both the Foreign Relations Committee and the press as an attempt to steal the limelight.[2]

[1] E. W. Kenworthy, "Senate Panel Will Conduct Broad Inquiry on Vietnam," *New York Times*, 4 Feb 66.

[2] Mark Watson, "McNamara and Wheeler Agree to Closed Hearings," *Baltimore Sun*, 5 Feb 66; "U.S. and South Vietnamese Leaders Meet at Honolulu," Department of State *Bulletin*, 28 February 1966, p. 302; James Reston, "Ships Passing in the Night," *New York Times*, 9 Feb 66; Joseph McCaffrey, 6:00 P.M. News, WMAL-TV, Washington, D.C., 4 Feb 66, Radio-TV-Defense Dialog.

If that was the case, the attempt failed, for as the hearings progressed they generated so much interest that NBC News decided to televise them in full. On the first day of the inquiry, 5 February, the Administrator of the Agency for International Development, David E. Bell, fielded a number of hostile questions from the committee about South Vietnamese refugees, allegedly unjustified official optimism, and the need for political and social reforms within the South Vietnamese government. Retired U.S. Army Lt. Gen. James Gavin followed on 8 February with a warning that U.S. policy in South Vietnam was alarmingly out of balance. Although precipitate withdrawal would be unwise, he said, the United States should restrict its operations to defensive enclaves until some political solution became possible. On the tenth, the former U.S. Ambassador to the Soviet Union, George Kennan, stated that there was "more respect to be won in the opinion of the world by a resolute and courageous liquidation of unsound positions than in the most stubborn pursuit of extravagant and unpromising objectives." Kennan recommended withdrawal from South Vietnam as soon as that step could be taken without damage to American prestige or to "the stability of conditions in the area."[3]

The administration's case received staunch support on 17 February, with the appearance of former U.S. Ambassador to South Vietnam Maxwell D. Taylor. Tracing the history of American policy in Southeast Asia, Taylor explained that by employing its troops, the United States hoped to achieve sufficient success on the battlefield to compel the enemy to accept an independent, non-Communist South Vietnam. In seven hours of testimony the next day, Secretary of State Dean Rusk emphasized that "toughness is absolutely essential for peace. . . . [If] we do not make it quite clear . . . where we are going to make our stand, then the prospects for peace disappear."[4]

The hearings received mixed reviews. The *St. Louis Post-Dispatch*, the *New York Times*, Walter Lippmann, and those commentators and newspapers that were usually critical of the Johnson administration's Vietnam policies praised the inquiry as an inestimable service to the American people that had helped deflate official rhetoric on the war. Others were less satisfied. Admonishing Fulbright to exercise restraint, the *Baltimore Sun* regretted the senator's apparently uncritical acceptance of the antiwar positions expounded by one of the committee's more outspoken members, Senator Wayne Morse of Oregon. To the *Sun*, Morse combined "a strong belief in his own righteousness with a rejection of conflicting opinion." The *Philadelphia Inquirer* printed under a large headline charges by Senator Hugh Scott of Pennsylvania that Fulbright had arranged the sequence of witnesses before the committee in a manner calculated to embarrass the president and that the hearings were giving aid and comfort to the enemy. Columnist David Lawrence quoted a comment by the Chairman of the Senate Armed Services Committee, Senator Richard Russell: "Every dissent prolongs the war. It makes no

[3] The hearings are contained in U.S. Congress, Senate, Committee on Foreign Relations, *Supplemental Foreign Assistance, Fiscal Year 1966*, 89th Cong., 2d sess., pp. 92, 227, 335.

[4] Ibid., p. 438. Quote from Rusk's testimony is on p. 629.

difference if some poor, half-baked beatnik . . . or some senator of the United States . . . demands that we get out. . . . This undoubtedly confirms Ho Chi Minh in his basic concept that the United States just wouldn't have the patience to wage a very lengthy war."[5]

In the end, the Johnson administration concluded that more had been gained by the inquiry than lost. It thought so well of the outcome that a few weeks after the hearings it tried to convince the government of France to play tapes of the meetings on the state-controlled television network in order to encourage, as one State Department official put it, "objective reporting on Vietnam in France." The French rejected the suggestion emphatically.[6]

Regardless of who won the debate, the administration remained concerned about criticism from Congress and the news media and attempted to forestall it. During the first week of March, with the president's knowledge and possibly at his behest, Undersecretary of State George Ball—who was generally known to have recommended that the United States withdraw from South Vietnam and who was on that account probably more influential with critics of the war than other members of the administration—contacted a number of prominent dissenters in an attempt to win either their support or their silence. At a series of lunches with columnists Walter Lippmann and James Reston and with Senators George McGovern, Abraham Ribicoff, and Joseph Tydings, Ball argued that "it was perfectly clear that under the present circumstances we had no option to do anything other than what we were doing. We faced a situation and not a theory; we had to see the war through, and at the price of substantial increased effort." The time had passed, he told Walter Lippmann in particular, "when any of us could afford to attack the policies now being pursued in South Vietnam."[7]

The president's efforts to forestall criticism of the war were complicated by the actions of the South Vietnamese. The United States had worked incessantly to improve the image of its ally, but the best of public relations programs had proved incapable of compensating for an unstable government and for troops who performed unpredictably in battle. In the months since February 1965, when Nguyen Khanh had withdrawn from South Vietnamese politics, a distressing series of coups and countercoups had occurred, leading in June to the accession of Air Vice Marshal Nguyen Cao Ky as premier and General Nguyen Van Thieu as chief of state. Although the two seemed more capable of holding the government together than had their predecessors, the problem of South Vietnam's image remained so heavily upon President Johnson's mind that he admonished them

[5] "Vietnam on TV," *St. Louis Post-Dispatch*, 9 Feb 66; ". . . and at the Senate Inquiry," *New York Times*, 9 Feb 66; Walter Lippmann, "Confrontation With China," *New York Herald-Tribune*, 15 Feb 66; "Inquiry on Vietnam," *Baltimore Sun*, 9 Feb 66; Saul Kohler, "Scott Says Fulbright 'Rigs' Senate Hearings on Vietnam," *Philadelphia Inquirer*, 18 Feb 66; David Lawrence, "Fulbright Hearings Criticized," *New York Herald-Tribune*, 15 Feb 66.

[6] Memo, John M. Leddy for the Secretary of State, 8 Mar 66, sub: Abortive Effort To Encourage Objective Reporting on Vietnam in France, FAIM/IR.

[7] Memo, George W. Ball for the President, 3 Mar 66. sub: Reports of Recent Conversations on South Vietnam; see also ibid., 1 Mar 66, sub: Lunch With George McGovern and a Group of Dissenting Senators. Both in FAIM/IR.

during the Honolulu meeting to study the criticisms levied by both the Fulbright committee and the *New York Times*. In that way, he said, they would understand the pressures he had to face and could conduct their government accordingly. He even advised Ky to act more "like a politician instead of just a general."[8]

For a time, Johnson's strictures appeared have some effect. Toward the end of January the Ky regime inaugurated a daily military briefing for the press, giving rise to hopes that the South Vietnamese might at last begin to take responsibility for telling their side of the war. Meanwhile, on the battlefield, as B–52 strikes kept the enemy off balance, American and South Vietnamese troops began entering enemy base areas regularly. South Vietnamese casualty rates also declined while those of the enemy appeared to increase. Observing those advances, *U.S. News & World Report* could only conclude, "The important thing is this, in the past few months, . . . the U.S. has prevented the Reds from winning the war, and is on the way to winning."[9]

Engagement at A Shau, March 1966

American confidence in South Vietnamese progress nevertheless began to decline on 9 March, when a North Vietnamese regiment attacked a South Vietnamese Special Forces camp at A Shau, some 45 kilometers southwest of Hue near the Laotian border. The enemy forced the garrison's defenders, 17 U.S. advisers and 360 South Vietnamese and Montagnard irregulars, to abandon all but one corner of the base. Late on the afternoon of the battle's second day, receiving word that ammunition was short and that food and water were gone, the commander of the III Marine Amphibious Force, Lt. Gen. Lewis Walt, ordered a fleet of helicopters to evacuate A Shau. When the rescuers arrived, the irregulars panicked, mobbing the aircraft and forcing the helicopters to back off empty. Although some survivors hid in the jungle to be rescued in scattered groups the next day, of 484 fighting men and civilian dependents in the camp only 186 returned. Five of the seventeen U.S. advisers were missing and presumed dead.[10]

Although noting that traitors among the camp's defenders had apparently betrayed the installation, early news accounts of the battle said little about problems. Some reports focused on the heroism of Maj. Bernard Fisher, USAF, who landed on the camp's enemy-controlled air strip to rescue a downed fellow flyer, an act for which Fisher later received the Medal of Honor. Others quoted one of the U.S. advisers who had praised his South Vietnamese and Montagnard com-

[8] Memo, Bill Moyers for U. Alexis Johnson, 16 Feb 66, sub: Conversation Between Johnson, Ky, and Thieu, FAIM/IR.

[9] Fact Sheet, Favorable Military Factors, Feb 66, Westmoreland History, bk. 4, tab A–9, CMH; Msg, Saigon 2598 to State, 19 Jan 66, DDI News from Vietnam file; "Turn for Better Seen in War," *U.S. News & World Report*, 7 Mar 66, p. 31.

[10] Ibid.; Msg, MACV to the SECDEF, 16 Mar 66, III MAF files, Office of Marine Corps History.

The Camp at A Shau

rades because "if they hadn't fought real hard, we'd have been overrun."[11]

The full story of what had happened emerged only on 14 March, four days after the fall of the camp, when Scripps-Howard reporter Jim Lucas revealed that the commander of the American detachment at A Shau, Capt. John D. Blair, had emerged from the jungle cursing the camp's South Vietnamese defenders as Viet Cong sympathizers and charging that only the Montagnards had fought to repel the enemy. According to Blair, one entire South Vietnamese company had gone over to the enemy. "If I could get my hands on Chung Wei [a South Vietnamese Army lieutenant]," he said, "I'd kill him."[12]

On the same day Lucas' story appeared, CBS News broadcast a report by correspondent Jack Laurence of an interview with Lt. Col. Charles House, the commander of the Marine helicopter squadron that had gone into A Shau. House confirmed that the irregulars had panicked. "So many people wanted to get out," he said, "they hung on the cables—almost pulled the helicopters into the [land-

[11] Ray Herndon [UPI], "Viet Cong Horde Crushes Stand of Green Berets," *Philadelphia Inquirer*, 10 Mar 66; "Viet Cong Overrun Isolated G.I. Post After Day's Battle," *New York Times*, 10 Mar 66; "Through Deadly Enemy Fire Pilot Lands To Save Comrade," *New York Herald-Tribune*, 11 Mar 66; Robin Mannock, "Wave By Wave They Came," *Washington Star*, 11 Mar 66.

[12] Jim Lucas, "A Shau Chief Charges Vietnam Treachery," *Washington Daily News*, 14 Mar 66.

ing] zone. And it was . . . a hell of a thing to have to do: some of them had to be shot in order to maintain control." Laurence asked whether that weighed on House's conscience. The Special Forces advisers, replied House, had opened fire in order to avert total chaos. "I was a little concerned," he continued. "I thought these people would shoot back, but they didn't. . . . They just turned their backs as they fell off and that was it."[13]

The reports caused immediate concern at the White House, prompting a Defense Department investigation that substantiated most of the details released by the news media. In Saigon, MACV information officers attempted to play down Blair's remarks as the product of great stress, and Blair himself subsequently told newsmen that some South Vietnamese irregulars had fought better than expected. As for Colonel House, the III Marine Amphibious Force awarded him the Navy Cross (second only to the Medal of Honor) for valor but then presented him with a letter of reprimand for his comments to the press, an action that had the effect of forestalling further promotions.[14]

Confrontation With the Buddhists, March–June 1966

Reports on the engagement at A Shau had hardly begun to fade before attention turned to a new problem confronting the South Vietnamese, one that was rapidly developing into a major political crisis. The central character was the commander of the I Corps Tactical Zone, Lt. Gen. Nguyen Chanh Thi, a powerful, charismatic leader who had long ruled South Vietnam's northern provinces virtually independent of the Saigon government and who appeared such a threat to the regime that rumors of an impending coup began to circulate every time he traveled to the capital. So cocksure was Thi that he sent troops twice during March to operate in the Demilitarized Zone despite orders from Saigon to refrain from entering the area.[15]

Backed by South Vietnam's Armed Forces Council, Ky decided to relieve Thi of command for insubordination. He requested Ambassador Lodge's concurrence, but Lodge refused. President Johnson was receiving "very fine" votes in Congress, Lodge said. Not only were critics of the administration's policies resorting to more and more picayune and untenable objections, the American people were

[13] Msg, NMCC to MACV, 14 Mar 66, 5th Special Forces Command Reporting file, Battle for A Shau, 69A729, box 14, WNRC.

[14] Msg, NMCC to MACV, 14 Mar 66; Msg, MACV to SECDEF, 16 Mar 66, III MAF files, Marine Corps Historical Center; Interv with Brig Gen Marion E. Carl, Investigative Officer for the House affair, 1969, Office of Marine Corps History; "U.S. Investigating Report of Treachery by the Vietnamese," *New York Times*, 16 Mar 66; [AP], "Yank Praises Nung Bravery in A Shau Fight," *Chicago Tribune*, 28 Mar 66.

[15] Msg, Saigon 3260 to State, 9 Mar 66, Westmoreland History, bk. 4, tab E-1, CMH; Msg, Saigon 3206 to State, 4 Mar 66, FAIM/IR; Westmoreland, *A Soldier Reports*, p. 169.

at last becoming used to Ky's government. If Thi responded by overthrowing the regime, the work of years would be lost.[16]

When Ky nevertheless decided to proceed, the State Department instructed Lodge to tell the press that the development was a routine change of command. That ploy deceived no one. Within hours, Neil Sheehan of the *New York Times* was telephoning Lodge for more information. Confronted by Sheehan's allegation that "a terrible row" was going on among the generals and that the United States was obviously facing a new era of instability in South Vietnam, the ambassador improvised a rationale. The removal of Thi, he told the reporter, was a big step toward stability, an act of strength and effectiveness, and a sign that the South Vietnamese could govern themselves. For the first time since the departure of the French, the central government had made a move against warlordism, imposing its will upon a strong and popular military commander in the clearest manner possible. Far from being a sign of weakness, Lodge said, the dismissal of Thi should be reported around the world as an act of courage.[17]

Lodge's prediction might have come to pass but for South Vietnam's militant students, who began demonstrating in Da Nang shortly after Ky dismissed Thi. The protest gained momentum over the next several days as the radical arm of the Buddhist movement, known for some time to have been plotting an antigovernment campaign, joined the disturbances. Forming a loose coalition called the Struggle Movement and protesting official corruption, inflation, and the government's disregard for the political rights of the people, the two groups carried the disorders to Hue, where they closed the schools and mobbed the radio station. Communist agitators joined in, attempting to magnify the unrest by egging the demonstrators on.[18]

Despite the increasingly anti-American tenor of the demonstrations and the possibility that the Communists were infiltrating the Struggle Movement, the State Department instructed Ambassador Lodge to maintain a neutral stance and to avoid making any comment to the press that might either diminish Ky's authority or give the demonstrators reason to believe the United States was working against them. In the same way, American officials were to avoid giving the impression that the U.S. mission and the South Vietnamese government considered all the agitators Communists or draft dodgers. Statements of that sort would merely deepen the protestors' anti-American bias and leave the United States vulnerable if the Ky government indeed fell.[19]

Given those instructions, the U.S. mission in Saigon would have preferred to say as little as possible about the crisis to the news media on grounds that the matter was a South Vietnamese problem best handled by South Vietnamese

[16] Msg, Saigon 3269 to State, 9 Mar 65, Westmoreland History, bk. 4, tab E-2, CMH.

[17] Msg, State 2653 to Saigon, 9 Mar 66; Msg, Saigon 3286 to State, 19 Mar 66; Msg, State 2673 to Saigon, 10 Mar 66; and Msg, Saigon 3288 to State, 10 Mar 66. All in FAIM/IR.

[18] CIA Information Cable, 23 Apr 66, Gard Papers, CMH; MFR, Westmoreland, sub: Meeting at Chu Lai, 24 Mar 66, Westmoreland History, bk. 5, tab B-1, CMH; Jeffrey Clarke, Advice and Support: The Final Years, 1965-1973, ch. 7 (hereafter cited as The Final Years), CMH MS, CMH.

[19] Msg, State 2862 to Saigon, 29 Mar 66, and Msg, State 2877 to Saigon, 29 Mar 66, both in FAIM/IR.

Buddhist Monk Blocks the Path of a Tank

spokesmen. Yet the Buddhists and students were cultivating the press assiduously, plying the Saigon correspondents with news releases and policy statements on all aspects of the situation while officials of the Ky regime maintained silence. Under the circumstances, the mission had little choice but to fill the vacuum. As Zorthian noted in response to questions from the U.S. Information Agency, unless the embassy made some attempt to provide perspective, correspondents would out of necessity turn to unreliable sources.[20]

Although a few inflammatory editorials appeared in the United States—the *New York Post*, for example, charged that the Johnson administration was supporting a hated military clique that persecuted Buddhists—most of the comments appearing in the press were restrained in tone. Agreeing with the State Department's line that the demonstrations were a cause of concern but appeared to pose little immediate threat to the Ky government, Neil Sheehan wrote that the bulk of the populations of Da Nang and Hue had nothing to do with the protests and appeared indifferent to Thi. Wesley Pruden of the *National Observer* accepted Lodge's view that the relief of Thi was a sign that the Ky regime might be matur-

[20] Msg, State 2743 to Saigon, 16 Mar 66, FAIM/IR; Msg, Saigon 1465 to USIA, 8 Apr 66, PPB 9, Press: Jan–Jun 66, 69A6507, box 7, WNRC.

ing. Dan Rather of CBS News noted that devout Buddhists were a minority of South Vietnam's population and that radical Buddhists constituted a minority within a minority. Robert Keatley of the *Wall Street Journal* observed that the relief of Thi had little significance in itself as far as the Buddhists were concerned and served merely as a pretext for demonstrations. What the Buddhists sought, Keatley added, was "the abolition of the military government in Saigon and their technique is, simply stated, blackmail the Saigon leadership: either surrender power or face violent and bloody street demonstrations."[21]

As the disorders continued and intensified, reporters covered each new demonstration in detail, dutifully recording allegations against the government, even the slogans on the demonstrators' signs. When three thousand predominantly Buddhist South Vietnamese Army troops joined the protests, marching through Da Nang behind a brass band, the reporters wrote of possible mutiny, terming the incident the most ominous development in the crisis. General Thi exacerbated the situation by ensconcing himself in Hue, where he issued a series of sarcastic statements to the press.[22]

On 24 March Ambassador Lodge warned Ky that continued disturbances without some corresponding effort on the part of the government to restore order would have a drastic effect upon American public opinion. During the first weeks of March, while demonstrations were occurring at Da Nang and Hue, the United States had lost 228 men killed, 850 wounded, and 2 missing. In the face of continued South Vietnamese "foolishness," he warned, the American people would never tolerate such heavy U.S. casualties.[23]

Although it leaked word that planning was under way to accelerate a return to constitutional government, the Ky regime apparently lacked enough of an internal consensus to make a formal declaration on the subject. Failing that, and in the absence of an official response to Buddhist and student grievances, demonstrations continued unabated. While conceding that Communists might have infiltrated street demonstrations, the *New York Times* editorialized on 4 April that a change of government in Saigon was clearly necessary. "The question is not if," the paper said, "but when—and how—a representative civilian regime can be established." Columnist James Reston meanwhile asserted that there was no cohesive national spirit in South Vietnam because there was no nation. The country was, he said, "a tangle of competing individuals, regions, religions, and sects, dominated by a single group of military warlords representing different regions,

[21] Msg, State 2743 to Saigon, 16 Mar 66; "It Happened Before in Saigon," *New York Post*, 17 Mar 66; Neil Sheehan, "7000 in Hue Call for Civilian Rule," *New York Times*, 15 Mar 66; Wesley Pruden, "Viet Regime Fires a General, Kills a Merchant," *National Observer*, 21 Mar 66; CBS Evening News, 25 Mar 66, Radio-TV-Defense Dialog; Robert Keatley, "South Vietnam Faces Political Crisis Again," *Wall Street Journal*, 17 Mar 66.

[22] Charles Mohr, "3000 Troops Lead Vietnam Protest Against Regime," *New York Times*, 3 Apr 66; [AP], "Ky Protests Spread Into Viet Capital," *Baltimore Sun*, 29 Mar 66; "Youths Taunt Marines Over Ky," *New York Herald-Tribune*, 31 Mar 66; [AP], "Anti-US Rally Held in Saigon," *Baltimore Sun*, 31 Mar 66; Msg, Saigon 3362 to State, 16 Mar 66, FAIM/IR.

[23] Msg, Saigon 3483 to State, 24 Mar 66, Westmoreland History, bk. 5, tab B-4, CMH.

an army without a country presiding over a people who have been torn apart by war and dominated and exploited . . . for generations."[24]

In an appearance on the ABC News program "Issues and Answers," the commandant of the Marine Corps, General Wallace M. Greene, attempted to dispel the gloom by stating that the situation at Da Nang was calm and that the demonstrations were "well ordered and semi-religious."[25] By 8 April the State Department's reports from the field were nevertheless far more pessimistic than anything that had thus far appeared in the American press. Government authority in the region north of Da Nang was almost nil, one message from Saigon warned. It had been replaced by "a rebellious combination of Buddhist control, student agitation, police and civil servant complicity, mob participation, undisciplined armed forces personnel in units and individually, and a numerically unknown but undeniable and increasing Viet Cong influence."[26]

As the disturbances grew, the U.S. mission became concerned that the press was paying little attention to anything but the crisis. Although the enemy had suffered sharp defeats and heavy casualties elsewhere in South Vietnam, officials of the U.S. mission told the State and Defense Departments, almost nothing had appeared in print about those engagements or about the fact that the enemy had increased infiltration through the Demilitarized Zone. To turn newsmen's attention away from the political situation, Washington agencies ought to highlight both of those subjects in briefings and press conferences.[27]

The State and Defense Departments carried out the suggestion, prompting a few stories on enemy infiltration, notably a front-page article in the *Washington Post* by Ward Just, who stated that North Vietnamese troops were "streaming into the Northern I Corps area in numbers . . . 'very close' to an invasion." The bulk of the news media's coverage nevertheless continued to concentrate on the disturbances, relating that Ky had sent three battalions of South Vietnamese marines on 5 April to occupy the air base at Da Nang and that troops of the rebellious 1st Division had responded by setting up armed roadblocks near the base. The headline—" 'Rebel City' Setting Up Defenses" claimed the *Washington Post*—made it seem as though all of Da Nang was about to explode.[28]

To the dismay of American officials, stories also began to appear suggesting that the crisis was affecting the conduct of the war. On 4 April Jack Foisie of the *Los Angeles Times* wrote that there had been "a noticeable decline" in the number of South Vietnamese military operations in the five northern provinces while U.S. operations had continued as usual. The next day, Seymour Topping of the *New York Times* observed that the campaign against the Viet Cong had begun to falter. Shortly thereafter, Andrew Hamilton of the *New York Herald-Tribune* noted

[24] Msg, State 2893 to Saigon, 30 Mar 66, FAIM/IR; "Vietnam: War Within War," *New York Times*, 4 Apr 66; James Reston, "Myths and Realities in Saigon," *New York Times*, 3 Apr 66.

[25] Richard Eder, "Top Marine Says Da Nang Is Calm," *New York Times*, 4 Apr 66.

[26] Msg, Saigon 3817 to State, 8 Apr 66, FAIM/IR.

[27] Msg, Saigon 3647 to State, 1 Apr 66, FAIM/IR.

[28] Msg, State 2923 to Saigon, 1 Apr 66, and Msg, State 2945 to Saigon, 2 Apr 66, both in FAIM/IR. Jack Foisie, " 'Rebel City' Setting Up Defenses," *Washington Post*, 5 Apr 66.

in a lengthy news analysis that some Department of Defense officials had become concerned that discord among the South Vietnamese might prompt the enemy to attack and annihilate a sizable American force to achieve a propaganda victory. The rioting, *Time* observed, was "a senseless and dangerous self-indulgence" for a country faced with possible extinction at the hands of an enemy.[29]

Publicly, the Johnson administration at first denied that anything was wrong. The riots were having little if any effect on either military operations or civilian programs, George Ball stated on the 11 April edition of "Face the Nation." The struggle between Ky and the Buddhist-student coalition was an almost necessary by-product of an emerging consciousness of politics, "simply a conflict as to the form of government which is best for the South Vietnamese people in order that they may continue the fight and resist the aggression from the North."[30] Privately, however, administration officials tended to agree with the press. The disturbances, Secretary Rusk told Prime Minister Thanat Khoman of Thailand, were having a serious impact on operations in the I Corps Tactical Zone.

The administration's public line held until the morning of 12 April. Then, aware that the Defense Department would shortly have to announce that for the first time American casualties for the previous week had exceeded those of the South Vietnamese, Arthur Sylvester acknowledged at a Pentagon briefing that the riots had indeed begun to disrupt the conduct of the war. South Vietnamese forces were mounting fewer operations against the enemy, Sylvester said, and a strike by Buddhist dock workers at the port of Da Nang was slowing delivery of bombs to American aircraft, reducing the number of tactical air sorties by about one-third.[31]

Sylvester's announcement came at a time when correspondents in South Vietnam were smarting after a recent attempt by U.S. military police to bar them from the scene of the rioting in Saigon. One of the MPs had allegedly even pointed a pistol at AP correspondent Peter Arnett. Although the U.S. mission had immediately withdrawn the jurisdiction of the military police over American newsmen, making reporters responsible for their own safety and subject to South Vietnamese law, the episode left a number of correspondents more distrustful than ever of official intentions. That mood increased when word began to leak that the bomb shortage Sylvester had attributed to striking Buddhist dock workers was in fact the result of a shortfall in production in the United States.[32]

When headlines began to play up the shortage, prompting congressmen to

[29] Jack Foisie, "Deadly Vietnam Contrast," *New York Journal American*, 4 Apr 66; Seymour Topping, "Crisis in Saigon Snags U.S. Effort," *New York Times*, 5 Apr 66; Andrew Hamilton, "U.S. Sees Viet Chaos Aiding Reds," *New York Herald-Tribune*, 11 Apr 66; "The Storm Breaks," *Time*, 14 Apr 66, p. 28.

[30] Msg, State 1733 to Bangkok, 2 Apr 66, FAIM/IR; Eric Wentworth, "Viet Fighting Unimpaired, Ball Asserts," *Washington Post*, 11 Apr 66.

[31] Max Frankel, "Vietnam Turmoil Now Slows War, Washington Says," *New York Times*, 12 Apr 66.

[32] [AP], "MP's in Saigon Draw Guns on U.S. Newsmen," *Chicago Tribune*, 9 Apr 66; Interv, author with Barry Zorthian, 22 May 79, CMH files; Arthur Krock, "Riddles of Our Vietnam Policy," *New York Times*, 19 Apr 66. For the extent of the bomb shortage, see Msg, Sharp to Westmoreland, 19 Apr 66, Westmoreland Papers, CMH.

charge that the government was misinforming the American people and that the United States appeared to be overextended in South Vietnam, General Wheeler cabled Westmoreland in dismay that sensation-seeking reporters were threatening the entire American buildup. While disavowing any attempt to conceal the truth from government officials, he wanted no more sensitive information falling into the hands of the press. "We Americans can do anything superbly," he said, borrowing a phrase from U. Alexis Johnson, "except keep our mouths shut."[33]

For a time the crisis in the northern provinces appeared to abate. On 10 April, reasoning that enough of a point had been made, Premier Ky withdrew his troops from Da Nang Air Base. Four days later, President Thieu announced that elections would occur within five months to select delegates for a constitutional convention. The Buddhists then agreed to allow the military coalition to remain in power until the civilian government framed by that constitution could take control. American military advisers meanwhile reported that, except within the I Corps Tactical Zone headquarters and the 1st Division, the Buddhists and students appeared to be making little headway in gaining support from the South Vietnamese armed forces.[34]

General Westmoreland took up the advisers' theme at a background briefing for the press on 30 April, when he emphasized that the disturbances were having at best a marginal impact upon the war. In the Mekong Delta, he said, the 7th Division had perhaps missed a chance to inflict heavy casualties on the enemy because one of its regiments had been in Saigon protecting the government. In the north, although South Vietnamese forces in the Da Nang region remained largely preoccupied with politics, most of the rebellious units of the 1st Division had returned to duty in the wake of a threat by the Military Assistance Command to withdraw U.S. advisers. The Viet Cong had meanwhile failed to take advantage of the situation, either because they were preoccupied with preparing an offensive for the post-monsoon season or because they hoped to gain political rather than military advantages from the disturbances.[35]

On the day Westmoreland gave his briefing, word arrived from Col. Nguyen Van Hieu, the chief of staff of the II Corps Tactical Zone and a highly respected officer, that despite the calm, all was not what it seemed. Hieu said that there was a very apparent deterioration of morale within South Vietnamese forces and less enthusiasm for offensive action than six months before. As a result, the army was in grave danger. The commander of the II Corps zone, Maj. Gen. Vinh Loc, had avowed in public that he favored neither Thi nor Ky, leaving his officers and enlisted men wondering, "If the commanding general is neither for the government nor for the Struggle Forces, who are we fighting and dying for?" Hieu went on to assert that the army was being systematically subverted by its Buddhist

[33] Joseph McCaffrey, "News Seven," WMAL-TV, Washington, D.C., 13 Apr 66, Radio-TV-Defense Dialog; Msg, Wheeler JCS 1974-66 to Westmoreland, 13 Apr 66, Westmoreland Papers, CMH.

[34] Clarke, The Final Years, ch. 7.

[35] Msg, MACV 15015 to CINCPAC, 2 May 66, Westmoreland History, bk. 6, tab A-1, CMH.

Police Encircle the Buddhist Institute

chaplains, who on several occasions had told various units to prepare to lay down their arms because the war was being fought for the good of the United States rather than that of the Vietnamese people. The situation was far more serious than it had been at any time in the previous two years, Hieu said, particularly since U.S. forces apparently did not recognize what was happening.[36]

Although the Chief of the Joint General Staff, General Cao Van Vien, confirmed most of Hieu's allegations, Westmoreland remained convinced that the situation was hardly as bad as it seemed. When U.S. casualties during the first week in May again exceeded those of the South Vietnamese, prompting criticism in the American press and inquiries from General Wheeler, Westmoreland responded that most South Vietnamese casualties resulted from enemy initiatives and that the Communists had launched few attacks of late. There were only slightly fewer South Vietnamese operations than usual, he said, and even the 1st Division, for all the Buddhist sympathies of its troops, seemed to be carrying out operations faithfully. Seeing no immediate crisis, Westmoreland flew to Hawaii on 12 May

[36] Msg, Saigon 4401 to State, 6 May 66; Memo, Zorthian for Lodge, 2 May 66, sub: Report of Conversation Between Brig. Gen. John F. Freund and Senior ARVN Officer; Memo, Freund for Westmoreland, 2 May 66. All in Westmoreland History, bk. 6, tabs G–4, B–2, and B–1, respectively, CMH.

to consult with Admiral Sharp. Ambassador Lodge meanwhile departed for consultations in Washington.[37]

With both Lodge and Westmoreland out of the country, Ky made his move. On the morning of 15 May, under the direct command of General Vien, South Vietnamese marines flew to Da Nang to secure the city's radio station, mayor's office, and military installations. Four days later, while Lodge and Westmoreland remained out of the country to avoid being linked with any developments, Vien forcibly ejected the dissidents from their pagodas in Da Nang. That move caused a series of confrontations with troops still loyal to Thi, but established effective government control over the city.[38]

As the Buddhists and their supporters lost ground they turned fanatical. A number of bonzes set themselves afire. Others issued rabidly anti-American statements, taking the line that the United States was responsible for what had happened and should intervene against Ky. Mobs burned the U.S. Information Agency library in Hue and ransacked the American consulate. That much of the violence was staged for the benefit of the American news media became obvious later in May, when the Buddhists on one occasion poured fresh gasoline on the corpse of a burning nun to provide better pictures for newsmen who had missed the start of the immolation.[39]

The violence nevertheless appeared to have the effect upon American public opinion that the Buddhists desired. On 20 May, shortly after Westmoreland returned to Saigon, General Wheeler informed him that questions were beginning to rise in Congress and the public about whether the South Vietnamese were serious in their efforts to maintain independence. "In all logic," Wheeler said, ". . . one cannot expect the American people to suffer indefinitely the continuation of this truly sickening situation." Although an optimist on most matters, he continued, "I think I can feel the first gusts of the whirlwind." Even if the contending factions achieved a semblance of solidarity and common purpose, the United States would have to recognize that it had "lost irretrievably and for all time" the support of some Americans. "Regardless of what happens of a favorable nature," Wheeler said, "many people will never again believe that the effort and sacrifices are worthwhile."[40]

Gallup polls at the time supported Wheeler's evaluation. Fifty-four percent of those queried said that if the strife evolved into large-scale warfare the United States should withdraw from South Vietnam. Another 48 percent said that they thought the South Vietnamese would never be able to establish a stable government. The American public, Secretary Rusk told Ambassador Lodge, appeared

[37] Msg, Wheeler JCS 2644-66 to Westmoreland, 12 May 66, and Msg, Westmoreland MACV 1529 to Wheeler, 12 May 66, both in Westmoreland Papers, CMH.

[38] Clarke, The Final Years, ch. 7; Westmoreland, *A Soldier Reports*, pp. 169-76.

[39] MFR, 20 Jun 66, sub: MACV Commanders Conference, 5 June 66, Westmoreland History, bk. 7, tab B-1, CMH. Msg, State 3697 to Saigon, 30 May 66, and Msg, Saigon 5149 to State, 31 May 66, both in FAIM/IR.

[40] Msg, Wheeler JCS 2837-66 to Westmoreland, 20 May 66, Westmoreland Papers, CMH.

to have concluded that the situation in South Vietnam was out of control and that the United States could do nothing.[41]

Westmoreland and Lodge attempted to allay the concern rising in Washington. The situation had been blown out of all proportion in the minds of the Johnson administration and the American people, Westmoreland told Wheeler, by the reports of correspondents attempting to make names for themselves by composing spectacular news stories on the crisis. Referring to the Gallup poll, Lodge told President Johnson that he was certain those who wanted to abandon South Vietnam were merely responding to insufficient information supplied by newspaper and television reports. "One television fireside chat," he told Johnson, "by you personally—with all your intelligence and compassion—could tip that figure over in one evening."[42]

Both Lodge and Westmoreland believed that the press accepted without question every assertion the Buddhists made. Yet as the crisis continued, the Saigon correspondents became increasingly wary of Buddhist claims. Richard Critchfield of the *Washington Star,* for example, pointed out that an antigovernment demonstration on 1 May, supposedly involving members of labor unions, in fact contained many of the same hired teenagers who had previously participated in Buddhist riots. Denis Warner noted that although the demonstrations in Saigon were in the hands of moderates and appeared relatively spontaneous, the Struggle Movement's propaganda in Hue appeared virtually indistinguishable from that of the Hanoi regime. Anyone who remembered the Viet Minh's tactics in 1945 would recognize that "every action, every manifesto, every decision taken in Hue seemed familiar." Quoting an authority on the National Liberation Front, Warner added that the Viet Cong had long ago penetrated the ranks of both the Buddhists and the students and were using the disturbances to compensate for recent military setbacks. "In Dalat, a small mountain resort where nearly everyone knows or recognizes everyone else," he said, "the demonstrators who wrecked the radio station were not recognized. They appeared not to be familiar with the layout of the town and when their task was finished they simply disappeared into the jungle."[43]

The reassertion of government authority in Da Nang cut the heart from the Buddhist movement. Although the Buddhists attempted to recoup on 7 June by blocking main roads in Da Nang, Hue, Quang Tri, and Qui Nhon, Ky sent units of the National Police to restore order. On 15 June he committed a four-battalion task force to reestablish full control over the last holdout positions in Hue and on 23 June sent other troops to seize the Buddhist Institute in Saigon, eliminating the last dissident stronghold.[44]

[41] Msg, Wheeler JCS 2844–66 to Westmoreland, 23 May 66, Westmoreland History, bk. 6, tab D–11; Msg, State 3575 to Saigon, 22 May 66, for Ambassador from Rusk, FAIM/IR.

[42] Msg, Saigon 4952 to State, 25 May 66, for President from Lodge, FAIM/IR. See also Msg, Westmoreland MACV 4070 to Wheeler, 22 May 66, and Msg, Westmoreland MACV 4081 to Wheeler, 22 May 66, both in Westmoreland History, bk. 6, tabs D–5, D–7, respectively, CMH.

[43] Denis Warner, "How Much Power Does Tri Quang Want?" *Reporter*, 5 May 66, p. 11.

[44] Clarke, The Final Years, ch. 7.

On 29 June, after the Buddhist crisis had ended, Ambassador Lodge admitted to President Johnson that, despite his earlier remarks, "In fairness to the press, and in fairness to our own effort, it must be known that the attitude of the press towards the Buddhists did undergo a marked change. Undoubtedly, this was due partly to their own observation, but I hope that the ceaseless backgrounding done here and in Washington played a part."[45]

Although the press had indeed turned against the Buddhists, the damage appeared to have been done. Gallup polls indicated at the beginning of June that 48 percent of those interviewed believed that the United States had made a mistake in sending troops to fight in South Vietnam, an increase of 11 percent since March. The figures were hardly indicative of a desire on the part of the American people to retreat from South Vietnam—many of those who considered the war a mistake might have wanted to end it by making an all-out effort to destroy the enemy. General Wheeler's apprehension that a whirlwind of disapproving public opinion might overtake the Johnson administration nevertheless appeared to have some foundation.[46]

The problem went deeper, however, than General Wheeler or anyone else in the Johnson administration realized. A comparison of public opinion trends during the Korean War with those of the Vietnam years would later indicate that public approval for each conflict, irrespective of press or television coverage, fell inexorably, in logarithmic progression, 15 percentage points each time U.S. casualties increased by a factor of 10 (100 to 1,000, 1,000 to 10,000, and so on). As American troops had moved eagerly to the offensive during the first months of 1966, in other words, they had incurred enough casualties to start the process. Although it would take more and more casualties to continue it, time had begun to gain on the United States and to lengthen for the enemy.[47]

[45] Msg, Saigon 5830 to State, 29 Jun 66, FAIM/IR.

[46] Erskine, "The Polls: Is War a Mistake," *Public Opinion Quarterly* 34 (Spring 1970): 134–50.

[47] The most thorough exposition on this subject is in John Mueller, *War, Presidents, and Public Opinion* (New York: Wiley, 1973). For a concise statement on the subject, see Mueller, "A Summary of Public Opinion and the Vietnam War," in Braestrup, *Vietnam as History*, app.

12

Political Attrition

Opposition to the war became more and more vocal in the United States toward the end of 1966. Believing the conflict a hindrance to the struggle of black Americans for racial equality, a prominent civil rights organization, the Southern Christian Leadership Conference, called during April 1966 for serious examination of whether the United States should withdraw from South Vietnam. Shortly thereafter, Senator Fulbright warned in a widely publicized speech that the United States was falling victim to the same fatal overextension of power and mission that had destroyed ancient Athens, Napoleonic France, and Nazi Germany. During May some eleven thousand protesters took to the streets in Washington, D.C., to show support for congressional candidates pledged to opposing the war; antidraft demonstrations occurred outside military installations across the United States; and slogan-chanting youths who denounced the role of American business in supporting the war briefly disrupted trading on the New York Stock Exchange. A Marxist professor of history at Rutgers University, Eugene D. Genovese, meanwhile made headlines by proclaiming that Communist ascendancy in South Vietnam was not to be feared but welcomed and that the United States would never experience racial harmony at home until it stopped making war upon the colored nations of the world.[1]

As antiwar rhetoric rose, a babble of conflicting advice and opinion on what the president should do next in South Vietnam assailed the Johnson administration. The Director of the Hudson Institute, Herman Kahn, advocated continuing on the same course in South Vietnam because Johnson's policies were the only ones capable of preserving American credibility around the world. Harvard University professor Henry A. Kissinger warned that American military tactics should change from large-unit operations against the enemy's main forces to securing populated areas. The military editor of the *New York Times*, Hanson W. Bald-

[1] For a brief survey of the antiwar movement's activities during this period, see Lester A. Sobel, ed., *Facts on File Yearbook, 1966* (New York: Facts on File, Inc., 1967), 26: 164, 184, 302.

win, advised the president either to go all out to win the war militarily or withdraw American troops from South Vietnam. Among the figures who opposed the war, University of Chicago professor Hans Morgenthau argued that the president should do everything he could to extricate U.S. fighting men from South Vietnam without losing face. Historian Arthur Schlesinger, Jr., advised Johnson to install a civilian government in Saigon and to reconvene the Geneva Conference to hammer out a settlement. White men, Schlesinger said, could never win a war on the Asian mainland.[2]

While activists demonstrated and intellectuals debated, the American public at large continued on its own course. Gallup polls reported on 6 July that although antidraft protests were occurring on college campuses, the vast majority of college students, both male and female, nevertheless favored the draft in principle even though college students in poor standing with their universities were subject to conscription. In the same way, although some 40 percent of the Americans interviewed in a September Gallup poll disapproved of the way the president was handling the war, that number included both those who wanted to slow the fighting and many others who urged that U.S. forces become more aggressive. By a margin of two to one, indeed, the American public favored rather than opposed continued involvement in South Vietnam. As pollster Louis Harris commented at the time, the dialogue was "not really between 'doves' and 'hawks,' but rather over what might be the most effective way to win our limited objectives and end the fighting."[3]

The Johnson administration sensed the American public's mood. Observing that spectacular military progress would be unlikely during 1967 because the South Vietnamese were adopting a caretaker mentality as the American effort grew, William P. Bundy warned that the public's impatience might well undermine the president's chances for reelection in 1968. Without clear signs of substantial progress in South Vietnam, he said, the Republican Party's candidate would be able to present clear and convincing alternatives to the policies Johnson had thus far espoused. The debate that followed would divide the American public, sapping the ability of the newly elected administration to carry on the war and convincing the enemy that the United States would almost certainly yield to Communist pressures in due course.[4]

Preserving the Public Image of the War, 1966

To demonstrate American resolve, the Johnson administration attempted throughout 1966 and 1967 to discredit its critics and to promote national unity

[2] "What Should We Do Now?—Five Experts Give Their Answer," *Look*, 9 Aug 66.

[3] George Gallup, "Most Students Like Present Draft Policy," *Washington Post*, 6 Jul 66; William Lunch and Peter Sperlich, "American Public Opinion and the War in Vietnam," *Western Political Quarterly*, Sep 69, p. 27; George Gallup, "Opinion Split on Viet Policy," *Washington Post*, 21 Sep 66; Louis Harris, "Confidence in Johnson on War Back to 42 Percent," *Washington Post*, 20 Sep 66.

[4] William P. Bundy, Working Paper, 21 Dec 66, sub: 1967 and Beyond, FAIM/IR.

whenever the opportunity arose. When the noted British philosopher Bertrand Russell announced that he would shortly convene an unofficial tribunal of leading international literary and political figures to investigate whether the United States had committed crimes against humanity in South Vietnam, the State Department took pains to discredit the inquiry by ensuring that a number of prominent individuals such as the president of Kenya and the emperor of Ethiopia disassociated themselves from the tribunal. In the same way, the State and Defense Departments instructed Lodge and Westmoreland to cultivate those visitors to South Vietnam who were well disposed to the war in order to enhance their pro-administration viewpoints. Among those who received special treatment were correspondent Howard K. Smith and author John Steinbeck. President Johnson himself occasionally took to the offensive. Confronting his critics head-on during a July 1966 campaign tour through the American midwest, he said, "There are people in our country who denounce air strikes against . . . North Vietnam, but . . . remain strangely silent when the communists in the South turn . . . mortars on an American hospital or blow up a busload of farmers. . . . I just wish they would ask themselves if their standard of judgment is really fair."[5]

In conjunction with the administration's efforts, the Military Assistance Command and the U.S. mission in Saigon strove continually to temper news stories that tended either to reflect poorly upon the South Vietnamese government or to embarrass President Johnson. During July 1966, for example, Westmoreland learned that a newsman from CBS was about to broadcast allegations that Prime Minister Nguyen Cao Ky was receiving kickbacks from the Saigon racetrack. He immediately instructed Bankson to contact the CBS bureau in Saigon to put the story into perspective. Bankson complied, emphasizing that the funds in question were being used for charitable purposes. He even prevailed upon the South Vietnamese government to open to CBS ledgers purporting to show how the money had been spent. Unconvinced, CBS broadcast a report that took Ky's guilt for granted, but Bankson's intervention created enough doubt to keep the matter from becoming a major scandal.[6]

The U.S. mission employed the same sort of approach later in 1966, when the Associated Press alleged that from 25 to 40 percent of the American aid reaching South Vietnam was showing up on the black market or being otherwise illegally diverted. The State Department proposed that the Military Assistance Command issue a formal rejoinder denying the charge, but Lodge and Westmoreland disagreed. Recognizing that any statement on the subject would have to refer to the corruption rampant within the South Vietnamese government, they chose instead to background the press on the dimensions of the problem and

[5] The main documentation detailing the effort to discredit the Russell tribunal is in FAIM/IR, in the Pol 27-12 Viet S files for 1966-1967. See also Msg, McNamara Defense 7222-66 to Westmoreland, 25 Nov 66, Westmoreland Papers, CMH; Msg, State 3919 to Saigon, 16 Jun 66, for Lodge and Westmoreland from Bill Moyers, FAIM/IR; William C. Wyant, Jr., "Remarks by Johnson in Midwest Depict War Critics as Unwilling To Back U.S. in Time of Crisis," *St. Louis Post-Dispatch*, 26 Jul 66.

[6] Westmoreland Historical Briefing, 10 Aug 66, Westmoreland History, bk. 8, tab B; [AP], "CBS Reports Ky Gets $15,000 a Week Racing Payoff, Called 'Charity' Fund," *Washington Post*, 29 Jul 66.

on the steps they were taking to find a solution. Within days, news stories based on the interview appeared in the press, all of them noting that sources in Saigon recognized the problem but discounted the allegation that up to 40 percent of American aid was disappearing. U.S. assistance to South Vietnam, the stories quoted those sources as saying, had succeeded at least temporarily in its primary task, halting the threat of destructive, runway inflation—something that could never have been done if that much property had been stolen.[7]

Whatever the merits of the argument, reliable information on the extent of corruption in South Vietnam was still largely nonexistent, giving official rebuttals delivered with authority and conviction at least as much validity in the public's eyes as newsmen's spectacular charges. Less easy to repair was another source of irritation to the Johnson administration, the recurring allegation by both the press and the Communists that U.S. and South Vietnamese military operations were slaughtering large numbers of civilians.

The Civilian Casualties Question Resurfaces

The issue had long been the subject of controversy. Morley Safer, for example, had used the theme in his report on the burning of Cam Ne. Official spokesmen attempted to divert the press to enemy atrocities with suggestions that newsmen examine those incidents to allow their readers and world public opinion to see how the Communists worked, yet the effort availed little. Not only were American activities more accessible to the press than enemy operations, Communist atrocities were also relatively commonplace in South Vietnam, leading newsmen to presuppose that the world at large was well aware of the enemy's terror tactics. The press therefore tended to dwell on events that seemed more worthy of special notice, where, for example, U.S. fire accidentally leveled a friendly village or killed innocent civilians.[8]

Noncombatant casualties first became an important political issue during the Fulbright hearings, when charges began circulating in Congress and the press that Operation MASHER/WHITE WING had produced six civilian casualties for every Viet Cong. The U.S. mission in Saigon responded that the numbers were exaggerated and that many civilian deaths were the result of Viet Cong tactics which drew noncombatants into the fighting. Besides recruiting civilians of both sexes as young as six years old to be intelligence gatherers, mission officials said, the enemy used civilians as proselytizers, porters, and assassins. The Communists also fired on American and South Vietnamese forces from populated areas in hopes of drawing return fire and causing an incident.[9]

[7] Hugh A. Mulligan, "Saigon's 'PX Alley' Offers Stocks of Glittering GI Goods," *Washington Post*, 16 Nov 66. Msg, State 85357 to Saigon, 15 Nov 66, and Msg, Saigon 11056 to State, 16 Nov 66, both in FAIM/IR. R. W. Apple, Jr., "Vast U.S. Aid Loss in Vietnam Denied," *New York Times*, 18 Nov 66.

[8] Msg, Saigon 3218 to State, 5 Mar 66, FAIM/IR.

[9] Msg, State 2749 to Saigon, 17 Mar 66, and Msg, Saigon 3594 to State, 30 Mar 66, both in FAIM/IR.

The Saigon Black Market

Secretary of Defense McNamara and General Wheeler both testified on civilian casualties before the Fulbright committee, but they found themselves hampered by a general lack of accurate information on the extent of the problem. As Ambassador Lodge noted in a cable to the State Department at the time, the Military Assistance Command was unable in many cases to determine who should be classed as a noncombatant. Were litter bearers and porters or people who planted mines for the enemy or grew crops exclusively for the Viet Cong to be considered civilians? Given the enemy's practice of removing the weapons of his dead, how were the black-clad corpses left behind after a battle to be identified as Viet Cong? Hundreds of structures and sampans were destroyed each day by air strikes and artillery, yet who was to know whether anyone was inside or if that person was an innocent civilian?[10]

The inadequacy of South Vietnamese and American record keeping complicated the problem. Prior to February 1966 the South Vietnamese armed forces maintained no statistics on civilian casualties and had difficulty counting even military wounded. The Military Assistance Command did require so-called backlash reports whenever civilians were killed or wounded by U.S. troops, but the

[10] Msg, Saigon 3594 to State, 30 Mar 66.

reports were compiled for only the most obvious incidents, when, for example, an artillery round fell short or jet bombers struck the wrong target. As for hospital records, those of the South Vietnamese Ministry of Health were almost useless until 1967, when American bookkeepers took charge, and those kept on civilians treated in U.S. military hospitals were generally too vague to reveal the source of an injury.[11]

So tangled did the problem seem that Ambassador Lodge told Secretary Rusk he saw no way around the impasse and had no wish to waste precious energies counting casualties when the war needed to be won. "It is hard here in Saigon to understand why there is so much distress in the U.S. about accidental killings by our side," he told William Bundy, "and apparently so little indignation about the fact that every time an American is killed here in Saigon [in a bomb throwing or terrorist attack] about twenty or thirty innocent Vietnamese women and children go down with him." Civilians had died during World War II, yet there had been no hue and cry about accidental killings because, on that occasion, "the extraordinary communist propaganda machine aimed at non-communist opinion, which is emerging today as just about the most successful communist activity, was not working against us."[12]

Westmoreland agreed with Lodge. When the State Department directed the U.S. mission to do all it could to provide accurate statistics for the sake of official credibility, the Military Assistance Command undertook a district-by-district survey of the problem but emphasized in its messages to Washington that designing an accurate system for counting civilian casualties was next to impossible. Refusing to become involved in what it termed "a statistical numbers game," the command continued to rely upon the wisdom and good sense of field commanders and on the rules of engagement that it had designed to protect civilian lives.[13]

The State Department nevertheless decided more could be done to improve American public opinion on the subject. At the beginning of August it dispatched Ambassador W. Averell Harriman to Honolulu with instructions to prod the military into developing a plan to spotlight Communist brutalities. The commander in chief, Pacific, notified Westmoreland of the development, suggesting that MACV's collection of photographs depicting enemy atrocities might be of use in attracting the press to the subject. Westmoreland responded that while the command was prepared to do more to publicize enemy atrocities, the Public Affairs Office was already releasing news and photographs of war crimes to the press, making extra coverage by the Saigon correspondents highly unlikely.[14]

[11] Msg, MACV 6769 to SECDEF, 3 Mar 66, DDI Statistics file; Msg, Saigon 3594 to State, 30 Mar 66. See also Guenter Lewy, *America in Vietnam* (New York: Oxford University Press, 1978), pp. 442f.

[12] Msg, Saigon 3594 to State, 30 Mar 66; Msg, Saigon 3404 to State, 19 Mar 66, for Bundy from Lodge, FAIM/IR.

[13] Msg, MACV 10144 to SECDEF, 31 Mar 66, DDI Statistics file. See also Msg, State 2749 to Saigon, 17 Mar 66, FAIM/IR.

[14] JCSM 4625-66, C/S CINCPAC to Westmoreland, 5 Aug 66, and Msg, Westmoreland MACV 6906 to CINCPAC, 10 Aug 66, both in Westmoreland Papers, CMH.

There was little, in fact, that Westmoreland could have done at that time to distract the press from civilian casualties, if only because mishaps were continually occurring.On 1 July, for example, American fighter bombers accidentally struck a friendly village, killing seven civilians and wounding fifty-one. A week later U.S. aircraft bombed and strafed a unit of South Vietnamese civilian irregulars, killing fourteen and wounding nineteen. Other incidents followed, culminating on 9 August, when F-100 fighters attacked a village in the heavily populated IV Corps Tactical Zone after receiving permission from South Vietnamese authorities. Sixty-three noncombatants died and another eighty-three were wounded. In all, misdirected artillery fire, armed helicopter attacks, equipment failures, and improperly conducted air strikes took more than eighty lives during July and August 1966 and wounded more than 250 persons, most of them civilians. To make matters worse, on the night of 11 August U.S. Air Force fighters mistakenly attacked the U.S. Coast Guard cutter *Point Comfort*, killing two American seamen and wounding five.[15]

Recognizing that many of the incidents were the result of a failure by American military officers and South Vietnamese civilian officials to follow MACV's procedures for controlling U.S. firepower, the American news media seized upon the accidents as an indication that many U.S. officers had become negligent in the way they fought the war. While admitting that mistakes were bound to happen in any conflict, Alex Dreier of ABC News charged that there was too much of a margin for error in South Vietnam. A *New York Times* editorial shortly thereafter added that the bomb tonnage being dropped on South Vietnam each week exceeded that dropped on Germany at the height of World War II. The social structure of the countryside was a shambles as a result, with the Communists the only beneficiaries. In a separate article *New York Times* correspondent Charles Mohr observed, "When harm comes to civilians or property, an inquiry is made. The usual official answer is that 'if it happened, it is against policy.' But critics doubt the value of an official policy that is widely ignored." He quoted an unidentified American military officer as saying, "I never saw a place where so many military orders are disobeyed as in Vietnam."[16]

Mohr's comments, along with other news stories on the subject, aroused considerable attention in Washington, where General Wheeler, Deputy Secretary of Defense Cyrus Vance, and other officials of the Johnson administration were meeting to discuss the problem. The group concluded that most of the incidents had taken place because a large number of experienced officers had just returned to the United States after serving one-year tours in South Vietnam, leaving behind new men who were as yet unfamiliar with MACV's rules of engagement. A number of South Vietnamese civilian officials, insensitive to the needs of their people, also seemed willing to approve every American request for an

[15] Westmoreland Briefing for the Press, 20 Aug 66, DDI MACV Backgrounders file.

[16] Alex Dreier, ABC News, 16 Aug 66, Radio-TV-Defense Dialog; "Civilian Casualties in Vietnam," *New York Times*, 21 Aug 66; Charles Mohr, "U.S. Acts To Save Vietnam Civilians," *New York Times*, 17 Aug 66.

Smoke From a Petroleum Storage Facility *near Hanoi hit by U.S. bombers.*

air strike, whether or not civilians were at risk. Notified of the group's conclusions, Westmoreland responded that he agreed and that he had already taken steps to reindoctrinate the men of his command in procedures to prevent what he called "the misapplication of friendly fire."[17]

At a meeting with the Saigon correspondents called both to emphasize MACV's concern for civilian casualties and to answer the questions appearing in the press, Westmoreland outlined some of the steps the command was taking to bring the problem under control. After presenting an analysis of the human and material failures involved in the incidents, he said that all air strikes in the IV Corps Tactical Zone would henceforward be approved by the office of the South Vietnamese corps commander, with civilian authorities at the province level excluded from the approval process except in emergencies, when troops were in direct contact with the enemy. Other measures included the establishment of a board of inquiry to review all of MACV's rules of engagement and the addition of a qualified South Vietnamese Army observer to forward air control flights whenever possible. Throughout the presentation Westmoreland and the other briefing officers made it clear that many things went wrong in war and that the incidents in question, as Westmoreland put it, "did not suggest any policy or procedural inadequacies." Neither Westmoreland nor the reporters present at the briefing made mention of the fact that U.S. forces in South Vietnam used their firepower with extreme liberality. That the shells, bullets, and bombs thus expended might sometimes miss their targets and hit civilians appears to have been everyone's foregone conclusion, an unavoidable fact of war.[18]

The Air War in North Vietnam Escalates

Westmoreland's presentation helped reduce the criticism surrounding the incidents in question, but the news media's interest in civilian casualties

[17] Msg, Wheeler JCS 4484–66 to Westmoreland, 17 Aug 66, and Msg, Westmoreland MACV 7176 to Sharp, 19 Aug 66, both in Westmoreland Papers, CMH.

[18] General Westmoreland Press Briefing, 24 Aug 66, DDI MACV Backgrounders file; Westmoreland Diary, 1 Sep 66, Westmoreland History, bk. 8, tab D.

American POWs Paraded Through Hanoi

continued, fueled by the air war in North Vietnam. That war had been going on for nearly two years but had rarely if ever encroached upon heavily populated areas until June 1966, when President Johnson approved strikes against petroleum storage facilities located in North Vietnam's Hanoi-Haiphong area. Analysts at the State Department argued against the raids on grounds that the United States had always claimed it would never escalate the war unilaterally and that there were some 1,000 people per square mile in the target area where before there had never been above 130. President Johnson nevertheless decided to approve the attacks, believing that the destruction of the enemy's fuel depots would shorten the war by impeding the flow of men and materiel into South Vietnam.[19]

Conducted between 27 and 30 June and accompanied by a careful public relations campaign, the attacks sparked antiwar demonstrations around the world but caused few public relations problems in the United States. While the *Washington Post* commented that the raids were long overdue, Harris polls reported that the American people approved by a margin approaching five to one. The Hanoi government, for its part, attempted to attract sympathy by claiming that U.S.

[19] Memo, Robert E. Patricelli for the Undersecretary of State, 6 Apr 66, FAIM/IR.

aircraft had indiscriminately bombed and strafed residential and economic areas, causing human and material losses to its people. The Military Assistance Command countered the allegation by pointing out that the strikes had been on target and that U.S. forces had taken every precaution to avoid damage to civilians. Hanoi itself diverted public attention in the United States by releasing photographs showing an angry mob of North Vietnamese jeering a parade of captured American pilots—a blatant violation of the 1949 Geneva Conventions which prohibited the exposure of prisoners of war to public curiosity and abuse.[20]

There matters stood until early December, when the United States, after several postponements because of bad weather, inaugurated Operation ROLLING THUNDER 52, a program of air attacks against targets within a ten-mile radius of Hanoi. Exceeding in concentration even the June and July strikes, the attacks stood out sharply against the decline in U.S. air operations in North Vietnam that had occurred during October and November.[21]

Recognizing that the operation would be controversial, General Wheeler instructed Westmoreland to refrain from depicting the strikes as a substantial increase in the level of the air campaign against North Vietnam. In part the product of the Johnson administration's continuing desire to avoid inflaming either pro- or antiwar sentiment in the United States, the request was due mainly to Wheeler's own conviction that, as he told Westmoreland, "any time we undertake a slightly different or increased initiative, it is characterized by those opposing U.S. policy as 'escalatory.' " Since "*escalation* has become a dirty word, such charges, true or false, impose further inhibitions here against moving ahead to win this war."[22]

The State and Defense Departments agreed, instructing the Military Assistance Command to point out in post-strike briefings for the press that the types of targets involved were all in categories struck previously. Should the enemy claim that civilian lives had been lost, information officers were to respond that, as in the past, the United States had striven to avoid hitting those scattered and small populated areas that existed in the vicinity of targets. To ensure that official spokesmen would be on strong ground in making that claim, the Joint Chiefs of Staff instructed the commander in chief, Pacific, to exercise extraordinary precautions to avoid civilian casualties. Only experienced, carefully briefed pilots thus participated in the strikes, which occurred only in weather that permitted positive visual sightings of both the target and the delivery of ordnance.[23]

[20] "Oil Targets," *Washington Post*, 30 Jun 66; Louis Harris, "Bombing Raises LBJ Popularity," *Washington Post*, 11 Jul 66; Bernard Gwertzman, "Ball Sees Little Chance of Russia, China Action," *Washington Star*, 30 Jun 66. The photographs in question appeared in the *Washington Post* on 1 July 1966, pages 1, 16. See also [AP], "TASS Says Hanoi Parades U.S. Pilot," *Washington Post*, 30 Jun 66.

[21] Memo, Robert H. Wenzel, S/S-O, for Benjamin H. Reid, 3 Mar 67, sub: ROLLING THUNDER Patterns in Late 1966, FAIM/IR.

[22] Msg, Wheeler JCS 6926-66 to Westmoreland, 11 Nov 66, Westmoreland Papers, CMH; Memo, William P. Bundy for Secretary of State, 25 Aug 66, sub: Proposed ROLLING THUNDER Program and Escalation Generally, Chron files, CMH.

[23] Msg, Joint State/Defense 83718 to Saigon, 12 Nov 66, and Msg, JCS 7735 to CINCPAC, 11 Nov 66, both in FAIM/IR.

The first attacks occurred on 2 and 5 December with more following on the thirteenth and fourteenth. The main targets were a large railroad marshaling yard at Yen Vien, 5.5 nautical miles northeast of Hanoi, and a huge vehicle depot at Van Dien, 5 nautical miles to the south. The North Vietnamese government reacted routinely to the first raids, but on 13 December Radio Hanoi announced that American jets had escalated the war by attacking Hanoi's suburbs and residential areas. Shortly thereafter, United Press International reported that the Soviet news agency TASS was making the same claim.[24]

In keeping with a policy of avoiding reaction to statements broadcast by Radio Hanoi, the U.S government at first refused to comment on the charges. Later, when the TASS report began to circulate, the State Department urged the Department of Defense to issue a communique refuting the implication that U.S. aircraft had escalated the war by deliberately attacking Hanoi. Defense complied, releasing word late on 13 December that "the only targets scheduled for attack in the Hanoi area during the last twenty-four hours were military targets which had been previously struck."

The statement might have satisfied the press but for the word of western travelers in Hanoi who confirmed that residential areas had indeed suffered damage. Lacking conclusive evidence to prove or disprove the claim, the Military Assistance Command refused to confirm or deny the reports and pointed out that there were no indications bombs had fallen on anything but military targets. The State Department took the same approach but immediately encountered reporters determined to establish whether any bombs had fallen within Hanoi's city limits, a development that the press considered tantamount to escalation. Unable to define Hanoi's boundaries, the State Department's Office of Public Affairs took a day to produce a map which showed clearly that the city's limits excluded all of the targets struck during the raids.

At least one reporter disputed the assertion, pointing out that during the period of colonial rule the French had defined Hanoi's boundaries broadly enough to include the targets, but by then much of the press was concentrating on the possibility that pilot error had damaged Hanoi. State Department spokesmen confirmed on 14 December that errors were always possible in war but immediately put the affirmation on background lest it become some sort of official admission of guilt. The next day, two days after the controversy had begun, State Department spokesmen allowed publicly for the possibility that houses near targets might have been damaged, but Deputy Secretary of Defense Cyrus Vance negated the effect of the admission by telling reporters shortly thereafter that available evidence provided "no basis for the allegation that any U.S. bombs fell on Hanoi." Vance added that much of the damage in question might have been caused by Communist surface-to-air missiles falling back to earth after missing their targets. Only on 22 December did the Johnson administration state unequivocally that

[24] This section is based on Directorate of Defense Information, Chronology of Government Statements and Comments, December 13, 14, 15, and 16 [early 1967], DDI file 33-a (NVN). See also Phil G. Goulding, *Confirm or Deny* (New York: Harper & Row, 1970), pp. 52–92.

it could not completely rule out the possibility of an accident. In an apparent attempt to placate world public opinion, it added, "If, in fact, any of our ordnance caused civilian injuries or damage, we regret it." Privately, the administration issued orders prohibiting American pilots from striking within ten miles of Hanoi.[25]

While official spokesmen formulated their position, the American news media turned ROLLING THUNDER 52 into a *cause celebre*. Although commentators were willing to accept that bombs had fallen on civilians in North Vietnam, they objected to the fact that, as the *Chicago Tribune* pointed out, the Communists appeared to have been "more truthful than the Washington news managers, who resorted to a series of denials and evasions and only confessed the facts after they had been found out." The *Kansas City Star* told its readers that even the argument blaming most of the damage to Hanoi on spent enemy missiles was an excuse based on a technicality. "World opinion will judge," the journal charged, "that raids within five miles of a population center would inevitably imperil civilians." Alexander Kendrick of CBS News added that the entire controversy was grounded in President Johnson's poor credibility and might have been avoided completely if the administration had conducted an open and aboveboard news policy. David Brinkley of NBC News agreed. "For two days and more," he said, "the communist charge . . . went around the world with no clear word of denial or explanation from Washington, and it is doubtful that the denial will ever completely catch up with the original report."[26]

The military saw the matter in a different light. "We were just starting to put some real pressure on Hanoi," Admiral Sharp wrote Wheeler in frustration on Christmas Eve of 1966. "Our air strikes on the rail yard and the vehicle depot were hitting the enemy where it was beginning to hurt. Then, Hanoi complains that we have killed a few civilians, hoping that they would get a favorable reaction. And they did. . . . Not only did we say we regretted it if any civilians were killed but we also stopped our pilots from striking within ten miles of Hanoi. Hanoi has been successful once again in getting the pressure removed. They will be encouraged to continue their aggression, hoping to outlast us."[27]

The Salisbury Affair

The outcry over the bombing of Hanoi appeared to be running its course when the first of a series of news reports filed directly from North Vietnam appeared

[25] Quote from Memo for the Secretary of State, 30 Dec 66, sub: Updated Chronology of Public Statements on Air Strikes in the Hanoi Area, FAIM/IR; Msg, Sharp to Wheeler, 24 Dec 66, Westmoreland Papers, CMH.

[26] "Managed News Again," *Chicago Tribune*, 28 Dec 66; "Vietnam and the Crisis of Credibility," *Kansas City Star*, 16 Dec 66; "The Kendrick Report," CBS News, 15 Dec 66, Radio-TV-Defense Dialog; "Huntley-Brinkley Report," NBC-TV, 16 Dec 66, Radio-TV-Defense Dialog.

[27] Msg, Sharp to Wheeler, 24 Dec 66, Westmoreland Papers, CMH.

in the 26 December edition of the *New York Times*. Written by the *Times'* assistant managing editor, Harrison E. Salisbury, the articles contradicted official American assertions about ROLLING THUNDER 52.[28]

Salisbury had first applied to enter North Vietnam six months earlier but had received no response until December, when the Communists abruptly decided to validate his visa. On his arrival in North Vietnam, he interviewed the country's premier, Pham Van Dong, and toured a number of towns and villages where North Vietnamese spokesmen said American bombers had harmed civilians. The reports that followed took up that theme, implying that the United States was striking civilian areas regularly. According to Salisbury, the town of Nam Dinh, an industrial complex some seventy-five kilometers south of Hanoi, had been the target of U.S. bombers for over sixteen months, even though its mayor swore it was a textile-producing center of little military significance. Describing "block after block of utter desolation," Salisbury claimed that the city's population of ninety thousand had been reduced to less than twenty thousand because of evacuations. Thirteen percent of its housing—the homes of 12,464 people—had been destroyed, at an expense of 89 civilians killed and 405 wounded. The reporter added that, besides destroying much of Nam Dinh, American bombers had hit the nearby Bao River dike six times, with many more near-misses.[29]

Describing the strike on the Van Dien vehicle depot, Salisbury compared what he called "the ground-level reality of United States bombing" to the version dispensed by U.S. official spokesmen. Van Dien was, he said, a large, open area situated just east of North Vietnam's Highway 1, with light buildings and compounds that "may or may not have been a truck park." The destruction there was far more general than MACV's news releases had implied, extending along both sides of Highway 1 and covering an area of "probably a mile or so on both sides of the highway." Among the structures destroyed was the Vietnam-Polish Friendship Senior High School, "lying . . . three quarters of a mile from the presumed United States target." Salisbury added that the North Vietnamese believed the United States was using raids on military objectives to disguise a policy of attacking civilian targets.[30]

In subsequent articles the reporter reiterated those themes, observing that air raids during October had destroyed every house and building in Phu Ly, a town fifty-five kilometers south of Hanoi. In the same way, American aircraft had leveled the countryside in the vicinity of the 17th Parallel. All of those attacks had been to no avail, Salisbury said, for despite the violence of the U.S. bombing program, North Vietnam repaired its roads and facilities with such remarkable

[28] Harrison E. Salisbury, "Visitor to Hanoi Inspects Damage Attributed to American Raids," *New York Times*, 26 Dec 66.

[29] Salisbury, "Raids Leave Blocks Razed, Fail To Cut Lines to Hanoi," *New York Times*, 27 Dec 66; Msg, State 111162 to All American Diplomatic Missions, 31 Dec 66, sub: Articles by Harrison Salisbury on North Vietnam, DDI North Vietnam (33) file.

[30] Salisbury, "Raids Leave Blocks Razed, Fail To Cut Lines to Hanoi."

Harrison Salisbury *(Photograph taken in 1984.)*

speed that supplies and war materiel continued to move along the nation's highways and railroads with relative ease.[31]

Salisbury's dispatches set off a furor in the United States, where opponents of the war accepted them at face value while prowar advocates urged even heavier bombing. Although Senators Richard Russell and George Smathers of the Senate Armed Services Committee urged President Johnson to forget world opinion and bomb Hanoi flat, many congressional leaders expressed embarrassment over the revelations. Congressman John Moss of California, for example, said he was irritated by the confusing way in which the Defense and State Departments had handled the bombing story and promised an investigation. Senator Vance Hartke of Indiana, a critic of the war, called for a halt to all bombing, whether in North or South Vietnam, and for the curtailment of large-unit ground operations in the South, all for the sake of preserving civilian lives. Senator John Pastore of Rhode Island was unwilling to go as far as Hartke but also advocated an immediate, unconditional halt to the bombing of North Vietnam to keep the conflict from widening into a thermonuclear war.[32]

The same sort of split appeared in the press, where conservative reporters backed the military while many of the rest expressed dismay. George Hamilton Coombs and Fulton Lewis III of Mutual Radio News, for example, considered the question of North Vietnamese casualties overblown. Coombs said that if the U.S. Air Force had intended to bomb Hanoi or any other city, "there wouldn't be enough standing to see from either ground or air." Lewis pointed out that while civilian casualties in North Vietnam were the result of pilot error or malfunctioning equipment, the enemy in South Vietnam waged an unrestricted terror campaign that purposely failed to discriminate between soldiers and civilians. Max Lerner of the *New York Post*, on the other hand, observed that the only way to make Pentagon releases square with Salisbury's reports without

[31] Msg, State 111162 to All American Diplomatic Missions, 31 Dec 66; Memo, Maj Gen John C. Meyer, USAF, for the Asst Sec Def (Legislative Affairs), 30 Dec 66, sub: Evaluation of Bombing of National Route One, DDI Operations file.

[32] Jed Stout, "Hanoi Area Bombing Inquiry Set," *Washington Post*, 30 Dec 66; Joseph Sterne, "Cease Bombing, Pastore Urges," *Baltimore Sun*, 30 Dec 66; Henry L. Trewhitt, "Viet Bombing Stirs Furor," *Baltimore Sun*, 28 Dec 66.

Salisbury's Photo of Damaged Civilian Areas

assuming deception by the government was to accept either a high degree of bombing inaccuracy by the Air Force or carelessness on the part of air crews. "Either is possible, or perhaps both," Lerner said. "But if so, we ought to be told." Walter Cronkite of CBS News was more caustic. Noting that Salisbury's revelations had widened the credibility gap, he added that there was no more flagrant example of the administration's poor public relations than the Pentagon's announcement that American bombers "were *not* attacking population centers in North Vietnam."[33]

Although the United States had never pursued a policy of bombing strictly civilian targets, U.S. reconnaissance photographs dating back to October 1966 showed clearly that civilian structures in Nam Dinh and Phu Ly had been damaged. Since Salisbury had seen enough of that destruction, as one State Department officer noted, "to lend credence to the wider assertions that are made concerning civilian casualties," the Defense and State Departments had little choice but to admit publicly that some damage had occurred.[34] They immediately

[33] George Hamilton Coombs, Mutual Radio News, 28 Dec 66, and Fulton Lewis III, Mutual Radio News, 28 Dec 66, both in Radio-TV-Defense Dialog; Max Lerner, "The Bombings," *New York Post*, 28 Dec 66; CBS Evening News, 27 Dec 66, Radio-TV-Defense Dialog.

[34] Msg, State 111162 to All American Diplomatic Posts, 31 Dec 66.

attempted to counteract any adverse reaction that might develop by pointing out that it was impossible to distinguish between damage caused by bombs and damage resulting from falling antiaircraft shells and spent missiles. When briefing the press on the measures the United States had taken to avoid civilian casualties, they also emphasized that all targets had been carefully selected and that in some cases the size of the bombs dropped had been reduced to preserve civilian lives. In letters to concerned congressmen and foreign dignitaries, the Office of the Assistant Secretary of Defense for Public Affairs noted as well that if some of Salisbury's facts were indeed correct, many others had been drawn from North Vietnamese propaganda pamphlets without attribution.[35]

Salisbury's failure to reveal his sources in his first reports ultimately told the most against him. Taking North Vietnamese allegations at face value, the reporter questioned whether Nam Dinh contained any military targets at all, implying that if they existed they were relatively insignificant. In fact, as Pentagon sources later announced, the city housed a petroleum storage facility, an important railroad marshaling yard, and a thermal power plant—installations so necessary to the enemy that he defended them with one of the heaviest concentrations of antiaircraft weaponry in North Vietnam. In the same way, the *Times* printed photographs by Salisbury purporting to show the ruins of the Roman Catholic cathedral at Phat Diem, the implication being that the United States was even bombing churches. Aerial reconnaissance pictures nevertheless showed clearly that the real cathedral had not been touched. Although Salisbury alleged in one dispatch that the countryside in the vicinity of the 17th Parallel had been devastated by American bombs, he had, in fact, never visited the area and had obviously received his information from the North Vietnamese. Worst of all, official spokesmen charged, he had delayed for two days before revealing that most of the statistics he had used in describing the destruction at Nam Dinh had been drawn directly from a North Vietnamese propaganda pamphlet entitled "Report of U.S. War Crimes in Nam Dinh City."[36]

The rebuttals had some effect. The *Washington Post* cast doubt on Salisbury's reliability as soon as it learned that the reporter had used propaganda in composing his dispatches. The *New York Times* itself appears to have had second thoughts. Besides allowing an article by Hanson W. Baldwin disputing Salisbury's conclusions to appear on page one, the paper came very close to disavowing the reporter. In an editorial entitled "The Tragedy of Vietnam" it rejected as false "the sweeping denunciations and false conclusions many Americans seem to have drawn from the statistics of civilian deaths and the pictures of destruction reported from Hanoi last week." Although stating that it remained critical of both the bomb-

[35] William P. Bundy, Material To Be Passed to Mr. George Brown, 30 Dec 66, Chron files, CMH; Neil Sheehan, "U.S. Concedes That Bombs Hit Civilian Areas in North Vietnam," *New York Times*, 27 Dec 66; Richard Fryklund, "Every Care Taken To Spare Civilians, Pentagon Says," *Washington Star*, 28 Dec 66; Goulding, *Confirm or Deny*, p. 63; Ltr, Goulding to Honorable Ogden Reid, House of Representatives, 30 Dec 66, DDI Air Incidents/Policy file.

[36] Msg, State 111162 to All American Diplomatic Posts, 31 Dec 66.

ing and the Johnson administration's public affairs policies, the paper added that its concerns were quite different "from saying there is even a shred of evidence . . . that the United States is deliberately bombing civilian targets. . . . The targeting restrictions in the North have been so precise and definite that the military feel some American pilots have given their lives because of them."[37]

Official rebuttals and the *Times'* retrenchment notwithstanding, Salisbury's allegations gained a wide audience throughout the United States. Interest remained high, in part, because the Defense Department never knew what Salisbury was going to say until it appeared in print and therefore could make clarifications and rejoinders only after the reporter had fanned the flames and gone on to other subjects. In part, it was also due to the fact that the Johnson administration continually gave ground on the reporter's allegations. While early in the controversy official spokesmen would concede only grudgingly that American bombs might have injured a relatively small number of civilians, by its end they were admitting, for example, that during one strike on the Yen Vien railroad yard three bombs had fallen on the target and forty outside.[38]

There were, of course, reasons for keeping detailed information of that sort from the press. The release of reconnaissance photographs or the regular tallying of how many bombs were on target could conceivably have given the enemy too much information about American capabilities. Yet by concentrating on the fact that only military targets were involved and by emphasizing the measures the United States was taking to minimize civilian casualties without spending any time on mistakes, U.S. offical spokesmen were really telling only one side of the story and preparing the ground for the controversy that was certain to arise when both sides became clear. Had the Johnson administration done a better job of explaining the nature of its bombing, Phil Goulding later stated, there might never have been an outcry.[39]

In all, Salisbury wrote some fourteen reports from Hanoi and another eight from Hong Kong summarizing his trip and his conclusions. When he returned to the United States he appeared in televised interviews, testified before Congress, and undertook a speaking tour. Confronted on several occasions by questions about his failure to identify his sources, he admitted to having made a "rudimentary error" but passed it off as "not very consequential."[40]

MACV's Statistics Questioned, March 1967

The debate on the effectiveness of the air war continued into 1967. Prompted by Salisbury's dispatches from Hanoi, opponents of the war charged that

[37] "The Tragedy of Vietnam," *New York Times*, 2 Jan 67; "Salisbury 'Casualties' Tally With Viet Reds," *Washington Post*, 1 Jan 67.
[38] [AP], "U.S. Admits Heavy Damage to North Viet Civilian Areas," *Baltimore Sun*, 22 Jan 67.
[39] Goulding, *Confirm or Deny*, pp. 90f.
[40] Howard G. Paster, "Salisbury Believes Hanoi Gates Open," *Editor and Publisher*, 21 Jan 67.

the bombing had done nothing to curtail Communist infiltration into South Vietnam and was serving mainly to stiffen the enemy's will to resist. An advocate of the war, syndicated columnist Joseph Alsop, contended on the other hand that the Military Assistance Command was doing all it could to cut off any move "to bring the boys home" by concealing the fact that enemy infiltration had declined sharply in the previous year. Caught between the two positions, the Johnson administration could not provide a clear picture of enemy strength because MACV's statistics failed to tally with those of intelligence agencies in Washington. Thus it began to cast about once again for some means of strengthening its political position.[41]

General Westmoreland offered little solace to those who sought to prove a decline in enemy infiltration in order to prove that the war was going well. When questioned by General Wheeler on the subject, he pointed out that the air campaign had become a significant hindrance to the movement of enemy supplies into South Vietnam but had done almost nothing to curtail the infiltration of combat units, most of which traveled at night. As for the failure of MACV's statistics to agree with those of Washington agencies, long periods of time, he said, were necessary before the command could assess infiltration rates correctly since the capture of pertinent documents and prisoners of war often lagged months behind the date of an infiltrator's arrival in South Vietnam. Although the command continued to look for the sort of decreases in infiltration that the Johnson administration wanted, intelligence continued to verify that the enemy was doing all he could to flesh out his larger units.[42]

The Military Assistance Command and the Defense Department convened a special intelligence conference in Honolulu during February 1967 in an attempt to harmonize the command's estimates with those of other agencies, yet that effort rapidly ran afoul of the Johnson administration's determination to show progress. Shortly after the conference concluded, as the antibombing campaign continued in the United States, General Wheeler informed Westmoreland that the president had requested a white paper for possible release to the press illustrating some of the positive results bombing had produced. In addition, Secretary of Defense McNamara had suggested that Westmoreland hold a special news conference on the subject. Almost as an afterthought Wheeler added that Westmoreland might note in both the white paper and the session with the press that battalion-size attacks by enemy units had undergone a noticeable reduction during the previous year—an obvious indication, he said, that the combination of air action in North Vietnam and ground operations in the South had impaired the enemy's ability to mount large-unit operations.[43]

Westmoreland fulfilled the president's and McNamara's requests but said nothing about enemy battalion-size attacks on the grounds that the information

[41] Msg, Wheeler CJCS 947–67 to Westmoreland, 2 Mar 67, and Msg, Wheeler CJCS 15494–67 to Westmoreland, 2 Mar 67, both in Westmoreland Papers, CMH.

[42] Msg, Westmoreland MAC 1297 to Wheeler, 6 Feb 67, Westmoreland Papers, CMH.

[43] Msg, Wheeler CJCS 15494–67 to Westmoreland, 2 Mar 67.

Washington analysts had cited to Wheeler was based on incomplete preliminary reports. Far from decreasing, he told Wheeler, enemy battalion-size attacks had increased dramatically during the previous year, going from ten in January 1966 to twenty-five in January 1967.[44] Alarmed, Wheeler immediately asked Westmoreland to withhold all information on that development from everyone but those with an absolute need to know. "If these figures should reach the public domain," he warned, "they would, literally, blow the lid off of Washington."[45]

Wheeler described the reasons for his concern in a lengthy memorandum to Westmoreland. Although under the old system of counting, only 45 major enemy attacks had occurred during 1966, under the new one that number had risen to 174. In addition, it now seemed that U.S. and South Vietnamese forces had made some 385 contacts with enemy battalions during 1966, most of them because of enemy initiatives. The implications of those figures were obvious and dangerous, Wheeler said. Large-scale enemy attacks had been used as a major element in assessing the direction of the war for the president, the secretary of defense, the secretary of state, Congress, and, to some extent, the American news media. Since military planners had few other straightforward yardsticks with which to measure the tempo of organized enemy resistance, those statistics had, in fact, been emphasized. Assuming that battalion- and larger-size attacks were a reliable measure of the enemy's ability and desire to take the initiative, the new figures meant that

> despite the force build up, despite our many successful spoiling attacks and base area searches, and despite the heavy interdiction campaign in North Vietnam and Laos [enemy] combat capability and offensive activity throughout 1966 and now in 1967 has been increasing steadily, with the January 1967 level some two-and-one-half times above the average in the first three months in 1966. The comparison of battalion contacts resulting from friendly versus enemy initiatives . . . adds weight to this conclusion.[46]

Wheeler went on to question the process that had been used to arrive at the new statistics, suggesting that since the overall number of attacks remained the same under both the old and new systems, the MACV staff had changed the definition of battalion attack to make it more inclusive than in the past. As he saw it, "The crux of the matter is intent; that is, was the enemy offensively seeking contact or was he reacting to our offensive actions? In this context, who fires the first shot has no bearing." Warning that any attempt by the MACV staff to inflate the size of the war in order to justify higher troop levels could only result in "trouble for us all," Wheeler concluded that the entire matter would have to be reviewed by a special operations and intelligence team. "I cannot go to the president," he said, "and tell him that, contrary to my reports and those of the other chiefs as to progress of the war in which we have laid great stress upon

[44] Msg, Westmoreland MAC 2344 to Wheeler, 10 Mar 67, and Msg, Westmoreland MAC 2450 to Wheeler, 14 Mar 67, both in Westmoreland Papers, CMH.

[45] Msg, Wheeler CJCS 1810-67 to Westmoreland, 9 Mar 67, Westmoreland Papers, CMH.

[46] Msg, Wheeler CJCS 184-67 to Westmoreland, 11 Mar 67, Westmoreland Papers, CMH.

the thesis [that] you have seized the initiative from the enemy, the situation is such that we are not sure who has the initiative in South Vietnam.''[47]

Westmoreland responded that while he welcomed the reviewing team, the statistics in question had resulted from the recent Honolulu intelligence conference and were considered the best available. Far from reflecting any inclination on the part of the Military Assistance Command to make the war seem bigger than it was, the new procedures gave results entirely consistent with the nature of the conflict. For if the total of battalion-size attacks had increased from 10 in January 1966 to 25 in January 1967, the total number of attacks in general, the same under both the old and new systems, had increased by the same proportion, more than doubling from 62 to 139 during the same period. That the enemy would generally exercise the initiative in launching battalion-size attacks was to be expected, Westmoreland concluded—especially in a war of the sort being fought in South Vietnam.[48]

Although the MACV chief of intelligence, Maj. Gen. Joseph A. McChristian, later avowed that the reviewing team upheld his procedures and statistics, Westmoreland in the end appears to have yielded to Wheeler's wishes. He reported to Admiral Sharp on 22 March that his staff and the reviewers had resolved all disputed issues by developing yet another set of definitions and formulas for assessing the enemy's combat initiatives. The practical result of the compromise was that nothing changed; the count of enemy battalion-size attacks during 1966 remained forty-five.[49]

Although Wheeler's objections to MACV's statistics appeared to be mere semantic quibbling, they were indicative of a malaise that by mid-1967 had worked its way into the American war effort. Westmoreland recognized it but seemed powerless to do anything. A full two months before MACV's record keeping came into question, he wrote in his diary, ''There is . . . an amazing lack of boldness in our approach to the future. We are so sensitive about world opinion that this stifles initiative and constantly keeps us on the defensive in our efforts to portray ourselves as a benevolent power that only acts in response to an initiative by the enemy. Therefore we become victims of our own propaganda and subject to political attrition.''[50]

Information officers in both South Vietnam and the United States would have agreed with Westmoreland's private assessment. Convinced that the war was

[47] Ibid.

[48] Msg, Westmoreland MAC 2450 to Wheeler, 14 Mar 67, Westmoreland Papers, CMH.

[49] Major General Joseph A. McChristian, *The Role of Military Intelligence, 1965–1967*, Vietnam Studies (Washington, D.C.: U.S. Army Center of Military History, Government Printing Office, 1974), p. 128; Msg, Westmoreland MAC 2715 to Sharp, 22 Mar 67, Westmoreland Papers, CMH. The statistic of 45 is in Office of the Assistant Secretary of Defense for Systems Policy and Information, Southeast Asia Statistical Summary, Table 2, CMH files. For details of the change in methodology referred to by Westmoreland, see Trip Report, DIAAP-42A, Maj Williams, 59708, 10 Apr 72, as cited in U.S. District Court Southern District of New York, General William C. Westmoreland v. CBS, 82 CIV. 7913 (PNL), *Plaintiff's Counter-Statement of Undisputed Material Facts, Annex B-Important Documentation Support of Plaintiff's Opposition to Defendant's Motion*, p. B398.

[50] Westmoreland Diary entry, Westmoreland History, bk. 13, tab A.

going as well as could be expected, they saw no need to hide anything and attempted to steer a middle course that respected the military's legitimate concern for security but also allowed the facts to speak. Early in 1967, for example, they concluded that the practice of characterizing American combat casualties as light, moderate, or heavy where single engagements were concerned had become a major threat to official credibility and that a return to the old policy of releasing numbers would be better. Although the Defense Department announced the names of U.S. missing, dead, and wounded daily and issued cumulative totals weekly, they reasoned that many persons in the United States and throughout the world believed that the U.S. government was attempting to conceal its casualties.[51]

The reports also tended to mislead the press. In the measurement used by the command, if a unit lost up to 5 percent of its members, casualties were announced as light. Losses from 6 to 15 percent were moderate and those above that heavy. Understood in context, the expressions were generally descriptive of a day's combat. Yet since most reporters had little knowledge of the size of the units involved in particular actions, eight-column headlines reporting heavy U.S. casualties sometimes appeared when only a platoon had been involved and ten men wounded. The result was a needless distortion of the war.[52]

Observing those trends and hearing the complaints of reporters in Saigon who believed the military was indulging in a cover-up, Colonel Bankson and the director of public information for the Office of the Commander in Chief, Pacific, Col. Willis Helmantoler, USAF, concluded that continued observance of the rule would destroy MACV's ability to deal with the press. They took General Westmoreland and Admiral Sharp aside during a reception honoring Sharp in Saigon and convinced the two that a change was necessary. At the same time Arthur Sylvester's successor as assistant secretary of defense for public affairs, Phil G. Goulding, notified Secretary McNamara that a problem of major proportions existed and that "no single step could do more to prove our credibility than to change this system and use numbers of some kind each night."[53]

Despite great concern at the Military Assistance Command that the revelation of casualty figures for individual actions would aid the enemy, the Defense Department decided to accept Goulding's suggestion. On 7 March Westmoreland notified the Saigon correspondents that the command would begin announcing casualty figures when significant operations occurred, if the existence of those operations had been revealed and disclosure would pose no danger to the units involved. The MACV Office of Information was to be the sole point of release for those figures, the only exception being an eyewitness account, in which case the reporter was to withhold unit designations lower than brigade, regiment, and group.[54]

[51] Memo, Phil G. Goulding for the SECDEF, 18 Feb 67, sub: Casualty Reporting, DDI Casualties file.

[52] Ibid.; OASD PA News Release, Fact Sheet-Casualties, 9 Mar 67, DDI Casualties file.

[53] Memo, Goulding for the SECDEF, 18 Feb 67, sub: Casualty Reporting, DDI Casualties file; Interv, author with Rodger Bankson, 13 May 80, CMH files.

[54] Memo, Capt J. N. Williams, USN, Acting Special Assistant for Southeast Asia, for Goulding,

As headlines, news stories, and editorials across the United States applauded the change, the MACV Office of Information sought further to improve relations with the press by issuing a revision of MACV Directive 360-1, which governed public information policies within the command. For the most part a restatement of rules already in effect, the new directive attempted to eliminate or smooth over potential sources of misunderstanding. Because disputes had arisen in the recent past over who should pay for medical treatment when a correspondent was wounded in the field, it reminded reporters that they themselves would be responsible if their employers refused to pay. In the same way, since reporters continued to complain about the length of time it took for news to make its way from the field to Saigon, the directive explained that the decision on whether to declassify a story was rarely what took the time. Instead, the gathering, transmitting, and checking of information from widely scattered locations when official lines of communication were saturated necessarily slowed the flow of information. Since newsmen also complained that the Military Assistance Command was releasing exaggerated statistics on enemy casualties, the directive conceded that the enemy-killed figures released each evening were inflated by duplication and error but noted that enemy soldiers killed by artillery or air strikes or who died of wounds were rarely counted, making the figure conservative in the long run.[55]

The M16 Rifle Controversy

Although the Saigon correspondents refused to concede that the body count was correct, they sympathized with the information officers' efforts and reciprocated by giving the Military Assistance Command the benefit of the doubt in many cases where they could have been far more critical. A case in point occurred during late 1966 and early 1967, after the United States decided to arm all U.S. troops in South Vietnam with a new automatic rifle, the M16. When rumors began to circulate in early 1967 that the weapon was jamming in combat and costing lives, a flurry of news stories appeared, but there was very little sustained critical comment of the sort that had occurred a year earlier when Morley Safer had revealed the burning of Cam Ne.

Television coverage of the problem, more extensive than that of the print media, exemplified the attitude of the press. First reports stated in a factual manner that there was some dissatisfaction with the rifle in South Vietnam and that at least one young marine had charged in letters home that practically every one of the American dead in a battle near the Demilitarized Zone had been found beside a jammed M16. When a special subcommittee of the House Armed Ser-

27 Feb 67, sub: Casualty Reporting; Msg, MACV 7835 to USARV, 7 Mar 67. Both in DDI Casualties file. OASD PA News Release, Fact Sheet-Casualties, 9 Mar 67.

[55] Msg, Defense 8642 to CINCPAC, COMUSMACV, 13 Mar 67, DDI Casualties file; MACV Directive 360-1, Public Information Policies and Procedures, 29 Mar 67, ans. A and E, CMH files.

vices Committee began an investigation into the charges, more comment occurred, but almost none of it by correspondents based in Saigon.[56]

Typical of the reporting that did originate in South Vietnam were two pieces by CBS correspondent Murray Fromson. In the first, after interviewing a young marine who expressed little confidence in the weapon, Fromson switched to the Marine commander in South Vietnam, Lt. Gen. Lewis Walt, for what became, in effect, a rebuttal. Observing that morale would decline sharply if the troops believed they were armed with an inferior rifle, Walt described the M16's advantages in jungle combat and stated flatly that the rifle was "the finest weapon our Marine Corps has ever been armed with."[57] In the second piece, Fromson attempted to investigate the rifle's performance. He presented more negative comments by young marines but then interviewed an Army ordnance expert who confirmed that the M16 indeed jammed if cleaned improperly but noted that the Army would shortly introduce a chrome-plated barrel to remedy the problem. At the end of the report, Fromson filmed a test in which an M16 was compared with one of its predecessors, the M14. After both weapons had been buried in a pile of sand, a marine retrieved and fired each. The M14 jammed immediately while the M16, equipped with a standard dust cover, functioned properly. Fromson concluded from that limited evidence that the M14 was "less dependable than the M16, exactly the opposite of what many marines report from the field."[58]

Although the reporter's conclusion was technically correct, there were nevertheless problems with the M16. According to evidence taken by the House Armed Services Committee, the propellent used in the rifle's ammunition had been altered without corresponding changes in the rifle's design, a condition that increased the weapon's rate of fire and caused it to jam. Because of the new propellent, the M16 also required more frequent cleaning than other rifles—a nuisance minor in itself but major in light of the fact that sufficient stocks of cleaning equipment had failed to reach some units in the field. As a result, the House Armed Services Committee heard allegations that at least a few marines had improvised cleaning rods out of old clothes hangers.[59]

All that information provided excellent material for exposes in the press, yet little of it appeared in any consistent manner, either because reporters, out of goodwill for the military, believed the official line or because they had no wish, as General Walt had warned, to undermine the confidence of U.S. troops in the basic infantry weapon. Added to those motivations may well have been rumors heard by at least a few reporters that when the first M16s to arrive in South Vietnam had gone to Army units, Marine junior officers had attempted to bolster

[56] NBC Nightly News, 10 May 67, and CBS Evening News, 22 May 67, transcripts in CMH files; Interv, author with Maj Gen Winant Sidle, 5 Jun 73, CMH files.

[57] Transcript, Murray Fromson, CBS Evening News, 12 May 67, copy in CMH files.

[58] Ibid., 23 May 67, copy in CMH files.

[59] U.S. Congress, House, Committee on Armed Services, *Hearings Before the Special Subcommittee on the M-16 Rifle Program*, 91st Cong., 1st sess., 1967. See also ibid., *Report of the Special Subcommittee on the M-16 Rifle Program*, no. 26, 19 October 1967, pp. 530f, hereafter cited as *Report on the M-16 Rifle Program*.

the morale of their own men by claiming that the rifle was actually inferior to the older M14s the marines continued to carry. Thus, when the marines found themselves receiving M16s and began to complain of malfunctions, some reporters dismissed the story as the misfiring of internal Marine propaganda.[60]

Inexperience on the part of some newsmen may also have been a reason. Covering a 1st Infantry Division operation named SHENANDOAH II on 5 October 1967, CBS correspondent Don Webster discovered a pile of damaged M16 rifles stacked outside a headquarters tent. Since three Americans had been killed and twenty-seven wounded during an engagement the previous day, Webster inquired about the weapons, to be told that all had been damaged by artillery, helicopter gunship, or enemy fire. The reporter accepted the story. Although he transmitted a report to CBS charging that two of the Americans killed had carried enemy AK47s in preference to their own weapons, he made almost nothing of the real story, the pile of rifles. As then Special Assistant for Southeast Asia Col. Lucius G. Hill, Jr., informed Goulding, the explanation given Webster in the field would never have held up with a more skeptical newsman because it raised other questions. When each of the M16s was damaged by American or enemy fire, Hill asked, "was one of our soldiers using it at the time? Does this mean that our artillery and gunships fired at our soldiers? Or does it mean that the M-16 was discarded for some reason and was then hit by fire? . . . These questions would be hard to answer in light of . . . [the division's] version." Although Bankson's successor as MACV Chief of Information, Brig. Gen. Winant Sidle, quietly cautioned Webster about making grandiose conclusions on the basis of interviews with one or two disgruntled riflemen, the Defense Department decided, on Hill's recommendation, to say nothing more on the matter.[61]

Shortly after Webster's piece appeared, the House Armed Services Committee released the results of its investigation into the M16. Charging that the failure of Army officials to correct deficiencies before the rifle was sent to South Vietnam "bordered on criminal negligence," the report sparked another flurry of comments in the press but failed to generate much long-term interest. At least one correspondent, Bob Erlandson of the *Baltimore Sun*, attempted to resurrect the issue in December by revealing that a Marine battalion in South Vietnam had replaced 286 out of 445 M16s because of premature wear in the barrels, but the story failed to catch hold. Other stories appeared on the subject during 1967 and into 1968, but the issue never again became a matter of much public relations concern to the Army.[62]

That the news media failed to make much of the issue may have worked to the detriment of U.S. forces in South Vietnam, if only because the lack of an out-

[60] Interv, author with correspondent Frank Faulkner, 1974, CMH files.

[61] Interv, author with Sidle, 5 Jun 73; Daily Staff Journal, ACofS, G-3, 2 Oct 67, 69A6597, box 8, WNRC; Don Webster, CBS News Report, 7 Oct 67, Radio-TV-Defense Dialog. Msg, MACV 33265 to OASD PA, Defense, 10 Oct 67, and Memo, L. G. Hill for Goulding, 19 Oct 67, sub: Webster's M-16 Story, both in DDI Press Flaps file.

[62] *Report on the M-16 Rifle Program*, p. 537. MFR [Dec 67], sub: Erlandson's M-16 Report, and Msg, CG III MAF to CG FMFPAC, 1 Dec 67, both in DDI Press Flaps file.

cry allowed deficiencies in maintenance and support for the weapon to go uncorrected. On 8 February 1968, at the height of the Tet offensive, the Deputy Commander of the U.S. Army, Vietnam, Lt. Gen. Bruce Palmer, notified all of his subordinate commanders that a recent M16 rifle review panel had cast "serious doubt" upon statements by various commanders that maintenance of the weapon was up to par and that all the troops had proper training. Nineteen percent of the soldiers interviewed had never zeroed in their weapons, Palmer said, and 35 percent had yet to receive formal training in the rifle's use.[63] Twenty-three percent reported that there were no weapons inspections in their units; 63 percent said their ammunition and magazines were almost never inspected. Although the Army had required the replacement of the mainspring in all M16s and all U.S. component commanders had reported repairs completed, the panel had discovered that many of the weapons had yet to be touched. Commanders at all levels had reported that M16 malfunctions were insignificant, but 38 percent of the soldiers interviewed had reported malfunctions in the previous four months, mainly the failure of the weapon to extract spent shells. Meanwhile, although depot stocks of cleaning equipment were adequate, troops in the field continued to complain that cleaning materials were unobtainable. A few soldiers had even resorted to using diesel fuel and insect repellent in place of bore cleaner. Commenting that the situation, especially the inaccurate official reporting, was intolerable, Palmer ordered all units under his command to correct deficiencies in the weapon within two weeks and to see to it that each soldier received two additional hours of refresher training on the maintenance and firing of the rifle.[64]

Westmoreland Goes to the United States, April 1967

If the press gave the Army the benefit of a doubt on a strictly military matter such as the M16, it nevertheless mistrusted the Johnson administration and continued to watch carefully for signs that the president was attempting to embroil the military in domestic politics. Westmoreland and Zorthian sought to avoid trouble on that front by playing down the political justifications of the war required of them, but with dissent in the United States rising as the war lengthened, they found themselves being drawn, despite their better judgment, into the political arena. Twice during 1966, for example, President Johnson had suggested that Westmoreland give speeches in the United States to bolster prowar sentiment, and twice the general had declined in an attempt to keep his command above politics. By early 1967 that course was no longer possible. When Johnson asked Westmoreland in February to go to New York to address the annual meeting of the AP Managing Editors Association, the president phrased the summons in

[63] Since every rifle is slightly different, each soldier has to "zero in," or adjust, his weapon's sights on a firing range to compensate for variances in drop and horizontal drift.

[64] Msg, Lt Gen Bruce Palmer, DCG USARV, to Maj Gen Hay et al., 8 Feb 68, sub: M16A1 Rifle, Westmoreland Papers, CMH.

Westmoreland Briefs Johnson in Washington

such a way that Westmoreland decided he had little choice but to comply. He took comfort in the fact that the trip might correct some of the misinformation he felt the press was spreading and assigned Bankson to write the first draft of the speech.[65]

The general's misgivings proved correct. During the speech he observed that the Communists had failed to understand the role that debate played in American democracy. Seeing every antiwar protest as evidence of crumbling morale in the United States, they thus tended to harden their resistance at the cost of American and South Vietnamese lives. As an afterthought, adverting to the burning of the American flag during an earlier antiwar demonstration in New York City's Central Park, he added that he and his men were "dismayed . . . by recent unpatriotic acts here at home."[66]

The speech caused an uproar in Congress and the press. Senator Fulbright charged that the Johnson administration was equating dissent with treason. Senator George McGovern of South Dakota agreed. Commenting that the president and his military leaders were attempting to lay the blame for their failures on

[65] Westmoreland, *A Soldier Reports*, p. 225; Interv, author with Rodger Bankson, 13 May 80, CMH files.
[66] Ibid. For the text of the speech, see "Westmoreland Address," *New York Times*, 25 Apr 67.

their critics, he charged that the Johnson administration had brought Westmoreland home to stifle criticism of the war. Far from aiding the enemy, McGovern continued, dissenters had exposed the contradictions, the falsehoods, and the resulting credibility gap which surrounded administration policy. The *Chicago Daily News* meanwhile asserted that "Dissent Is Not Treason"; the *New York Post* termed the speech a form of domestic psychological warfare; and Walter Lippmann suggested that the justification of administration policy was the province of politicians, not generals. Lippmann added that the Communists would undoubtedly find in Westmoreland's mission unprecedented confirmation that President Johnson knew he had failed to unite the country behind his policies, a fact that could only be of great comfort to them.[67]

As the outcry developed, fed in part by news that U.S. fighter bombers had struck targets within Hanoi's city limits for the first time since December 1966, almost as many congressmen and editorial writers defended Westmoreland as opposed him. The *Washington Star* considered the uproar "an astonishing display of nonsense." The *Washington Post* observed that the government of a free society might be unable to punish dissent but was under no obligation to refrain from reply and rebuttal. The *Denver Post* said that Westmoreland's comments about the effect antiwar demonstrations had on troops in the field "ought to stimulate reappraisal by many of those who protest against the U.S. presence in Vietnam." Although disagreeing with Westmoreland's remarks, Senator Mike Mansfield of Montana stated that the general had as much right to express his views as those who opposed administration policy.[68]

Westmoreland's speech before a joint session of Congress on 28 April served further to dampen criticism. Although there was speculation in both Congress and the press that the general had come to Washington to request additional troops and that his visit was meant to prepare the nation for enlargement of the war, many of his critics were impressed. Senator Mansfield called his review of U.S. strategy "soldierly." Senator Eugene McCarthy of Minnesota granted that it was an "objective appraisal." Senator Robert Kennedy of New York called it "a fine presentation."[69]

If Westmoreland's trip improved the Johnson administration's public relations, it nevertheless worked ultimately to the detriment of the MACV information program. As Goulding's Special Assistant for Southeast Asia, Col. Winant Sidle, observed, prior to the trip each of Westmoreland's Saigon backgrounders had been followed by a rash of favorable news stories, some of them reproducing

[67] Andrew J. Glass, "Senators Blast War Widening, Dissent Curbs," *Washington Post*, 26 Apr 67; "Dissent Is Not Treason," *Chicago Daily News*, 26 Apr 67; Walter Lippmann, "The Intervention of the General," *Washington Post*, 27 Apr 67.

[68] "Stifling Dissent," *Washington Star*, 27 Apr 67; "Meeting Dissent," *Washington Post*, 27 Apr 67; "Westmoreland's Challenge to Critics," *Denver Post*, 25 Apr 67; Joseph R. L. Sterne, "Mansfield Chides Dissenters," *Baltimore Sun*, 27 Apr 67.

[69] Richard Lyons, "General Is Cheered on the Hill," *Washington Post*, 29 Apr 67; John Herbers, "Congress Expects War Escalation," *New York Times*, 29 Apr 67; Memo for General Wheeler, 1 May 67, sub: Congressional Reaction to General Westmoreland's Speech, Westmoreland History, bk. 16, tab 1.

the general's comments almost word for word. After it, all that changed. Suspecting that Westmoreland had become a tool of the Johnson administration, newsmen replaced their favorable coverage with more skeptical appraisals. In that sense, by yielding to Johnson's wishes, Westmoreland had subjected himself and his information program to much the same sort of "political attrition" he had so decried earlier in the year.[70]

[70] Interv, author with Sidle, 5 Jun 73.

13

The Benefit of a Doubt

Although Westmoreland's trip to the United States was of some temporary benefit to the Johnson administration, major public relations problems remained in South Vietnam, most notably where the South Vietnamese government and armed forces were concerned. Continuing newspaper reports from Saigon charged that the South Vietnamese armed forces were failing to do their part in the fighting and that the ineptitude of the country's officials was retarding the effort to win the peasantry to the side of the government. The United States resorted to a number of expedients over the years in an attempt to remedy the problem, but time and again had to face the fact that it could accomplish little until the South Vietnamese took action.

Typical was an article by Ward Just that appeared on 6 November 1966 in the *Washington Post*. Describing conditions in Quang Nam Province, where the city of Da Nang was located, Just noted that although the province chief, who had received a Ph.D. in political science from Michigan State University, was honest and intelligent, he was a native of North Vietnam who had little in common with the people he supposedly governed. After only six months on the job he had taken a six-week sabbatical to lecture in the United States, suspending important social reforms until he returned. According to Just, the U.S. marines had become the only functioning government in Quang Nam, providing security and training local forces. The reason they were so important could be seen every Friday afternoon, when Marine units returned to their bases from the field "bone tired and dragging their butts," while freshly shaved and neatly dressed South Vietnamese junior officers headed toward the bright lights of Da Nang for the weekend. So pervasive had the marines' influence become, the reporter concluded, and so weak was that of the South Vietnamese government, that qualified observers believed the province would revert to Communist control within two weeks of the Americans' departure.[1]

[1] Ward Just, "Pacifying a Province," *Washington Post*, 6 Nov 66.

Generals Westmoreland and Wheeler Meet Ambassador Bunker

A center of resistance to the Saigon regime during the Buddhist crisis, still seething with discontent, Quang Nam was hardly typical of South Vietnam as a whole. Yet Just's conclusions had enough validity to disturb the Johnson administration. Benjamin Read, the executive assistant to the undersecretary of state, told Undersecretary Nicholas Katzenbach that any attempt to shore up the South Vietnamese by sending more Americans would undoubtedly create conditions throughout South Vietnam similar to the ones Just had described. "If we are willing to put in ten per cent because we are impatient," he said, " . . . will not our accelerated impatience prompt us later to put in twenty per cent, fifty per cent, one hundred per cent?"[2]

In the belief that the effort to win the peasantry to the side of the government, the so-called pacification program, was essential to progress in South Vietnam, President Johnson had already taken steps that he hoped would remedy the problem. Notifying the U.S. mission in Saigon in mid-October that he wanted to see "marked improvements" in pacification within ninety days, he all but threatened to remove the program from civilian control. Since it was extremely doubtful that anyone could meet his demands within the time allotted, General Wheeler immediately instructed Westmoreland to prepare for the inevitable.[3]

Improving the Image of the South Vietnamese War Effort

While the civilian agency established within the U.S. mission to coordinate the ninety-day effort, the Office of Civilian Operations, struggled to control the far-flung program, Zorthian and Westmoreland began a new public relations campaign to counteract the news media's negative reporting of pacification. As Zorthian refined statistics on constructive aspects of the program for release to the press, Westmoreland notified all American unit commanders to be alert for occasions to emphasize positive developments. "It should be the goal

[2] For Read's comment, see Memo, Jack Rosenthal for Nicholas Katzenbach, 6 Nov 66, sub: Ambassador Lodge's Response to Our Pacification Cable, FAIM/IR.

[3] Msg, Wheeler JCS 6339-66 to Westmoreland, Sharp, 17 Oct 66, Westmoreland Papers, CMH.

of each component's command information program," he said, echoing the CONARC memorandum of 1965, "to so indoctrinate our servicemen that they will 'talk up' civic action . . . not only when talking to newsmen but in their letters and on their return to the United States." Whenever possible, "progress should be attributed to South Vietnamese efforts. Only thus can we hope to dispel the frequent assertion at home that the effort in Vietnam is largely an American operation."[4]

Although the Office of Civilian Operations did some consolidating, ninety days proved too little time to make the changes Johnson wanted. Shortly after arriving in Saigon on 25 April, Lodge's successor, Ambassador Ellsworth Bunker, therefore announced that the Military Assistance Command would assume charge of the program. He renamed it the Office of the Assistant Chief of Staff for Civilian Operations and Revolutionary Development Support (CORDS).

Although a few positive news stories appeared, the attempt to improve the image of the pacification program also had little effect. Much of the press continued to concentrate on the program's failures. Wendell Merick of *U.S. News & World Report*, for example, applauded the decision to give Westmoreland control of pacification but observed that progress would continue to be painfully slow. With only nine hundred revolutionary development teams available in July 1967 to handle the more than fourteen hundred hamlets, even if the United States and South Vietnam doubled that number every year, many years would pass before there would be enough teams to cover the whole country. Merick asserted that U.S. forces would have to take control of the war if the United States expected to get anything done and that many American officers had already adopted a "one job at a time" mentality, preferring to see strictly military operations take precedence over pacification. *U.S. News & World Report* took up the theme again on 31 July, in an article entitled "The Truth About the War in Vietnam." In some regions, the magazine said, pacification was in danger of total collapse. In others, goals were being lowered, and the program seemed to be making little headway even in the area around Saigon. The situation in the II Corps Tactical Zone, where 80 percent of the population lived in so-called secure hamlets, exemplified the problem. At least one American officer there insisted that control was possible only because U.S. troops were present. The enemy would return, he said, just as soon as they left.[5]

The inability of the South Vietnamese armed forces to pursue the war with any vigor provided material for derogatory news stories and complicated MACV's effort to foster a positive view of pacification. An April 1967 article by Associated Press correspondent Peter Arnett typified the problem. Observing that "all of

[4] Ltr, Westmoreland to All Commanders, 22 Oct 66, sub: Command Emphasis on Revolutionary Development/Civic Action, Westmoreland History, bk. 10, tab B-1. See also Mission Council Action Memo 125, sub: Minutes of the Mission Council Meeting of 26 Sept 66, Westmoreland History, bk. 9, tab B.

[5] Wendell Merick, "A Way Out for U.S. in Vietnam War," *U.S. News & World Report*, 17 Jul 67, p. 28; "The Truth About the War in Vietnam," *U.S. News & World Report*, 31 Jul 67, p. 40.

the present Vietnamese generals fought on the French side in the Indochina war, and the . . . stigma of having once belonged to a defeated army has never really been erased from the Vietnamese officer mentality,'' Arnett said that the introduction of U.S. troops into combat had caused many South Vietnamese officers to relax their efforts. As a result, South Vietnamese forces were fighting even less than they had two years earlier, when American units had entered the war to avert a Communist victory. The South Vietnamese Army, Arnett concluded, lacked the strength, unity, and morale it needed to defeat the Communists.[6]

General Westmoreland disputed claims of that sort. During 1966, he told his commanders, the South Vietnamese armed forces had expanded by 50 percent. They ran the government and provided most of the country's district and province chiefs, even though they possessed but the thinnest veneer of leadership. Convinced that more had to be done to put the efforts of the South Vietnamese in a better light, the general consulted with Zorthian during January 1967 to develop a public relations program for the South Vietnamese armed forces that would work in tandem with the one on pacification.[7]

Rejecting the sort of hard-hitting publicity campaign that might have drawn undue attention to itself, information officers decided on a more indirect approach. If U.S. commanders cultivated a continuing awareness of South Vietnamese efforts to improve, they reasoned, numerous opportunities to point out legitimate successes would arise. To that end, they instructed American officers in the field to compile lists of colorful or outstanding South Vietnamese personalities and operations and to steer reporters in those directions. Besides advising their South Vietnamese counterparts on techniques for attracting favorable news coverage, those officers were also to accompany reporters into the field when possible to single out well-run South Vietnamese operations for special news coverage.[8]

In support of that program, the MACV Office of Information intensified its advice to the South Vietnamese armed forces on proper public relations techniques and increased its efforts to train South Vietnamese information officers. It also included South Vietnamese accomplishments in the short radio and television news clips it produced for release in the United States and began developing material on the subject for use in Defense Department briefings. To publicize South Vietnamese advances in technical areas, information officers likewise made a special effort to assist newsmen and authors in developing articles for specialized magazines such as *Aviation Week*, *Radio & Electronics*, and *American Rifleman*.

The command's information officers also specified the themes that were to predominate in conversations between military officers and newsmen where the South Vietnamese armed forces were concerned. Avoiding derogatory comments,

[6] Peter Arnett, ''South Viet Army Lacks Strength, Unity, Morale,'' *Washington Star*, 21 Apr 67.

[7] MFR, 9 Feb 67, sub: MACV Commanders' Conference of 22 Jan 67, Westmoreland History, bk. 12, tab D–11; Westmoreland Diary, 16 Jan 67, Westmoreland History, bk. 12, tab C; MACV Directive 550–3, 23 Jan 67, sub: Public Awareness of RVNAF Operations and Activities, DDI RVNAF 26–a file, hereafter cited as MACV Directive 550–3.

[8] The source for this section is MACV Directive 550–3.

officers were to characterize the South Vietnamese Army as "a body of individuals in varying states of training and efficiency but, by and large, capably and single-mindedly prosecuting the war." They were to depict South Vietnamese commanders as "able, dedicated" officers and the individual South Vietnamese soldier as a first-class fighting man. Everyone was to stress that the South Vietnamese Army had greatly improved in the previous year and that the nation's technical training was "at a high state of development in many areas."

Although attempts to improve reporting of the South Vietnamese portion of the war continued, little in fact changed. Most U.S. officers in the field were too busy to spoon-feed reporters and left them to their own devices. South Vietnamese field commanders continued to shun publicity, perhaps reasoning, as they had during the Diem years, that higher headquarters would interpret favorable news stories as an attempt on their part to curry favor with the Americans prior to launching a coup.

That the South Vietnamese were doing as well as the Military Assistance Command claimed was also difficult for many newsmen to see. During nine of the first thirty weeks in 1967, U.S. casualties outnumbered those of the South Vietnamese, and during the six months ending in May 1967, American casualties even exceeded the number of South Vietnamese youths drafted into the army. At any given moment, between 9 and 20 percent of the South Vietnamese Army's regular force personnel and from 17 to 30 percent of its regional and popular forces personnel were listed as deserters. The Saigon correspondents wrote of South Vietnamese successes but nevertheless tended to consider the problems they saw as by far the more important story.[9]

Westmoreland recognized that there were problems. On one occasion, he wrote in his diary of a South Vietnamese airborne operation in which the troops dropped improperly, left the field too soon, and, as he put it, "trampled the crops of a lot of angry peasants."[10] Yet, like General Harkins and Ambassador Nolting in 1963, he believed that the South Vietnamese needed encouragement as much as criticism and therefore tried to impress upon everyone the need for toleration. There were, he continually pointed out, extenuating circumstances. The enemy's major formations were located in border areas where U.S. troops operated, while better than half of South Vietnam's maneuver battalions provided security for the common people in built-up areas, a task for which they were better suited than Americans but which provided little opportunity for the large engagements that produced heavy enemy casualties. In addition, the South Vietnamese were suffering from the usual dilution of strength that accompanied the rapid expansion of an armed force. As new recruits poured into infantry units, support structures sagged, leaving artillery, transportation, and medical facilities insufficient to sustain large mobile operations of the sort the United States conducted. As

[9] Msg, Wheeler JCS 6105 to Westmoreland, 2 Aug 67, and Msg, Westmoreland MACV 7180 to Wheeler, 2 Aug 67, both in Westmoreland Papers, CMH; Msg, Westmoreland MACV 7757 to Frank Bartimo, 18 Aug 67, DDI RVNAF 26-a file.

[10] Msg, Westmoreland MACV 7180 to Wheeler, 2 Aug 67.

for the desertion problem, Westmoreland believed that most departures were temporary, the product of homesickness or of a soldier's need to help his family harvest crops. The South Vietnamese government's efforts to enforce regulations strictly while improving the care of military dependents and increasing pay and food allowances had reduced desertions during the first six months of 1967 by 50 percent.[11]

Westmoreland's protestations to the contrary, the ineffectiveness of South Vietnamese troops was readily apparent to any newsmen willing to do research in the field. *New York Times* correspondent Tom Buckley, after accompanying an operation in Long An Province just south of Saigon, compared a South Vietnamese platoon to a U.S. unit operating nearby. The South Vietnamese were well dressed and freshly shaved while the Americans were unshaven and dirty from many nights in the field. The Americans discussed the South Vietnamese ''with idle, humorous contempt. Nothing the ARVN did could surprise them any more. 'We're not heroes,' [they said], . . . 'but we stay and fight if we have to. If there's trouble today, you just watch the ARVN's. They'll *didi mow.*' (The phrase is corrupt Vietnamese, known to every G.I. It means, approximately, 'bug out,' or run away.)'' Later in the story, Buckley reported that the South Vietnamese unit indeed disappeared as soon as firing commenced.[12]

By mid-1967 the South Vietnamese Army's lack of aggressiveness had become almost a given attribute of the war. In July South Vietnamese battalion-size operations were lasting only eight battalion-days while those of U.S. forces ran thirty-nine. Although differences in mission accounted for part of the discrepancy, and, if body counts are to be believed, South Vietnamese troops tended to kill about as many of the enemy per day of combat as the Americans, South Vietnamese maneuver battalions in the field actually made contact with the enemy only 27 percent as often as U.S. maneuver battalions. Since they held their own when they did fight, even when they lacked artillery and logistical support, their failure to be aggressive apparently resulted less from cowardice than from poor motivation and the refusal of officers to make effective use of their men. The problem received wide play in the press, where David Halberstam, for example, told of a conversation he had held with a North Vietnamese Army major who had defected to the South. Asked what he could do if given command of a South Vietnamese Army battalion, the major responded, ''I could command a division in North Vietnam. But a platoon here, even a squad, I could not do that. What can you do? They have no purpose.''[13]

The low regard in which many U.S. officers held South Vietnamese troops further complicated the situation. As far as improving the South Vietnamese military was concerned, one American general noted after the war, ''we were really

[11] Msg, Westmoreland MACV 7757 to Frank Bartimo, 18 Aug 67.

[12] Tom Buckley, ''The Men of Third Squad, Second Platoon, C Company, Third Battalion,'' *New York Times Magazine*, 5 Nov 67, p. 32.

[13] OASD PA Fact Sheet [mid-1967], sub: RVNAF Effectiveness, DDI RVNAF 26–a file; David Halberstam, ''Return to Vietnam,'' *Harper's*, December 1967.

quite indifferent. Most of us did not want to associate with them."[14] The result was that South Vietnamese troops received minor roles in U.S. operations for the sake of appearances while Americans did the fighting and won the headlines.

That American general officers occasionally voiced their dissatisfaction anonymously to the press only made matters worse. On 7 August 1967, for example, as complaints about the South Vietnamese Army's failure to take the initiative were beginning to rise in earnest in Congress, an unidentified American general gave an interview to R. W. Apple of the *New York Times*. "Every time Westmoreland makes a speech about how good the South Vietnam army is," that officer said, "I want to ask him why he keeps calling for more Americans. His need for reinforcements is a measure of our failure with the Vietnamese."[15]

Decrying the disloyalty of the officer Apple had quoted, General Wheeler questioned the Military Assistance Command on the article. Westmoreland responded that "it is inconceivable to me that any general officer in Vietnam would make such a statement. Any general who is serving here or who has made an honest appraisal as a result of a professional visit could not come to the conclusion that a need for reinforcements is a measure of our failure with the Vietnamese. Progress is not failure and by every measure there is increasing progress." As for Apple, Westmoreland noted that the reporter worked mostly from his own personal contacts, rarely resorting to the MACV staff for material. "I have watched Apple become more critical and more argumentative during recent months," he continued. "Barring some dramatic and irrefutable turn for the better here, we can expect him to continue to play the role of doubter and critic. He is probably bucking for a Pulitzer prize."[16]

Westmoreland's assertions of progress notwithstanding, by August 1967 concern was increasing in the United States that the war was in stalemate. During July a number of articles appeared suggesting that the United States and South Vietnam had lost the initiative because few large operations of the sort that had taken place during 1966 were occurring. Comparing the North Vietnamese Army to that of South Vietnam, the *Christian Science Monitor* noted that if the enemy was in check he still went undefeated. *Time* meanwhile warned that the war was hardly going as well as the Johnson administration had hoped and that the gains made during the previous eighteen months might shortly disappear. Editors everywhere injected derogatory comments about the South Vietnamese Army. *Newsweek*, in particular, observed that even if progress had occurred, South Vietnamese troops "all too often . . . displayed stupendous ineptitude, as well as a distressing reluctance to fight."[17]

[14] Douglas Kinnard, *The War Managers* (Hanover, N.H.: University Press of New England, 1977), p. 92.

[15] Msg, Wheeler JCS 6336 to Westmoreland, 8 Aug 67, Westmoreland Papers, CMH. See also Msg, Wheeler JCS 6105 to Westmoreland, 2 Aug 67, Westmoreland Papers, CMH.

[16] Msg, Westmoreland MAC 7576 to Wheeler, 12 Aug 67, Westmoreland Papers, CMH. See also Msg, Wheeler JCS 6336 to Westmoreland, 8 Aug 67.

[17] Tran Van Dinh, "A Look at the Vietnamese Armies," *Christian Science Monitor*, 15 Jun 67; "The War: Taking Stock," *Time*, 14 Jul 67, p. 20; "The War in the Delta," *Newsweek*, 14 Aug 67, p. 28. See

General Westmoreland refused to agree that the enemy had gained the initiative. If that were so, he told General Wheeler, the more than thirty large operations under way during August and the more than five thousand small-unit actions undertaken each day would have had more success finding him. Instead, the enemy remained elusive, accepting major combat along the Demilitarized Zone and in the Central Highlands, where sanctuaries were available and lines of communication short, but for the most part contenting himself with attacks on South Vietnamese operations in support of civic action programs. The enemy held the initiative only momentarily, Westmoreland said, when he harassed U.S. and South Vietnamese base areas, actions that were relatively inexpensive to him but that generated a maximum of publicity.[18]

To clarify what was happening, Westmoreland once more resorted to a public relations initiative. Making haste carefully, as he put it, to avoid charges that the military was once again waging an organized propaganda campaign, he took pains to schedule on-the-record news conferences for his major field commanders so that they could put the situation in each of South Vietnam's corps tactical zones in proper context. He also ordered the command to make an extra effort to move reporters to the scenes of important South Vietnamese actions and to keep a running tally of South Vietnamese successes for release to the press.[19]

As the campaign developed, Admiral Sharp suggested that the Military Assistance Command allow selected South Vietnamese officers to participate in briefings designed to stress the effectiveness of the war effort. Westmoreland was open to the idea but in the end demurred. If the South Vietnamese took part, he told Sharp, it would have to appear to be entirely their own idea. Otherwise, the United States would open itself to the charge that it was conducting a propaganda campaign.[20]

The increased interest of the Military Assistance Command in the effectiveness of South Vietnamese operations attracted the Saigon correspondents to the subject but failed to produce the sort of stories Westmoreland had sought. In mid-September Peter Arnett observed that South Vietnamese inefficiency and lack of will were costing American lives—in the ambushing of convoys on roads supposedly guarded by South Vietnamese troops and in the shelling of U.S. installations by squads of guerrillas who moved freely past sleeping South Vietnamese sentries. South Vietnamese troops took Saturdays and Sundays off while their allies and the enemy continued to fight. In the same way, American battalions often fought through the night while most South Vietnamese units quit at sundown if they could. Finally, Arnett said, many South Vietnamese operations were staged in areas where no guerrillas were known to be or where only small numbers were present.[21]

also Msg, Westmoreland MACV 7180 to Wheeler, 2 Aug 67, Westmoreland Papers, CMH.

[18] Msg, Westmoreland MACV 7180 to Wheeler, 2 Aug 67.

[19] Ibid.

[20] Msg, Westmoreland MACV 7430 to Sharp, 8 Aug 67, Westmoreland Papers, CMH.

[21] Peter Arnett, "South Vietnamese Army Fights Five-and-One-Half Day Week," *Washington Post*, 17 Sep 67.

Other reports in the same vein appeared, especially in *Newsweek*. On 25 September the magazine printed an article by its bureau chief in Saigon, Everett G. Martin, who charged that U.S. claims of South Vietnamese progress were misleading. He quoted an American colonel who had just returned to South Vietnam after three years: "everyone must admit that militarily we are better off than we were three years ago. With five hundred thousand U.S. troops, more planes, and more artillery, we should be. Wherever U.S. troops occupy the ground, security is better. But otherwise, I don't see any change. All the old problems are with us." Noting that the U.S. command seized on every action in which the South Vietnamese Army as much as did its duty as evidence of improvement, Martin leveled the usual charge that South Vietnamese troops avoided the enemy. In corroboration he quoted a second American. "Their military intelligence is better than it was," that official said, "so they can avoid contact more efficiently."[22]

One week later, *Newsweek* returned to the attack with an insulting article by another correspondent, Merton Perry, entitled "Their Lions, Our Rabbits." Although he admitted that some effective South Vietnamese military units existed, Perry nevertheless said that poor officers, poor pay, and a lack of motivation had rendered the bulk of South Vietnam's army unwilling or unable to perform even the limited task of protecting rural civic action teams. Some 360 pacification workers had been assassinated to that date in 1967, while South Vietnamese Army detachments idled nearby. Claiming that a few units had given up their combat roles completely to supply American troops with beer, prostitutes, and laundry services, Perry added that the South Vietnamese armed forces were as sick as the society that had begotten them—riddled with "factionalism, corruption, nepotism, inefficiency, incompetence, and cowardice."[23]

The Military Assistance Command decided against attempting to rebut the Arnett and Martin articles. Although information officers disagreed vehemently with the conclusion that the South Vietnamese Army was becoming increasingly ineffective, they believed that a frontal assault would only attract attention to the argument. As it was, Westmoreland told the Joint Chiefs of Staff, Arnett's bias against the South Vietnamese was so well known that none of the Saigon correspondents had even bothered to query the command on the reporter's allegations.[24]

Perry's article, on the other hand, prompted the South Vietnamese government to ban all sales of the offending issue of *Newsweek*. The magazine's bureau chief in Saigon, Martin, informed the U.S. mission that Perry's editors in New York had significantly altered the tone of the piece by eliminating qualifications and explanations from the manuscript prior to publication. Because of those revisions and because the South Vietnamese appeared to be prepared to retaliate fur-

[22] Everett G. Martin, "Vietnam: Last Chance," *Newsweek*, 25 Sep 67, p. 64.

[23] Merton Perry, "Their Lions, Our Rabbits," *Newsweek*, 9 Oct 67, p. 44.

[24] Msg, Westmoreland MAC 8875 to Gen John McConnell, Acting CJCS, 20 Sep 67, Westmoreland Papers, CMH; Defense Information Fact Sheet [Oct 67], sub: Newspaper Article by Peter Arnett, DDI RVNAF 26–a file.

ther against the magazine, Martin said that he had advised his employers not to protest the ban. He requested that the U.S. mission do the same. Ambassador Bunker complied with the request but privately placed little credence in Perry's innocence. The reporter "attended recently one of a series of small, informal dinners I have been having for the press," he told Rusk, "and gave me the impression of being thoroughly disenchanted with everything here and I doubt whether it is possible for him to report objectively."[25]

The Village of Ben Suc

If the attempt to improve the image of the South Vietnamese armed forces faltered, the effort to convince the press of the effectiveness of the pacification program was hardly more successful. A case in point occurred during January 1967, when the U.S. command launched Operation CEDAR FALLS in an attempt to deny the enemy use of the Iron Triangle, an infamous Viet Cong base near Saigon reputed to contain an enemy regional headquarters. Since the civilian population of the region, especially the inhabitants of the village of Ben Suc, willingly supported the Communists, the Military Assistance Command decided to remove everyone to refugee centers and to turn the entire area into a free strike zone where allied forces could fire without clearance or hesitation.[26]

With Harrison Salisbury's allegations about the air war in North Vietnam gaining wide play in the press, initial news reports of CEDAR FALLS tended to fall on the inside pages of most newspapers, where they drew little attention. None of the stories concentrated on the fate of the people of Ben Suc. The *New York Times* published a routine article that relayed word from U.S. officers in the field that the people would be given new land and homes and would be better off than before. *Newsweek* noted that American troops had done their best to make the relocation as painless as possible and quoted approvingly a statement by a U.S. officer that "Charlie's monkey business in the Iron Triangle is going to be through for good." Criticism began to appear in the press only in July 1967, when an article on the operation by correspondent Jonathan Schell appeared in the *New Yorker*. Less a report of events than a scathing attack on MACV's way of making war, the piece rapidly became a source document for the antiwar movement in the United States.[27]

Schell's account of the operation was a catalog of everything critics of the war considered wrong with the pacification program. Ben Suc had been a prosper-

[25] Msg, Saigon 8347 to State, 12 Oct 67, FAIM/IR. See also Msg, Saigon 7987 to State, 12 Oct 67, FAIM/IR file.

[26] Statement of Lt Col R. L. Schweitzer, G–5, 1st Inf Div [Aug 67], sub: Operation CEDAR FALLS, CEDAR FALLS file, CMH.

[27] "Viet Cong Village To Be Bulldozed," *New York Times*, 11 Jan 67; "The Iron Triangle," *Newsweek*, 23 Jan 67, p. 85; Jonathan Schell, "The Village of Ben Suc," *New Yorker*, 15 Jul 67, p. 28.

ous village inhabited by healthy peasants, the reporter said, but in 1965, when the United States began escalating the war, all had changed. Although U.S. pilots and artillerymen routinely sought the permission of South Vietnamese authorities before bombing or shelling populated areas, Ben Suc's province chief had been an outsider with little knowledge of the area and little time on the job. As a result, careless bombing had reduced the center of the village to rubble, and many villagers had been injured by indiscriminate U.S. and South Vietnamese artillery fire. U.S. psychological warfare teams had dropped leaflets warning that death was imminent for anyone who continued to support the Viet Cong and depicting American weapons with teeth and claws devouring those who failed to rally to the government. As CEDAR FALLS progressed, males between the ages of fifteen and forty-five were removed from their families and taken to the provincial police headquarters for interrogation. Some were singled out as Viet Cong for the slightest of reasons—careful grooming, better than average clothing, or lack of a government identification card. Many were abused by their South Vietnamese Army captors.

The women and children left behind were meanwhile forced to fend for themselves. Although the plan for the operation called for the villagers to bring along all their possessions, Schell charged that many were actually allowed to take only what they could carry. Families with homes near where a government truck was parked could take almost anything they wanted, but those living at a distance, without men to help, could carry only clothing, cooking utensils, and one or two bags of rice. That South Vietnamese soldiers assisted the people in loading the trucks Schell reported as a source of wonderment to the U.S. troops present. "You saw it," one American officer reportedly guffawed. "The Arvins loaded those trucks. We've never seen anything like it."

By the time the refugees had arrived at their destination, a barren field near the village of Phu Cuong, Schell continued, "They had lost their appearance of healthy villagers and taken on the passive, dull-eyed, waiting expression of the uprooted." Further trials awaited. Although the military plan that had brought them to Phu Cuong had worked with precision, from the moment they had come under the jurisdiction of pacification officials little had gone well. Shelter and facilities to receive the people were almost totally lacking because South Vietnamese social service agencies had been informed of the operation only hours before. The camp that grew up was thus a disjointed affair created by hundreds of uncoordinated decisions by the large number of overlapping organizations in one way or another responsible for pacification. Along with the loss of dignity entailed in those circumstances came a new title for the villagers. At Ben Suc the people had been called "hostile civilians" to signify to visiting reporters that they all supported the Communists and deserved relocation. At Phu Cuong they became "refugees," a twist of semantics designed to imply that "they were not themselves the enemy but . . . 'the people,' fleeing the enemy."

Schell ended the article by describing the U.S. Army's destruction of Ben Suc. After bulldozers cut swaths "across the backyard fences, small graveyards, and

Viet Cong Captured During Operation CEDAR FALLS

ridged fields of the village," U.S. Air Force jets pulverized the rubble a second time in hopes of collapsing tunnels under the village too deep to reach by other means. The overall impression he had received, he implied, was one of guilt compounded by mindless destruction—as if, "having once decided to destroy it, we were now bent on annihilating every possible indication that the village of Ben Suc had ever existed."

Yesterday's news, the article had little impact in the United States but caused an uproar at the Military Assistance Command. The command claimed that CEDAR FALLS had been one of the most successful U.S. operations of the war to date and called upon officers present during the operation to refute Schell. Their reports stated that far from being an innocent village peopled by prosperous peasants, Ben Suc had been an underground city honeycombed with tunnels going deep into the earth. Near the village soldiers had discovered a well-equipped enemy hospital stocked with a large variety of medicines as well as vast stores of rice—far more than could ever have been consumed by the villagers alone. The rebuttals denied that indiscriminate bombing and shelling had occurred but refused to rule out the abuse of prisoners that Schell had apparently witnessed, noting merely that incidents of the sort were bound to happen

Soldiers Load Rice Captured During CEDAR FALLS. *Note U.S. markings on bag.*

in war. To Schell's contention that good grooming and better than average dress had been the main criteria behind the detention of some villagers, the reports noted that several of the individuals in question had turned out to be the highest-ranking enemy political and propaganda cadre ever captured. In all, CEDAR FALLS had produced 711 Viet Cong killed; 537 enemy deserters; 210 prisoners of war; 578 captured weapons; 60,000 rounds of ammunition; and 2,240 grenades, mines, and booby traps. Further evidence that Ben Suc had been a Communist redoubt for years could be found in the fact that the operation had provided the Military Assistance Command with an intelligence bonanza: some 285 pounds of enemy documents and reports that revealed the Communist side of many earlier engagements and compromised upcoming enemy operations in the Saigon region.[28]

If Schell had distorted the strictly military portion of the operation, relying on unflattering stereotypes of the South Vietnamese Army and the mistaken notion that the population of Ben Suc was composed of mainly innocent, victimized peasants, his description of the confusion at Phu Cuong was more

[28] Statements of Lt Col R. L. Schweitzer [Aug 67], sub: Operation CEDAR FALLS, and sub: Comments Relating to the 15 July 1967 *New Yorker* Article, "The Village of Ben Suc," CEDAR FALLS file, CMH.

accurate. Although the Military Assistance Command contested the reporter's charge that South Vietnamese efforts at the refugee center had been inept, it said little about his contention that the camp had grown up in disorder because that was true. Although the American director of pacification for the region, John Paul Vann, had protested, II Field Force kept knowledge of CEDAR FALLS from everyone, even the South Vietnamese government, until two days before the operation. Little was thus done either to prepare for the refugees or to keep lines of authority and communications from becoming tangled.[29]

A running feud that developed over several days between Vann and Maj. Gen. William DePuy, commander of the U.S. 1st Infantry Division and the officer responsible for evacuating Ben Suc, typified the confusion. DePuy was convinced that the Office of Civilian Operations could never handle so sophisticated an operation. At the planning conference preceding CEDAR FALLS he told Vann that civilian managers would "fall on their faces" as soon as they were called upon to do anything and that the U.S. Army would as usual have to do the job itself. Further confrontations followed. Contending that the camp at Phu Cuong was a disgrace and that the United States would face severe criticism if the press found out, DePuy even attempted to take control of the refugee center. Only the word of the commander of II Field Force, Lt. Gen. Jonathan O. Seaman, who toured the facility and decided Vann had matters about as well in hand as could be expected, forestalled the move.[30]

Many of the problems that occurred at Phu Cuong would, indeed, have come about even if the operation had been planned in detail weeks in advance. For example, Schell made much of the fact that the peasants' pigs were allowed to root at will through the camp. In fact, the peasants refused to leave the animals in holding pens because they mistrusted the camp's administration. The problem came under control only after several weeks, when Tet holiday feasting reduced the swine population. In the same way, only after the evacuation was well under way did it become apparent that the least stable elements of the population of Ben Suc had fled the village for Phu Cuong first. Used to living in squalor, they clustered near the main entrance of the camp, giving a bad impression to all visitors who passed through that gate. In later relocation operations, the first villagers to arrive at holding areas were either separated from one another or given accommodations where they would be less likely to attract attention.[31]

By late 1967, indeed, the Military Assistance Command had rethought the entire concept of relocating villages with Viet Cong sympathies. Although Schell's article had little to do with the decision, about the time the piece appeared an official report critical of U.S. handling of refugees was leaked to the press, spark-

[29] Director, Region III, OCO, for Province Representative, Binh Duong, 20 Feb 67, sub: After Action Report, Operation CEDAR FALLS, CEDAR FALLS file, CMH.

[30] MFR, 14 Jan 67, sub: Complaints of Maj. Gen. William DePuy Relative to OCO/GVN Performance on Handling Refugees During Operation CEDAR FALLS, CEDAR FALLS file, CMH.

[31] Memo, Director, Region III, OCO, for Province Representative, Binh Duong, 20 Feb 67, sub. After Action Report, Operation CEDAR FALLS, CEDAR FALLS file, CMH.

The Refugee Camp at Phu Cuong

ing an investigation by a Senate committee on refugees chaired by Senator Edward Kennedy of Massachusetts. That discussion as well as information from Viet Cong prisoners of war and deserters confirmed what many U.S. pacification officials had long believed: disgruntled villagers forcibly separated from their homes and the graves of their ancestors provided an excellent source of recruits for the enemy. Since South Vietnam's refugee centers were already overflowing and the country's government obviously resented the forced relocation of its subjects by foreign troops, the command yielded. In concurrence with the South Vietnamese Joint General Staff it issued a combined campaign plan for 1968 that instructed military commanders to avoid displacing any more people than necessary. Although refugees continued to flow into the country's cities voluntarily, from that time onward the Military Assistance Command sought to keep South Vietnam's peasants on the land, where they could support themselves.[32]

[32] MFR, 17 Jan 67, sub: Meeting With Gen. Thieu, 13 Jan 67, Westmoreland History, bk. 12, tab C-5; Westmoreland, *A Soldier Reports*, pp. 303f; Operations Report, Lessons Learned, 25th Division, Period Ending 31 Oct 67, p. 41, CMH files; MACCORDS, The Refugee Operation: National Overview, Dec 67, Refugees 1967 file, CMH.

Robert Komer Takes Charge of Pacification

With the establishment of the CORDS organization in May there seemed some hope that the rivalries and crossed lines of communication that had hampered the pacification program in the past would disappear. President Johnson assigned as director of the program one of his ablest aides, Robert Komer. Working within the MACV organizational structure but possessing ambassadorial rank in his own right, Komer moved vigorously to streamline his organization and to gain for his agency the assets and recognition it needed to survive in an environment where strictly military pursuits usually prevailed.[33]

With Civilian Operations and Revolutionary Development Support (CORDS) a part of the Military Assistance Command, discussions began on whether the MACV Office of Information should take responsibility for the agency's public relations. The arrangement seemed logical. Westmoreland had authority over the release of all information on his command, and the MACV Office of Information was his representative. In addition, MACV's information structure reached to battalion level, employing several hundred officers and enlisted men throughout South Vietnam in information-related capacities. On the other hand, although CORDS' information office employed various psychological operations officers part-time in each of the country's provinces and corps tactical zones, it existed only at the command level and could muster but one true information officer, Elinor Green, who worked mainly in Saigon.[34]

Officials within the agency were nevertheless concerned about having their program explained by the military. As the chief of the agency's Refugee Division, George Goss, observed, the public spokesman for pacification needed to be a civilian "to impress the skeptics that we are striving to preserve the integrity of civil programs and operations." To have the MACV Office of Information represent CORDS, whatever the value of its resources, might result in the submersion of civil programs under those of the military or the confusion of military and civilian objectives in the minds of the American public.[35]

When the MACV Office of Information proposed as an interim measure combining CORDS' information assets with its own and appointing a single action officer within the command to coordinate news of pacification, officials within CORDS balked. Observing that the author of the proposal, Bankson, obviously misunderstood the importance of the CORDS reorganization, Komer's military assistant, Maj. Gen. George I. Forsythe, pointed out that almost half of the South Vietnamese Army and a large proportion of the U.S. mission's military and civilian advisory staff were engaged in pacification. Because of that, he said, the MACV Office of Information "should organize and man itself to devote a major effort

[33] ASD PA, Talking Paper, sub: U.S. Government Organization for Vietnam Support and Operations: In Vietnam, in Washington, DDI Pacification (Early Rev Dev) (1) file.

[34] Memo, MACOI for Chief of Staff, MACV, 29 Jun 67, sub: CORDS Information Program, folder 93, Press, 1967–68, DepCORDS Papers, CMH.

[35] Memo, George Goss, Chief, Refugee Division, CORDS, for ACofS, CORDS, 5 Aug 67, sub: CORDS Information Program, folder 93, Press, 1967–68, DepCORDS Papers, CMH.

(possibly something on the order of one-half of its effort) to the pacification program.''[36]

Forsythe's suggestion had little chance of acceptance. Although the Johnson administration and the U.S. mission in Saigon would have liked nothing better than to see the MACV Office of Information devote half its time to publicizing pacification, a long line of failed efforts in that regard bore testimony to the news media's fascination with American military operations and its unwillingness to change perspective. In August, Zorthian and the new MACV Chief of Information, Brig. Gen. Winant Sidle, worked a compromise. Stressing that Zorthian, a civilian, was minister-counselor for public affairs and would inject a civilian viewpoint into public affairs policies touching pacification, Sidle absorbed the program into the MACV Office of Information but also appointed a high-ranking civilian to be deputy chief of the agency's Public Information Division.[37]

Although supposedly in charge of public relations for civilian operations, Zorthian and Sidle quickly found that Komer was difficult to restrain. Known as Blowtorch at the White House because he applied heat regularly to his colleagues, Komer continually dabbled out of turn in his agency's public relations. Shortly after taking office, for example, he instructed one of his assistants to prepare a communique on upcoming South Vietnamese elections at the village and hamlet levels for delivery to the press shortly after the last returns were in. ''It could emphasize the impressive nature of these elections,'' he said, ''the high proportion of voters' participation, and end with a few well chosen remarks about the significance of this exercise in basic democracy as seen by the U.S. mission. Let me have a draft and I will jazz it up appropriately.''[38]

The memorandum found its way to Zorthian, who responded immediately. ''Puhleeze, Robert,'' he wrote, ''while we may look like block-heads, we do have a certain amount of professionalism.'' Pointing out that the command held wrap-ups for all South Vietnamese elections and had done so since May 1965 at least seven times, he added that all briefings of that sort were held on background since the U.S. mission left on-the-record briefings on South Vietnamese topics to the South Vietnamese government. ''We *will* back stop their effort with our own backgrounder by knowledgeable political section officers,'' Zorthian continued. ''. . . But I would hope to avoid, at all costs, giving you an opportunity to 'jazz it up appropriately.' You did that with the last report you issued . . . and we still have not recovered.'' He added that Komer needed to learn to work within channels—''Why don't you let us know when you are going to write letters to senators . . . which will be released and make news?''—and in conclusion referred to a recent article by Ward Just. Since Just ''says you are the 'key to pacification —tweedy, nervous, energetic, bright, enthusiastic, pipe-smoking, volatile,' why

[36] Memo, Maj Gen George I. Forsythe, Assistant Deputy for CORDS, for Komer, 16 Jul 67, sub: CORDS Information Program, DepCORDS Papers, CMH.

[37] Memo, Elinor Green, CORDS IO, for ACofS, 6 Aug 67, sub: CORDS Information Program, and Memo, Forsythe for Komer [Aug 67], sub: CORDS Information Program, both in folder 93, Press, 1967–68, DepCORDS Papers, CMH; Interv, author with Barry Zorthian, 13 Nov 80, CMH files.

[38] Memo, Komer for Mr. Calhoun, 26 May 67, CMH files.

don't you devote these considerable talents to straightening out the Revolutionary Development program. . . . You provide the substance and we'll provide the press relations.''[39]

Zorthian was emphatic for a reason. Although optimistic articles surfaced occasionally in the press—on 26 May *Life* reported Lt. Gen. Lewis Walt's pride in the progress of pacification in the I Corps Tactical Zone—the news media were becoming more outspokenly critical of the program, leaving little room for mistakes by officials as important as Komer. On 28 May, for example, the *New York Times* published a pessimistic article asserting that the South Vietnamese Army was undependable and that pacification was lagging. Two weeks later, *U.S. News & World Report* charged that, although there was room for optimism about the future, pacification had yet to show any progress. Edmund Stillman repeated the point in a *New York Times Magazine* article published on 18 June, adding that since the South Vietnamese government was ''incompetent and weak'' and Premier Ky ''a mere facade,'' the United States could never win the war and ought to withdraw.[40]

Pressure increased during August 1967, when the Chairman of the House Subcommittee on Government Operations and Information, John Moss, informed the Johnson administration that he would shortly release a report criticizing pacification. ''The subcommittee is concerned,'' he told Secretary Rusk, ''. . . with the failure of the government of South Vietnam to take substantive action, with conviction and determination, in a number of areas. . . . Specifically, we are deeply concerned about the lack of meaningful progress and reform in the lagging and floundering pacification program.'' American officials serving in South Vietnam, Moss said, needed to take a firmer stand in their dealings with the local government. Too often in the past they had attempted ''to avoid 'rocking the boat' '' rather than press for necessary reforms. ''The continuing support of the congress and the people of the United States is in large part dependent upon there being reasonable prospects for success,'' Moss added ominously. ''Without substantive GVN [Government of Vietnam] action in a number of areas . . . prospects for success are minimal, the advisability of continued United States involvement is questionable and could lead to a reassessment of the United States' position.''[41]

The State Department responded calmly to Moss' criticisms, noting that where unfinished business remained, the congressman's recommendations would receive serious consideration. Privately, President Johnson was agitated. Although the subcommittee had yet to release its report to the press, he instructed the U.S. mission in Saigon to compile information on the most recent developments in pacification for use as a rebuttal.[42]

[39] Memo, Zorthian for Ambassador Robert W. Komer, 31 May 67, sub: Your Memorandum of May 26, 1967, to Mr. Calhoun, CMH files.

[40] ''The Two Wars of Gen. Lew Walt,'' *Life*, 26 May 67, pp. 77–84; ''War at Crisis,'' *U.S. News & World Report*, 12 Jun 67, pp. 29–31; Edmund Stillman, ''The Short War-and the Long War,'' *New York Times Magazine*, 18 Jun 67, pp. 7f.

[41] Msg, State 27847 to Saigon, 27 Aug 67, FAIM/IR.

[42] Msg, State 28424 to Saigon, 29 Aug 67, FAIM/IR.

Komer answered the president's request by observing that since the subcommittee's report had yet to be published he was unsure of how to proceed. He suggested that someone in the White House contact the subcommittee staff to determine Moss' objections so that relevant facts and figures could be made public before the report appeared. "Just tell us what target to shoot at," he said, "and we'll let fly."[43]

When it became apparent that the staff was hardly likely to surrender its initiative in that manner, Komer took action on his own. In a letter to Moss, he noted that the congressman had spent little time on pacification during a recent trip to South Vietnam and requested an opportunity to satisfy the subcommittee's doubts before the report appeared. He then enumerated a number of areas where he believed improvements had occurred. An increasing proportion of the South Vietnamese Army was becoming directly involved in providing security for pacification. A newly developed Hamlet Evaluation System indicated that a larger portion of South Vietnam's population was coming under government control. During the first half of 1967, indeed, 1,037 villages and 4,616 hamlets had elected local governments. Although problems remained, Komer concluded, the pacification program had achieved results during 1967 far more substantial than ever before. "I am unaware," he said, "of a single category in which there has been overall regression."[44]

In an apparent attempt to go around Moss, Komer also wrote Congressman Ogden Reid, a member of the subcommittee who had remained in Washington while Moss traveled to South Vietnam. "It is really quite hard on us out here who are trying to do a job which needs doing," he said, "to get whacked about 'lagging and floundering.' You didn't come along on the last trip, but I can assure you that in their all too brief time in Vietnam the chairman and staff had mighty little time available to spend on pacification and heard little to warrant such a slam. What happened?"[45]

Reid passed the letter to Moss, who was outraged. Avowing in a letter to Komer that he and his staff had spent most of their time in South Vietnam discussing pacification, he termed Komer's letter to him "offensive" and the one to Reid "insulting." Throughout his time in South Vietnam, he continued, he had expressed strong concern about the lack of progress in pacification. "Though you cite the fact of elections in 1037 villages and 4616 hamlets, may I remind you there are over 2500 villages and 12500 hamlets in South Vietnam." Although unwilling to charge a total lack of progress, Moss told Komer that he considered his doubts well founded. They were the product, he said, of the many reorganizations pacification had experienced over the years, each accompanied by "new and glowing promises" that had gone largely unfilled.[46]

Komer sent Moss what he termed "a handsome apology" but continued to

[43] Msg, Saigon 4420 to State, 30 Aug 67, Moss Committee file, DepCORDS Papers, CMH.
[44] Ltr, Komer to Honorable John E. Moss, 29 Aug 67, Moss Committee file, DepCORDS Papers, CMH.
[45] Ltr, Komer to Honorable Ogden Reid, 29 Aug 67, Moss Committee file, DepCORDS Papers, CMH.
[46] Ltr, Moss to Komer, 21 Sep 67, Moss Committee file, DepCORDS Papers, CMH.

consider the congressman's report a "shameful" attempt to "belabor our Vietnam effort." In the end he dismissed what had happened as a bureaucratic power play. "Ambassador Bunker was less disturbed about this than I was," he told Westmoreland. " . . . The Moss technique is to cow the State Department and AID [the Agency for International Development] by such means, I'm told."[47]

Elections, September 1967

Whatever the accuracy of Komer's statement, his disagreement with Moss underscored the U.S. government's lack of credibility on pacification. During the first week of September, while the Moss affair was still developing, an opportunity arose to improve the image of the effort. During the Buddhist crisis of the previous year, the military leaders of South Vietnam had promised to begin drafting a constitution and to provide for a popularly elected government. They had kept their word. Between September 1966 and June 1967 they held elections to select a constituent assembly, village councils, and hamlet chiefs. On 4 September a fourth election was to be held to select a president and senate. Pitting candidates of national stature against one another for the first time, that event provided an excellent opportunity to demonstrate democracy in action in South Vietnam. Its impact on the Viet Cong, Hanoi, and world public opinion, so the reasoning went, was bound to be enormous—ample demonstration to all the world that the South Vietnamese government was beginning to claim the allegiance of its people.[48]

Overcoming the news media's skepticism was the challenge, but the U.S. mission in Saigon had already tried many of the techniques it would use. A year before, prior to the September 1966 election for a constituent assembly, the Joint U.S. Public Affairs Office had predicted that press coverage would be heavy, perhaps rivaling that given the off-year congressional elections in the United States, and that at least some reporters would arrive in Saigon with the intention of exposing election rigging by the South Vietnamese government. The U.S. mission had taken pains to warn the South Vietnamese of the importance of keeping the election fair. It had also gone to great lengths to ensure that the Saigon correspondents received a balanced picture, holding background briefings for the press and arranging for interviews and transportation to special events. As a result, although news stories had indeed alleged that many South Vietnamese had voted only to avoid later difficulties with the authorities, some five hundred newsmen from

[47] Handwritten note by Komer attached to General Westmoreland's copy of Memo, Komer for Ambassador Ellsworth Bunker, 27 Sep 67, Moss Committee file, DepCORDS Papers, CMH.

[48] Allen Goodman, *Politics in War: The Bases of Political Community in South Vietnam* (Cambridge, Mass.: Harvard University Press, 1973), pp. 40–63. Memo, Walt W. Rostow for the President, 8 Mar 67, sub: Major Themes You Should Leave Firmly in the Minds of Thieu and Ky, vol. 67, National Security Papers, LBJ Library.

South Vietnamese News Conference, September 1967. *Thieu is at microphone, Ky in right foreground.*

around the world had covered the election with few if any reports of irregularities appearing.[49]

Similar precautions preceded the September 1967 presidential elections. Concerned lest charges of election fraud mar the outcome, the U.S. mission kept close watch on the activities of the political candidates, especially Ky. When reliable reports indicated that Col. Nguyen Ngoc Loan, Ky's political ally and director of the National Police, had begun systematically summoning police and security officers from throughout South Vietnam to Saigon to instruct them on bribing and blackmailing officials who could influence the election, Ambassador Bunker put an end to the meetings by speaking privately with Ky. In the same way, when growing tensions between Ky and Thieu threatened to divide the leadership of South Vietnam, Bunker and Westmoreland, at President Johnson's request, met with the two generals. Noting that the president had repeatedly assured members of Congress that the elections would have no effect on the war, they told Thieu and Ky that their rivalry threatened to discredit both the election and the president of the United States in the eyes of the American public and Congress.

[49] Msg, Saigon 3754 to State, 17 Aug 66, and Msg, Saigon 6414 to State, 21 Sep 66, both in FAIM/IR.

A week later Ky and Thieu agreed to a division of power in which Thieu held the presidential office while Ky became vice president with control over the cabinet and the armed forces.[50]

During the presidential campaign the American news media at first took a highly skeptical view of events, actively seeking irregularities when Thieu and Ky's opponents began charging election rigging. When an aircraft carrying opposition candidates to a political rally landed in the wrong place, effectively canceling the rally, the newsmen thought they had the evidence they needed. Yet as the campaign developed, opposition candidates were able to pursue their activities with little if any interference. Although the South Vietnamese constitution required the screening of candidates to rule out those with Viet Cong sympathies, the government rescinded censorship of the South Vietnamese press, in effect allowing the candidates and their supporters to criticize Thieu and Ky with impunity. As a result, the skepticism of American newsmen diminished to such an extent that Ambassador Bunker later told President Johnson he considered the "balanced tone" of press reporting from Saigon helpful.[51]

The Johnson administration was nevertheless unwilling to leave anything to chance. When the South Vietnamese, at U.S. prompting, invited a panel of prominent American politicians, publishers, labor leaders, and clergymen to Saigon to monitor the elections, the White House issued detailed instructions for handling the group. Noting that "there is highest level interest in this mission," those guidelines recognized that there would be some contact between the panel and the press but stipulated that "a busy schedule should minimize opportunities for exposure to emotional and slanted attitudes." To that end, although the group was to have every opportunity to see what it wanted and to get an honest picture of the election, information officers were to avoid exposing it to a mass news conference and were to permit as little opportunity as possible for cynical reporters to poison the visitors' outlooks. Suggesting that the Military Assistance Command select individual programs and itineraries to suit the interests, professions, and ages of the observers, the guidelines also urged the command to choose escort officers carefully. "A good, gung-ho Jewish chaplain," for example, might accompany the President of the Synagogue Council of America, Rabbi Jacob P. Rudin.[52]

As it happened, the observers could hardly have been kept from the press or the press from them. Governor Richard Hughes of New Jersey carried on a running debate with reporters who contested his acceptance of the election. The U.S. senators in the party—Edmund Muskie of Maine, Bourke Hickenlooper of Iowa, and George Murphy of California—also discussed matters openly with newsmen. In the end, despite the influence of the press, the president's fears proved groundless. All of the panel members agreed that the election had been

[50] Msg, Saigon 28095 to State, 14 Jun 67; Msg, Saigon 28493 to State, 21 Jun 67; and Msg, Saigon 305 to State, 5 Jul 67, all in FAIM/IR. See also Msg, State 217671 to Saigon, 28 Jun 67, FAIM/IR.

[51] Msg, Saigon 2686 to State, 9 Aug 67, for President from Bunker, and Msg, Saigon 3824 to State, 23 Aug 67, for President from Bunker, both in FAIM/IR.

[52] Msg, State 27494 to Saigon, 26 Aug 67, FAIM/IR.

relatively honest. As General Westmoreland noted in a cable to Admiral Sharp, a few even gave the appearance of having changed their perspective on the war.[53]

Although the American news media for the most part agreed that the election had been fair—even usually critical *Newsweek* noted that Thieu and Ky had received only 34.8 percent of the total vote, enough for victory but far from the mandate possible if engineering had occurred—the election had little long-term effect on the image of the South Vietnamese government. The defeated candidates immediately charged that tampering had occurred, whereupon their supporters took to the streets in a series of wild protest demonstrations. The Thieu regime responded by arresting Truong Dinh Dzu, the runner-up in the election with 17.2 percent of the vote and the only candidate to campaign on a strong peace platform. The move was unfortunate, making Thieu and Ky seem ungracious winners when, in fact, even if all of Dzu's charges had been substantiated and all of the disputed ballots disqualified, Dzu rather than Thieu would have been the loser, dropping to fourth place among the candidates.[54]

By the third week in September the novelty of the election had worn off, and most of the Saigon correspondents had returned to their accustomed themes: American troops in combat, debates on the efficacy of the bombing, the ineptitude of the South Vietnamese Army, and the lack of progress in pacification. On the twenty-seventh, Walt Rostow at the White House cabled the U.S. mission in Saigon to request once more that Bunker, Westmoreland, and Komer ''search urgently for occasions to present sound evidence of progress in Vietnam.'' The United States, he said, ''must somehow get hard evidence out of Saigon on steady if slow progress in population control, pacification, VC manpower problems, economic progress in countryside, ARVN improvement, etc. Little comes through despite what we know to be most serious efforts out your way.''[55] The mission complied, making the theme of progress dominant in its public relations on pacification for the next three months, but General Sidle nevertheless had his doubts. Few reporters had been convinced that the United States was making headway with pacification, he told General Westmoreland, because the Military Assistance Command had oversold the program at a time when it should have been much more modest in its claims. From mid-1966 onward, the command had claimed pacification was succeeding, when, in fact, ''we did not truly know whether we were making progress or not.''[56]

The press, for its part, remained unwilling to bend where the performance

[53] Msg, Westmoreland MAC 8384 to Sharp, 5 Sep 67, Westmoreland Papers, CMH. See also Howard R. Penniman, *Elections in South Vietnam* (Stanford: Hoover Institution on War, Revolution, and Peace-American Enterprise Institute Publications, 1972), pp. 66, 76. Penniman was one of several prominent specialists who accompanied the panel as consultants.

[54] ''An Election, a Barrier, and Talk of Peace,'' *Newsweek*, 18 Sep 67, p. 27; Penniman, *Elections in South Vietnam*, p. 84.

[55] Msg, Walt Rostow to Ambassador Bunker, 27 Sep 67, NSC files, box 59, Country file on Vietnam, LBJ Library.

[56] Memo, Sidle for Chief of Staff, MACV [Sep 67], sub: Joint State/Defense Msg 45007, DDI Pacification Reporting file.

of the South Vietnamese was concerned, an attitude typified by the comment of an unnamed newsman during an informal press conference with Vice President Hubert Humphrey. On his first visit to South Vietnam during November 1966, Humphrey told a hastily assembled group of newsmen, "When you speak to the American people give the benefit of a doubt to our side. I don't think that's asking too much. We're in this together." The newsman turned to his companion to grumble, "Benefit of a doubt? Hell, what do they think we've been doing for the past six years."[57]

[57] James H. Broussard, Summary of Findings on American Press Assessments of Vietnam War, 15 Aug 80, CMH files; "Whose Benefit? Whose Doubt?" *Newsweek*, 13 Nov 67, p. 63.

14

Claims of Progress—and Counterclaims

Doubts about the way the United States was fighting the war began to rise in earnest within the Johnson administration toward the end of 1966. During September the intelligence community overwhelmingly endorsed reports that the bombing of North Vietnam had yet to create insurmountable difficulties for the enemy. In November, arguing that the benefits of bombing the northernmost portions of North Vietnam hardly justified the cost, Secretary of Defense McNamara proposed as an alternative the construction of a barrier of obstacles and electronic sensors along the length of the Demilitarized Zone. The proposal alarmed Westmoreland and the Joint Chiefs of Staff, who recommended instead an intensification of the bombing in North Vietnam. The enemy, they argued, would have little difficulty evading McNamara's line of obstacles and sensors. McNamara countered that the system would at least allow the United States to monitor enemy infiltration into South Vietnam more efficiently. President Johnson sided with his secretary of defense in May 1967 to the extent that he authorized construction of the barrier and discussions on whether to limit the bombing of North Vietnam to the region south of 20° north latitude.[1]

The Problem of Statistics

The questioning began to surface in public during August 1967, giving weight to allegations already circulating in the press that the enemy had matched the American buildup with one of his own and that the war was at a standstill. The Preparedness Subcommittee of the Senate Armed Services Committee provided the occasion. Committee chairman John Stennis of Mississippi favored

[1] Msg, Wheeler JCS 3903-67 to Westmoreland, 27 May 67, Westmoreland Papers, CMH; *Pentagon Papers*, 4: 7, 154–87.

Enemy Porters on the Ho Chi Minh Trail

escalation of the air war and sought to discredit those who argued that attacks on North Vietnam had little effect in the South. Meeting in executive session over several weeks, Stennis' subcommittee took testimony from the Joint Chiefs of Staff and most of the other senior officers directing the air war in North Vietnam. All favored mining the port of Haiphong and lifting the restrictions that had until then limited air attacks.[2]

Appearing on 25 August, the last day of the hearings, Secretary of Defense McNamara took the opposite position. He advocated a limited campaign that concentrated on bombing enemy infiltration routes. Communist forces in South Vietnam required only fifteen tons of supplies per day, McNamara said, an amount so insignificant that it would slip through even the most devastating bombing campaign. Those who argued in favor of intensified air attacks hoped to win the war on the ground in the South by breaking the enemy's will in the North. Yet lacking the heavy industry so essential to the economies of Western nations, the agrarian economy of North Vietnam could withstand almost any attack on its industrial base with only minor adjustments. In the same way, the people of the country were so inured to hardship, deprivation, and death that nothing short of all-out attacks on population centers would shake them. As for mining Haiphong, McNamara said, North Vietnam's supply requirements were so small that a mere 540 tons per day would suffice. The enemy could move that amount through Chinese ports to the north or directly from ship to shore across the country's beaches at night.

Leaked to the press, McNamara's testimony made headlines across the United States. The *New York Times* charged that after two-and-one-half years of escalation in South Vietnam and a buildup of 500,000 troops, the secretary's testimony showed that the military situation was little better than it had been when Americans first entered the war. "American escalation has been matched by the Communists," the paper said, "and the stalemate has merely . . . moved to a higher level of combat, casualties, and destruction."[3]

[2] For allegations of stalemate, see "War at Crisis," *U.S. News & World Report*, 12 Jun 67, p. 29; Tran Van Dinh, "A Look at the Vietnamese Armies," *Christian Science Monitor*, 15 Jun 67. This section is drawn from *Pentagon Papers*, 4: 197–204.

[3] "Generals Out of Control," *New York Times*, 1 Sep 67. See also "Hill Report: 'We Have Not Achieved War Objectives,'" *Washington Post*, 1 Sep 67; "A Better Way," *Christian Science Monitor*, 1 Sep 67.

The furor over the bombing and the continuing allegations of stalemate put the Johnson administration on edge, especially since General Maxwell Taylor and presidential adviser Clark Clifford had recently returned from a tour of Asian capitals with word that many foreign leaders supported American ends in South Vietnam but believed that the United States was losing the propaganda war. Proposals on ways to remedy the situation were circulating within the administration even before McNamara's testimony. General Wheeler believed that Westmoreland might hold a special briefing for the press to present a precise, factual case for progress over stalemate. The session might cover such items as the number of enemy base areas U.S. troops had neutralized in each corps tactical zone and how many miles of roads, railroads, and waterways they had opened. Westmoreland was willing to give the briefing, but Admiral Sharp took strong exception. "We have trapped ourselves," he said, "because of our obsession to quantify everything. If you can't put a number on it, it isn't worth talking about. . . . I suggest that we attempt to move away from the great dependence on demonstrating our results with numbers and concentrate on the less tangible but more important results of our operations."[4]

As Sharp had perceived, the Johnson administration's claims of progress were under attack in part because official statistics explaining the war failed to stand up to scrutiny, a fact that the Saigon correspondents took delight in demonstrating. Richard Harwood of the *Washington Post*, for example, noted in a 3 September article that "the war just doesn't add up." Earlier in 1967 the Military Assistance Command had informed President Johnson that the South Vietnamese armored regiment stationed in Saigon had shown a dramatic improvement in combat skills, killing 125 of the enemy during the previous year for every 10 fatalities of its own. The statistic was impressive, Harwood continued, but the performance of the squadrons composing the regiment hardly squared with expectations. In all of 1966 the 8th Armored Squadron had killed 1 enemy soldier; the 5th, 12; the 10th, 23; and the 9th, 148. Those units together had suffered only 14 killed in action. A single U.S. Marine battalion stationed near the Demilitarized Zone, Harwood said, had claimed more enemy dead and had suffered more killed in just three days of fighting during July.[5]

The Body Count

Harwood's comparison of U.S. marines under attack near the Demilitarized Zone with South Vietnamese armored units stationed in relatively quiet Sai-

[4] Msg, Sharp to Wheeler, 3 Aug 67, Westmoreland Papers, CMH. See also Memo, Philip Habib, EA, for Secretary Rusk, 5 Aug 67, sub: Report to the President on the Clifford/Taylor Mission, FAIM/IR. Msg, Wheeler JCS 7126 to Westmoreland, 30 Aug 67; Msg, Wheeler JCS 6105 to Westmoreland, 2 Aug 67; Msg, Westmoreland MAC 7180 to Wheeler, 2 Aug 67, all three in Westmoreland Papers, CMH.

[5] Richard Harwood, "The War Just Doesn't Add Up," *Washington Post*, 3 Sep 67.

gon was unfair but illustrative of the way both the Military Assistance Command and the press misused statistics. More pointed criticisms came from the Defense Department's Office of Systems Analysis, where doubters argued that "the losses we claim to have inflicted upon the enemy, even if accepted at face value, are not enough to break his will." The United States claimed to have killed or captured nearly half of the enemy's new recruits and over one-fourth of the men coming of military age in North Vietnam every year. Yet given the enemy's obvious willingness to sustain heavy casualties when the need arose, losses of that magnitude were sustainable indefinitely. The British had borne worse during World War I, absorbing a loss of half their men in a five-year age group. Even if all the figures were correct, the supplies and equipment U.S. forces claimed to have captured from the enemy equaled but one-third of the total enemy requirement, a loss the Communists could easily replace.[6]

In fact, MACV's statistics exaggerated the number of enemy killed in action and the amount of supplies captured. The quantity of supplies involved was so small that the point was hardly worth arguing, but the distortion of the body count was a different matter because it disguised the fact that in 83 percent of all engagements the enemy had the initiative. Fighting on his own terms, he broke contact when he pleased.

The enemy's preferred tactic, the analysts explained, was to use a small force, perhaps a squad, to surprise a larger American unit. Firing from concealed positions, the attackers inflicted as many casualties as possible and then moved out before artillery and air strikes could respond. As a result, although the United States and South Vietnam enjoyed at least a five-to-one kill ratio in large engagements where the enemy attacked a perimeter or foolishly stood and fought, ambushes—along with mines, booby traps, and sniper fire—accounted for three-quarters of all U.S. casualties, a fact that tended to flatten kill ratios. In the opinion of systems analysts, American field commanders were inevitably embarrassed by those losses, believing themselves rather than their tactics at fault. To protect both their own careers and those of their superiors, they compensated by padding their claims of enemy killed. "We know reliably," the analysts said, "through . . . interviews, personal acquaintances who have been in the field, and press interviews of soldiers and junior officers, of case after case of inflated body counts." The situation was further complicated by the fact that although field units automatically made allowances in their counts for bodies the enemy had concealed as well as for indirect kills by air strikes and artillery, the figures those units submitted were subjected to an "almost universal doubling . . . at battalion and again at brigade" headquarters as they passed up the chain of command. The result was that while the Military Assistance Command claimed 55,000 enemy killed in action during 1966 at best 19,500 could be accounted for. "The degree of probable delusion" revealed by that discrepancy was so great that it merited

[6] This section is based on OASD SA, Military Results and Initiative in Vietnam, 17 Oct 67, Thayer Papers, folder 38, Comparative Forces file, CMH.

Punji Stakes, A Major Source of Allied Casualties

national concern. "For even if the claims are accurate," the analysts concluded, "the enemy merely *may* be in trouble, and in fact we are overstating that."

Although more cautious, the State Department's analysts were reaching the same conclusions. Fred Green of the Bureau of Intelligence and Research told William P. Bundy that in many cases the statistics the Johnson administration wanted to use to justify claims of progress were either "highly questionable or incomplete." Noting MACV's claim that kill ratios favored the United States, Green observed that however heavy the enemy's casualties, his recruitment rate and level of infiltration were more than sufficient to compensate.[7]

General Westmoreland disputed the contention that American forces were exaggerating the body count and promised to investigate any instance where discrepancies between actual and reported kills were alleged. At the beginning of October he dispatched teams to each major headquarters in South Vietnam to monitor the body-counting process. Those teams upheld his point of view, reporting that by and large the count was honest.[8]

[7] Memo, Fred Green for William Bundy, 22 Sep 67, sub: Indicators of Progress, Thayer Papers, folder 86, CMH.

[8] Mission Council Action Memo 233, 6 Oct 67, sub: Meeting of October 2, 1967, MCAM file for 1967, CMH; MACV History, 1967, p. 967, CMH files.

Those findings may have satisfied Westmoreland, but the chief of the MACV Office of Information, General Sidle, recognized that both the press and critics of the war were bound to question the survey's results and to charge that the units involved had followed proper procedures only because official observers were present. Sidle therefore informed Westmoreland that he intended to deemphasize body counts and all other statistics in MACV's future briefings. Following Admiral Sharp's line of reasoning, he pointed out that both the command and the Office of Information had fallen into a routine of passing figures to the press without explaining the operation they were illustrating or showing how the action fit into the overall picture of the war. In the future, he said, information officers would concentrate on meaningful, newsworthy items, fleshing out otherwise dry statistics with details and interesting firsthand accounts of what had happened.''[9]

Countering the Negative Viewpoint of the Press

Although the press was hardly blameless, Sidle was convinced that much of the talk of stalemate and many of the doubts arising within the Johnson administration were the fault of the Military Assistance Command. A small group of reporters was indeed hopelessly biased against the war, he told Westmoreland, but the majority of those who questioned MACV's claims of progress were ''convinced that we have not been telling the whole truth, . . . that we tend to be overoptimistic, and therefore our talk of progress at the present must be taken with a large grain of salt.'' The press agreed that the command had made great strides against the enemy but faulted ''the entire governmental public affairs operation for overstating the case earlier.''

Sidle recognized that he would need time and patience to counter the negative attitudes of the Saigon correspondents. Many of the veteran reporters he might have relied upon to support the war and the military point of view—Jim Lucas, Marguerite Higgins, Jack Foisie, and others—were no longer in South Vietnam. In their place stood a group of more than 450 accredited news men and women that surged beyond 500 in times of crisis or special event. They represented over 130 news-gathering organizations: the *Jyllands Posten* of Denmark, *Joon-Gang Ilbo* of Korea, *Mainichi Shimbun* of Japan, the *Philippine Daily Star*, the *Valley Times* of Southern California, the *London Economist*, the American television networks, and Agence France Presse, to name a few. Some veteran reporters remained—Arnett, Faas, Mohr, Sully, Shaplen, Merick, and Deepe—but a number of them had been critical of American policy in the past. Although the others included many respected newsmen—Lee Lescaze and Stanley Karnow of the *Washington*

[9] This section is based on Memo, Brig Gen Winant Sidle, Chief of MACV Information, for Westmoreland, 11 Sep 67, sub: Improvement of the MACV Information Program, JUSPAO Papers, RG 84, 71A2420, box 7, P-1 Policy files, WNRC.

Post, Bernard Weintraub and Peter Braestrup of the *New York Times*, Elizabeth Pond of the *Christian Science Monitor*, and Robert Erlandson of the *Baltimore Sun*—many more had little direct knowledge of the Vietnam War and only limited experience in journalism. As always, fewer than one-third of all accredited correspondents were working reporters. The rest, as Bankson had found, were hangers-on or support personnel.[10]

In dealing with so polyglot and unwieldy a group, Sidle did what he could to improve official credibility but concentrated as well upon keeping the press within what he considered reasonable bounds. Because he believed that MACV's briefing team had, as he put it, "a chip on its collective shoulder," he replaced it with a group of officers more likely to get along with the press. He also began looking for talented lower-ranking officers to take the place of the generals who periodically briefed the press. In the eyes of the Saigon correspondents, lower-ranking officers had little personal stake in the programs they described, but generals had something to sell. To ensure that the briefers had as wide an audience as possible, he also made a special effort to have television news teams cover the sessions. If the networks proved unwilling, he told Westmoreland, he would have crews from the Armed Forces Television Network station in Saigon film the presentations for later transmission to Washington, where the Office of the Assistant Secretary of Defense for Public Affairs could release them.

Sidle likewise made a special effort to ensure that qualified information officers were always within easy reach of the press. When the Military Assistance Command moved its headquarters from Saigon to Tan Son Nhut Air Base, taking the Office of Information with it, he made certain that the small information office established in the public affairs building by Rodger Bankson contained a knowledgeable information officer at all times. Without that office, he told Westmoreland, the command would have lost whatever tenuous influence it had with its critics among the Saigon correspondents, few of whom would have taken the time to travel to Tan Son Nhut to verify their facts. With an office convenient to the press in Saigon, on the other hand, the command could easily correct errors and distortions before they were reported.

If Sidle believed in being near the press, he was nevertheless convinced that too much familiarity with newsmen could only make trouble. Thus he made it a point always to be present at Zorthian's weekly backgrounders. Reporters complained, recognizing that the presence of a general officer at the sessions served to dampen the proceedings, but he refused to relent. Rumor had it, he later observed, that Zorthian "gave away the family jewels" during the meetings. If so, he wanted the practice stopped.[11]

[10] MACOI, List of Accredited Correspondents, Nov 67, CMH files. For a picture of the press corps in Saigon at the end of 1967, see Peter Braestrup, *Big Story: How the American Press and TV Reported and Interpreted the Crisis of Tet in 1968 in Vietnam and Washington*, 2 vols. (Boulder, Colo.: Westview Press, 1977), 1: 1-51.

[11] Interv, author with Sidle, 5 Jun 73, CMH files. See also Memo, Peter Heller for Zorthian, 15 Sep 67, sub: Paper on Press Initiative, JUSPAO Papers, RG 84, 71A2420, box 7, Policy Guidelines file, WNRC.

Sidle was also concerned about the ability of television newsmen to extract unsavory information from enlisted men and junior officers in the field. Believing that information officers paid too much attention to the writing arms of the press and too little to television, he instructed his field representatives to ensure that television newsmen received not only logistical support and briefings when they arrived in an area but also the undivided attention of an official escort. With a knowledgeable officer on hand at all times, he reasoned, unexpected and possibly damaging incidents could be explained without delay. Otherwise, the reporter would draw his own conclusions from the comments of whatever enlisted men and company grade officers he happened upon. General Westmoreland put the matter more succinctly. A sizable number of the negative news stories that appeared on television in the United States were the fault of unthinking soldiers who either acted improperly on camera or made disparaging remarks, he observed in a memorandum to all major commanders. Neither of those problems would occur, he said, in the presence of a qualified military observer.[12]

Sidle believed that television reporters received substantial bonuses when the networks aired their films on prime time news programs and that negative stories appeared to have the best chance of selection. He thus had little confidence that his system of escorts would improve television coverage of the war and took several other steps that could apply equally to both the electronic and the print media. He told Westmoreland that when he learned of a case of biased or erroneous reporting, he planned to contact the information officer of the unit involved to determine whether all of MACV's prescriptions had been observed. If everything was in order, he would contact the newsman who had made the report "to straighten him out." In the meantime, information officers would keep records of all instances of exaggerated or erroneous reporting. If a correspondent continued to be irresponsible, the command would forward a list of his transgressions to Washington, where the Office of the Assistant Secretary of Defense for Public Affairs could deal directly with his employers. The MACV Office of Information would also develop a "hardhead list" of reporters it considered "worst cases." When errors occurred, Sidle said, his staff would attempt to deal with the individuals involved but would pass them to Westmoreland and Bunker if an approach from higher-level officials appeared more promising.

Sidle's program had been in effect little more than a month when television newsmen demonstrated that they could circumvent the system. On 9 October 1967, CBS News correspondent Don Webster, accompanied by cameraman John Smith, visited the scene of a recently concluded nighttime firefight involving units of the 1st Infantry Division. When the newsmen decided to separate to cover different parts of the battlefield, a military escort accompanied each, the division public information officer following Webster while an enlisted man assigned to the information office accompanied Smith. Observing an enemy body with an

[12] Msg, MAC 34348 to All Major Commanders, 19 Oct 67, sub: Television News Team Visits, box 7, 70A872, WNRC.

ear removed lying to one side, Smith persuaded his escort to duplicate the atrocity by severing an ear from another body nearby. The cameraman then furnished the knife and photographed the resulting incident. The rarest of news events, a violation of the laws of war filmed as it happened, the episode proved irresistible to CBS, which aired it on that evening's news. Don Webster narrated the segment as though the occurrence it depicted had been entirely spontaneous.[13]

Zorthian and Sidle

The commanding general of the 1st Infantry Division, Maj. Gen. John Hay, immediately arrested the enlisted man and sought authority to proceed against Smith. The U.S. embassy in Saigon nevertheless barred any move against the cameraman, apparently in the belief that the arraignment of a civilian reporter before a military tribunal would only increase criticism of the war. Although Webster was never implicated in the incident and received the support of his bureau chief, many of the Saigon correspondents, according to Sidle, considered his conduct in the affair unprofessional. In the end, both he and Smith were subpoenaed to testify at the trial of the enlisted man but left the country rather than appear. The soldier was convicted, reduced in rank, and fined.[14]

Sidle's attempt to restrain negative reporting was also hindered by the extreme spirit of competition that prevailed among some of the Saigon correspondents, a few of whom were constantly searching for loopholes in MACV's regulations. On 17 November 1967, for example, Maj. Gen. Bruno A. Hocmuth, U.S. Marine Corps, died under fire in a helicopter crash. The MACV guidelines urged reporters to withhold the names of the dead until next of kin could be notified, but UPI's bureau chief in Saigon, Eugene Risher, put word of Hocmuth's death on the wire as soon as he heard of it. Risher's action angered the rest of the Saigon correspondents, most of whom considered the move a breach of journalistic ethics. Risher argued, however, that the story involved a national figure whose death was so important that it overrode the next-of-kin guideline. He was the only bureau chief in South Vietnam, according to Sidle, who felt that way.[15]

[13] Msg, Westmoreland MAC 9545 to Harold K. Johnson, 11 Oct 67, Westmoreland Papers, CMH.

[14] Msg, MACV 42962, MACV Judge Advocate General, to General Counsel and ASD PA, 27 Dec 67, DDI 33B (NVN): POW's/Defections file; Interv, author with Sidle, 5 Jun 73.

[15] Msg, Sidle MACV 37556 to ASD PA, 16 Nov 67, and Msg, Defense 3364 ASD PA to Sidle, 21 Nov 67, both in DDI Press Flaps file.

In a cable to the Office of the Assistant Secretary of Defense for Public Affairs, Sidle noted that the Military Assistance Command had only two weapons to combat Risher's tactics. It could withdraw the reporter's accreditation or attempt to withhold all embargoed information from UPI until after its formal release in Saigon. Since sanctions against the entire UPI organization would almost certainly create a situation similar to the one that had resulted from Ambassador Taylor's action against Beverly Deepe in 1965, Sidle recommended disaccreditation. In the past, the only correspondent disaccredited for a journalistic infraction, Jack Foisie of the *Los Angeles Times*, had lost his privileges because of a blatant security violation, the revelation of a troop movement before the Military Assistance Command had given clearance for an announcement. Since Risher's case was a violation of trust rather than of security, his disaccreditation on those grounds, Sidle said, would set a worthy precedent. Establishing MACV's authority in cases where security violations were not at issue, it would put "real teeth" into the MACV guidelines.[16]

Zorthian agreed with Sidle's recommendation, but the Assistant Secretary of Defense for Public Affairs, Phil G. Goulding, refused to go along. Risher's premature announcement of Hocmuth's death was reprehensible, Goulding told Sidle, but hardly grounds for punitive action. There was a dictum accepted by both the American public and the press that certain figures were so prominent that news of them took precedence over other considerations. As to whether Hocmuth fit into that category, no accepted dividing line or definition existed, leaving room for an honest difference of opinion. Goulding believed that the reporter thus deserved the benefit of a doubt.[17]

Although Risher went undisciplined, the U.S. mission in Saigon did succeed in tempering the work of at least one offending reporter, going around a so-called hardhead to his editors. The incident occurred in December 1967, when Ambassador Bunker received word that the older brother of Jonathan Schell, Orville, was about to publish in the *Atlantic Monthly* an article critical of the way the United States and the South Vietnamese handled refugees. Confusing that article with another recently completed by Jonathan Schell that scored U.S. and South Vietnamese tactics in Quang Ngai and Quang Tin Provinces, Bunker immediately cabled Secretary McNamara. He suggested that since the editor of the *Atlantic* was Robert Manning, a former Assistant Secretary of State for Public Affairs, the magazine might agree to publish some response from the government along with the article. Bunker included a critique of the article on Quang Tin authored by an officer of the U.S. mission.[18]

McNamara responded two weeks later that the article in question was by Orville rather than Jonathan Schell. Published in the January 1968 edition of the

[16] Msg, Sidle MACV 37556 to ASD PA, 16 Nov 67.
[17] Msg, Goulding Defense 3364 to Sidle, 21 Nov 67, DDI Press Flaps file.
[18] Msg, Saigon 715 to State, 30 Dec 67, for McNamara from Bunker, FAIM/IR; Orville Schell, "Cage for the Innocents," *Atlantic Monthly*, Jan 68, p. 29; Jonathan Schell, "Quang Ngai and Quang Tin," *New Yorker*, 9 Mar 68, and "Quang Ngai and Quang Tin II," *New Yorker*, 16 Mar 68.

Atlantic, it was, he said, "less troublesome in tone and content" than that of Jonathan, "thanks in part to the editing of Bob Manning." The Defense Department had achieved that effect by showing Manning the mission's critique of the essay on Quang Ngai, even though it had little application to the article in question. McNamara added parenthetically that although the critique was impressive, he doubted its utility as a counter to the work on Quang Ngai because it showed the government of South Vietnam in a very bad light and supported the general thrust of the article if not the specifics.[19]

The Order of Battle Controversy

Of all the steps Sidle took to improve MACV's credibility, the attempt to deemphasize statistics was the least successful. The fluid nature of the fighting in Vietnam eliminated the fixed battle lines by which progress in earlier wars had been measured. Lacking such geographic evidence of success, the Johnson administration resorted time and again to numbers to demonstrate that its policies were working. Compounding Sidle's problem was the fact that the command had begun practices that were no longer beneficial but that had to be continued for the sake of credibility.

Earlier in the war, for example, information officers had briefed the press on the enemy's order of battle whenever the Military Assistance Command updated its figures in order to keep the new estimate from leaking unexplained to newsmen. By mid-1967, however, the command and the Central Intelligence Agency were caught up in a dispute over which had the more accurate figures. Both apparently agreed that the number of the enemy's hard-core regular troops remained the same and that the number of guerrillas should be increased from the 40,000 previously estimated to 90,000. They differed on the issue of enemy sympathizers, the so-called Self-Defense (part-time soldiers living in enemy-controlled areas) and Secret Self-Defense units (part-time clandestine organizations operating in government-controlled areas). The CIA wanted to include them in the new order of battle, but the Military Assistance Command refused. The two might have settled for a gentlemen's agreement in which each side kept its own books but for the fact that a briefing for the press would have to follow ratification of the estimate. As Sidle pointed out long afterward, an agreement on the figures seemed imperative because any disagreement would have leaked to the press where it would have become a threat to official credibility.[20]

In MACV's estimation, the enemy's Self-Defense and Secret Self-Defense

[19] Msg, State 10067 to Saigon, 18 Jan 68, FAIM/IR.

[20] Msg, JCS 7143 to Westmoreland, 30 Aug 67, Westmoreland Papers, CMH; Msg, Joint State/Defense 5803 to Saigon, 21 Oct 67, CMH files; Memo, OASD PA for Mr. Richard Moose, Executive Office of the President, 19 Oct 67, sub: MACV Press Briefing on Enemy Order of Battle, DDI Statistics file; Interv, author with Sidle, 5 Jun 73. For the definitions of Self-Defense and Secret Self-Defense Forces, see CICV, Monthly Order of Battle Summary, Jul 66, pp. 2-3, CMH files.

Forces had little effect on the war. "They operate entirely in their own hamlets," Westmoreland's deputy, General Creighton Abrams, told Wheeler. "They are rarely armed, have no real discipline, and no military capability. They are no more effective in the military sense than the dozens of other non-military organizations which serve the [Viet Cong] . . . in various roles."[21] The Central Intelligence Agency disagreed. Far from being mere "fellow travelers" or "fifth columnists" too young or too old to take part in combat, those units, according to the agency's analysts, contributed greatly to Communist capabilities and "not infrequently" went armed into combat. For the sake of credibility, they said, any briefing on the subject had to be candid.[22]

The command disputed that conclusion. If seventy to eighty thousand Self-Defense and Secret Self-Defense Forces were included in the enemy's order of battle, Abrams told Wheeler, overall enemy strength would appear to range between 420,000 and 430,000 men, a dramatic increase from the previous estimate of about 300,000. "We have been projecting an image of success over recent months and properly so," Abrams said. "Now, when we release the [larger] figures . . . newsmen will immediately seize on the point that the enemy force has increased about 120,000 to 130,000. All available caveats and explanations will not prevent the press from drawing an erroneous and gloomy conclusion as to the meaning of the increase. All those who have an incorrect view of the war will be reinforced and the task will become more difficult."[23] Abrams buttressed his argument by noting that Joseph Fried of the *New York Daily News* had already learned of the change from an anonymous source but was withholding the figures at the request of Sidle and Zorthian, who had convinced him that any announcement would be premature. General Westmoreland added his own weight to Abrams' argument in a subsequent message to Wheeler, noting that the addition of questionable figures to the overall estimate of enemy strength "distorts the situation and makes no sense."[24]

A tug-of-war ensued between the Military Assistance Command and the Central Intelligence Agency, culminating on 14 September in a compromise that gave the command most of what it wanted.[25] At a meeting of intelligence analysts representing all parties to the dispute, the Central Intelligence Agency agreed to exclude Self-Defense and Secret Self-Defense Forces from the order of battle. The command, for its part, slightly increased its estimate of enemy forces and

[21] Msg, Abrams MACV 7840 to Wheeler, Westmoreland, Sharp, 20 Aug 67, Westmoreland Papers, CMH.

[22] Memo, George A. Carver, Jr., Special Assistant for Vietnamese Affairs, CIA, Office of the Director, for Phil Goulding, ASD PA, 13 Oct 67, sub: Transmittal of Requested Comments, DDI Order of Battle file.

[23] Msg, Abrams MAC 7840 to Wheeler, Westmoreland, Sharp, 20 Aug 67, Westmoreland Papers, CMH.

[24] Msg, Westmoreland MAC 7859 to Wheeler, 20 Aug 67, Westmoreland Papers, CMH.

[25] Msg, Westmoreland MAC 8703 to Gen J. P. McConnell, Acting CJCS, 14 Sep 67, Westmoreland Papers, CMH. See also Richard A. Hunt, Pacification: Managing the "Other War," ch. 9, CMH MS, in CMH files.

adopted a range of figures rather than an exact number to describe their size.

The dispute nevertheless continued, centering on the press release that would announce the new estimate. On 27 September the Office of the Assistant Secretary of Defense for Public Affairs approved a draft briefing that included an extensive treatment of Self-Defense and Secret Self-Defense Forces, apparently on the theory that credibility required a thorough explanation of why those units were unmentioned in the order of battle. The Central Intelligence Agency supported the draft, asserting that without a full explanation newsmen would conclude that the United States was hiding a drastic increase in the size of the enemy's forces. Westmoreland, Komer, and the MACV staff, however, recommended the removal of any mention of Self-Defense and Secret Self-Defense Forces from the briefing on the grounds that an in-depth treatment of the subject would threaten all of the benefits that the briefing was supposed to promote.[26]

Seeking a compromise, the Office of the Assistant Secretary of Defense for Public Affairs early in October proposed dropping any mention of Self-Defense Forces in favor of an explanation suggested by the Military Assistance Command that unarmed "fifth columnists" and Viet Cong sympathizers would no longer figure into official tallies of enemy strength just as unarmed supporters of the South Vietnamese government had never been counted as part of the South Vietnamese armed forces.[27] The Central Intelligence Agency again took strong exception, warning that the command was playing a dangerous statistical game that might well backfire.[28]

In the end, the Central Intelligence Agency won the complete explanation it had sought but again compromised by accepting MACV's characterization of the Self-Defense Forces as largely innocuous. The White House approved the final briefing after making a number of changes to emphasize that although the command had underestimated the number of guerrillas present in South Vietnam in the past, guerrilla strength had probably declined over the previous year. Delivered on 24 November 1967, the briefing made little stir in the press, which largely accepted the new numbers as a matter of course.[29]

[26] Msg, ASD PA to MACV, 27 Sep 67, and OASD PA Talking Paper [Sep 67], sub: MACV Press Briefing on Enemy Order of Battle, both in DDI Order of Battle file. The Westmoreland-Komer recommendation is in Msg, Sidle MAC 9237 to Daniel Z. Henkin, 2 Oct 67, sub: Proposed Briefing on Enemy Order of Battle, DDI Order of Battle file.

[27] Msg, Goulding Defense 8311 to Sidle, 3 Oct 67, and Msg, Sidle MAC 9366 to Goulding, 6 Oct 67, both in DDI Order of Battle file.

[28] Memo, George A Carver, CIA, for Goulding, 13 Oct 67, sub: Requested Comments, DDI Order of Battle file.

[29] For all the changes the White House made, see OASD PA Memo, Col Lucius G. Hill for Daniel Z. Henkin, 3 Nov 67, sub: Status of Attached Press Briefing on OB, DDI Order of Battle file. The final briefing delivered to the press is in Memo, MACV OI for the Press, 24 Nov 67, sub: MACV Briefing on Enemy Order of Battle, DDI 33-C (NVN) Military Order of Battle file.

Demonstrating Progress, September–October 1967

Although the threat posed by the dispute over the order of battle gradually subsided, President Johnson was becoming increasingly concerned about his public relations. The press and television seemed dominated by news of election irregularities in South Vietnam, the possibility of a rift between Thieu and Ky, debates on the bombing, and stories on the South Vietnamese Army's ineptitude. Allegations that the war was in stalemate were also continuing. In the belief that the outcome of the war might be at stake and that winning the support of the American people was a critically important dimension of the fighting, Johnson directed the Military Assistance Command and the U.S. mission in Saigon to begin amassing statistics to demonstrate that the United States and South Vietnam were gaining on the enemy. Officials at all levels, he said, were to search diligently for occasions to present that evidence to the American public.[30]

The U.S. mission in Saigon responded to the directive with an elaborate public relations plan antithetical to most of what Sidle was attempting to accomplish. According to that plan, while Bunker, Westmoreland, and other senior officials held press conferences to compare the situation in 1967 with the one that had prevailed two years earlier, the Military Assistance Command would sponsor visits by newsmen to villages and hamlets that illustrated progress in the pacification program. Taking care to brief reporters on major American and South Vietnamese failures, the command and the U.S. embassy in Saigon would also address controversial subjects appearing in the press in a series of hard-hitting briefings designed to allay credibility problems. As those programs gained momentum, the public affairs office would accelerate the release of captured enemy documents that either revealed enemy failures or admitted by word or inference that the war was going poorly for the Communists. To build up the South Vietnamese, the Military Assistance Command would meanwhile institute a series of progress reports for the press designed to place current events in South Vietnam in proper perspective.[31]

The outline of the campaign received wide support within the Johnson administration with only a few officials expressing doubts, most prominently McGeorge Bundy, who had resigned from the White House staff to become the president of the Ford Foundation but who still kept abreast of issues. Asked in early October to suggest steps the United States might take to strengthen its public image, Bundy put his response into the context of the president's coming campaign for reelection. Terming the war "a long, slow business in which we cannot expect decisive steps soon," he warned Johnson that the administration would shortly have to defend its record on the war and that any public relations initiative undertaken for the sake of appearances alone would undoubtedly be characterized as political maneuvering in press reports. "There is a credibility gap," he said, "and

[30] Msg, Joint State/Defense 45007 to Saigon, 28 Sep 67, DDI Pacification Reporting file.
[31] Msg, Saigon 7867 to State, 7 Oct 67, CMH files.

it makes no difference that the press has done more to make it than we have." What the United States needed, he said, was for newsmen to begin to find progress in South Vietnam for themselves. One important reporter such as Joseph Alsop, favoring the government's point of view but reporting independently, was worth ten official spokesmen.[32]

Bundy's suggestion clashed with Johnson's judgment that public opinion required urgent reinforcing but helped inject a note of caution into the promotional campaign that developed. For although the Johnson administration remained dedicated to airing the best possible arguments for its position, it did at least decide against relying on great volumes of hard statistics to demonstrate its case. By concentrating on general trends to indicate the favorable direction of the war, it sought to avoid charges in the press that it was manipulating statistics and contriving analyses.[33]

The U.S. mission in Saigon also attempted to move the administration toward moderation. When the president decided to increase the flow of captured enemy documents to the press as a means of demonstrating that even the enemy believed the United States and South Vietnam were doing well, Zorthian moved immediately to ensure that only the most credible materials appeared and only a few at a time. When pressure from the White House continued for the release of more and more documents, Zorthian satisfied the demand by creating a staff of four Americans and three South Vietnamese to expedite the declassification process but still made certain that the program never reached saturation by releasing only those documents he considered truly significant. The result was that the press continued to take captured enemy documents seriously whenever they appeared.[34]

The information officers were less successful when dealing with Johnson himself. When the president discovered an extremely optimistic appraisal of the situation in the II Corps Tactical Zone by the corps commander, Lt. Gen. Stanley R. Larsen, he insisted that the commanders of the other corps tactical zones write similar reports so that he could use them in briefing the press. In the same way, when CORDS developed its system of Hamlet Evaluation Surveys, he quoted the reports in his conversations with newsmen despite warnings from pacification officials that the statistics were at best rough indicators of what was going on in South Vietnam. When he persisted, the best information officers could do was to take steps to ensure that all of MACV's reports using the statistics avoided inviting unwarranted conclusions. General Sidle meanwhile cautioned the Defense Department to avoid making extravagant claims for pacification, once

[32] Memo, McGeorge Bundy for the President, 7 Oct 67, sub: Vietnam—October 1967, FAIM/IR.

[33] Msg, Joint State/Defense 52950 to Saigon, 12 Oct 67; Msg, Joint State/Defense 58043 to Saigon, 21 Oct 67; and Msg, Joint State/Defense 75209 to Saigon, 27 Nov 67, all in CMH files.

[34] Memo, Deputy Ambassador Locke for Zorthian, 28 Sep 67; Memo, Zorthian for Mr. Calhoun, 15 Aug 67; Msg, Saigon 10921 to State, 11 Nov 67; Ltr, Zorthian to Otis E. Hayes, Dep Asst Director, USIA, 12 Dec 67. All in FAIM/IR.

more noting that the success of the program had been exaggerated in the past.[35]

The press, for its part, greeted the onset of the campaign with skepticism. When Zorthian inaugurated a series of interviews during October with the most senior members of the U.S. mission, William Tuohy of the *Los Angeles Times* applauded the cordiality of the meetings but noted that there was a pronounced difference between what the officials were saying and what the press had thus far reported to the American public. While officials claimed that the press criticized the South Vietnamese Army unfairly, he said, the press could see only twelve years of wasted American advice and training. For months and years officials had touted each succeeding pacification plan as the one that would turn the tide. Having heard it all so many times before, the press remained skeptical.[36]

Anticipating that sort of reaction from many reporters, Zorthian during mid-October began a search for some way to publicize the progress of the war that avoided the negative interpretations newsmen tended to impose upon even the most forthright official declarations. At first he considered producing a comprehensive television "Report to the Nation" in which Ambassador Bunker, General Westmoreland, and other top members of the U.S. mission would discuss various aspects of the war without the press participating. He decided against the idea because the broadcasts would have to use the American television networks, which would undoubtedly edit and package them to suit their own tastes. He next proposed a "Briefing Vietnam" program in which important mission officers, relying on captured enemy documents and visual aids, explained little-understood aspects of the war. The U.S. mission would control the direction and content of the program, with the television networks serving only as conduits for the final product. When the networks objected to becoming the passive instruments of government information officers, Zorthian abandoned the idea. Hoping for the best, he settled for a series of briefings to cover those aspects of the war that he believed needed publicizing.[37]

As the campaign gained momentum, President Johnson followed developments in South Vietnam avidly, searching for any occasion that would combat reports of stalemate. During the last days of October, when the enemy experienced a major defeat in a series of hard-fought engagements near Loc Ninh, a South Vietnamese district capital in the northern part of the III Corps Tactical Zone, he notified Westmoreland that he wanted the Military Assistance Command to make a suitable statement commending the South Vietnamese units involved in the battle if that was possible. Aware that the MACV Office of Information was

[35] Msg, MACV to CG, III MAF, et al., 21 Aug 67, CMH files; MFR, MACCORDS-PP, 16 Oct 67, sub: Reporting and Statistics; MFR, MACCORDS-PP, Lt Col F. G. Gosling, Chief, RDS Branch, 16 Oct 67, sub: Reporting and Statistics; Memo, Sidle for Chief of Staff, MACV [late 67], sub: Joint State/Defense Msg 45007. All in CMH files.

[36] William Tuohy, "Newsmen's View of Viet War Fails To Match U.S. Optimism," *Los Angeles Times*, 29 Oct 67.

[37] Msg, Saigon 8446 to State, 13 Oct 67; Msg, State 63607 to Saigon, 3 Nov 67; and Msg, Saigon 10850 to State, 10 Oct 67, all in FAIM/IR.

Troops Pinned Down Near Dak To

already emphasizing the South Vietnamese portion of the operation and that the battle had been largely an American affair, Westmoreland demurred. News accounts of the operation were reasonably well balanced, he told General Wheeler, making a "hard sell" inappropriate. The decision proved fortunate. A few days later Westmoreland learned that South Vietnamese units involved in the battle had looted the village of Loc Ninh and several nearby French plantations, a circumstance that might have come to light if the press had been following South Vietnamese aspects of the battle more closely. In the end, General Wheeler satisfied the president's request by taking a circuitous route. With the assistance of the Office of the Assistant Secretary of Defense for Public Affairs, he saw to it that a reporter supportive of U.S. policy, Orr Kelly of the *Washington Star*, received enough information to author a favorable article highlighting the South Vietnamese Army's role in the battle.[38]

[38] Msg, CJCS 9468 to Westmoreland, 5 Nov 67, Westmoreland Papers, CMH; Memo, Zorthian for Eugene M. Locke, Deputy Ambassador, 7 Nov 67, FAIM/IR; Msg, Westmoreland MAC 10547 to Wheeler, 6 Nov 67, Westmoreland Papers, CMH; Memo, Richard Fryklund, Dep ASD PA, for Wheeler [Nov 67], DDI Operations file; Orr Kelly, "Loc Ninh Emerging as a Significant Fight," *Washington Star*, 21 Nov 67.

The Battle of Dak To, November 1967

Shortly after the Battle of Loc Ninh ended, American forces discovered four North Vietnamese regiments massing for an attack on Dak To, a Special Forces camp located in the western highlands province of Kontum. The three-week battle that followed, involving some of the hardest fighting of the war to that date, provided the Johnson administration with what it hoped would be yet another opportunity to publicize progress.

The administration's ends were complicated by the fact that in the eyes of the press the battle was hardly a clear-cut victory for U.S. forces. At the height of the fighting, General Westmoreland complained to Abrams that the Saigon correspondents were distorting what was happening. United Press International and the Associated Press had both issued stories detailing how the enemy had surrounded and trapped U.S. units on a hill and how enemy fire had driven off fleets of helicopters, leaving scores of wounded Americans to wait more than three days for rescue. Part of the problem, Westmoreland said, was that the press appeared to be a day-and-a-half behind the action. "I suggest," he said, "that you make an all out effort to get the Saigon press in [to Dak To] and to put this action . . . in the proper context, refuting, if refutation is appropriate, that units have been surrounded and slaughtered."[39]

Abrams could do nothing of the sort. One of the few occasions during the war when massed enemy units stood their ground and fought, the battle was little less spectacular than the descriptions appearing in the press. Dak To, Abrams told Wheeler, was surrounded by mountains whose heavily jungled slopes had to be secured foot by foot. The enemy had fortified the tops of the peaks and ridges, in one case building three separate trenches that made complete circles around the top of a mountain. The fighting, as a result, had been tenacious. Although 75 to 80 percent of the American wounded were treated and returned to their units, several large groups of injured soldiers, pinned down by intense enemy fire, waited up to two days before helicopters could land to evacuate them. In the end, only one soldier died because of the delay, and the enemy pulled back from his positions, losing more than 1,600 men according to MACV estimates. Better than 300 Americans were nevertheless killed and over 1,000 wounded.[40]

The enemy's intentions in fighting so heavy and prolonged an engagement came under almost immediate scrutiny in both official circles and the press. General Abrams hypothesized that the North Vietnamese had hoped to inflict a maximum number of casualties upon American units to intensify antiwar sentiment in the United States. Vice President Ky suggested that the offensive had been an enemy attempt to discredit the newly elected Thieu-Ky regime. Ameri-

[39] Msg, Westmoreland JCS 10011 to Abrams, 21 Nov 67, Westmoreland Papers, CMH.

[40] Msg, Abrams MAC 11329 to Wheeler, 22 Nov 67, Westmoreland Papers, CMH; Admiral U. S. G. Sharp and General William C. Westmoreland, *Report on the War in Vietnam (As of 30 June 1968)* (Washington, D.C.: Government Printing Office, 1970), p. 139.

can pacification officials working in South Vietnam's heavily populated coastal provinces believed that the enemy had attempted to draw U.S. forces away from the coastal plain in order to regain some of the influence over the peasantry he had lost during the previous year. That the battle had indeed led to a drawdown of U.S. forces in the region, contributing to an increased sense of insecurity among the people, lent some credence to that interpretation.[41]

The news media replayed all of those themes, adding the pessimistic observation that if the battle had indeed been an American victory it might yet become a psychological defeat, if heavy U.S. casualties weakened the willingness of the American people to continue the war. Even if that will held firm, Lee Lescaze of the *Washington Post* observed, nothing of permanence had been gained. Although driven from the hills for a time, the enemy would rearm and regroup, to return at will to the bunkers and trenches that remained in the mountains around Dak To. At the time, neither official analyses nor newspaper commentaries considered the possibility that the enemy might have been planning a major offensive against South Vietnam's cities and had sought at Dak To to draw American troops into the hinterlands where they would be out of the way.[42]

Westmoreland's Visit to Washington, November 1967

Although General Westmoreland continued to believe that the press had sensationalized the battle, news coverage of the event raised only moderate concern in the Johnson administration, which had chosen that moment to play host once more to Westmoreland in Washington. Although the purpose of the meeting was supposedly to discuss how the United States could achieve maximum progress during the next six months, there appears to have been little doubt in military circles that the general was participating in a major public relations initiative. His presence in Washington created opportunities not only to promote the theme of progress in the war but also to attack critics of the administration's war policies and to bolster the president's sagging standing in the polls. On the day General Wheeler informed Westmoreland of the president's wishes, he also noted that the general ought to be prepared to appear on television.[43]

Ambassador Bunker accompanied Westmoreland to Washington, but the general was by far the more visible of the two, testifying before the House Armed Services Committee and delivering a major address to the National Press Club.

[41] Msg, Abrams MAC 11239 to Wheeler, 22 Nov 67, Westmoreland Papers, CMH; Msg, Saigon 11830 to State, 24 Nov 67, for the President from Locke, FAIM/IR; Msg, NHT 1535, Lt Gen William Rosson, MACV CofS, to Westmoreland, 10 Dec 67, Westmoreland Papers, CMH.

[42] Lee Lescaze, "Dak To Battleground the Enemy's Choice," *Washington Post*, 25 Nov 67.

[43] Msg, Wheeler JCS 9381 to Westmoreland, 3 Nov 67, and Msg, Wheeler JCS 9566 to Westmoreland, 6 Nov 67, both in Westmoreland Papers, CMH. The president's standing in the polls was at an all-time low, See Louis Harris, "Public Confidence in President Plunges to an All-Time Low of 23%," *Washington Post*, 23 Oct 67, and George Gallup, "46% Now Feel Viet War Is Mistake," *Washington Post*, 23 Oct 67; Msg, Wheeler JCS 9682 to Westmoreland, 10 Nov 67, Westmoreland Papers, CMH.

Both he and Bunker appeared on the NBC Television program "Meet the Press." On each occasion he stressed that American units were grinding the enemy down and that South Vietnamese forces were increasingly effective on the battlefield. As those trends continued, within about two years the United States would be able to begin phasing down the level of its commitment and to turn a larger and larger share of the fighting over to the South Vietnamese. "It is significant," he told the National Press Club, "that the enemy has not won a major battle in more than a year. In general, he can fight his large forces only at the edges of his sanctuaries. . . . His guerrilla force is declining at a steady rate. Morale problems are developing within his ranks."[44] Ambassador Bunker emphasized the same themes in his appearances, noting upon arrival in Washington that every prospect existed for the progress already taking place in Vietnam to accelerate in the coming year. President Johnson, for his part, took advantage of the moment to distinguish between responsible dissenters, those members of Congress who disagreed with his policies, and the antiwar movement, whose actions he described in a White House news conference as "storm trooper bullying" and "rowdyism."[45]

That the president was launching a major new public relations initiative was apparent to the press even before Westmoreland and Bunker arrived in Washington. Columnist Joseph Kraft commented sardonically that in bringing Westmoreland home, the president was obviously attempting to shore up support for the war. The general would arrive with a message of great progress, he predicted, expressed in numbers of enemy dead, weapons captured, and peasants newly brought under the control of the Saigon government. James Reston of the *New York Times* was also critical. Noting that Johnson was mobilizing his "big guns" to mount a major counterattack against critics of the war, he added that the offensive was already successful in at least one respect: "The doubters in the cabinet and sub-cabinet have shut up at dinner parties. . . . In fact, there is no longer any debate, let alone open dissent [within the administration] . . . only closed ranks and closed minds to anything but the official line."[46]

More criticism followed once Westmoreland and Bunker began to speak out. *Newsweek* and the *New York Times* cabled their representatives in Saigon to request material to refute official claims of progress. Ward Just of the *Washington Post*, in an article headlined "President's Hard-Sell on Vietnam," characterized the administration's public relations campaign as "nine days of statistics old and new, intelligence estimates revised and unrevised, of prediction, evaluation, opinion, conjecture, fact, rumor and logic—delivered from such varied pulpits as press clubs and television studios, airports and sidewalks." Equally dubious, Hedrick Smith of the *New York Times* reported that because of previously exaggerated

[44] Address by General W. C. Westmoreland to the National Press Club, 21 Nov 67, copy in CMH files. See also Msg, Westmoreland HWA 3455 to Abrams, 26 Nov 67, Westmoreland Papers, CMH.

[45] E. W. Kenworthy, "Johnson Retorts to Critics of War; Scores Rowdyism" *New York Times*, 18 Nov 67; Eric Wentworth, "Bunker Reports War Gain," *Washington Post*, 14 Nov 67.

[46] Joseph Kraft, "Westmoreland's Trip, Swing by LBJ Have Common Themes," *Washington Post*, 14 Nov 67; James Reston, "Washington: Communique From the Home Front," *New York Times*, 17 Nov 67.

official estimates of the war's progress there was a tendency on Capitol Hill and even among some officials of the Johnson administration to receive the new estimates with caution. Many members of the government remained concerned about the president's persistence in using doubtful statistics to gauge progress in South Vietnam.[47]

Just also addressed the question of progress. Since the Johnson administration believed that the American people would turn against the war if the United States failed to show progress, he said, the president was becoming increasingly vexed by his inability to communicate his sense that the war was being won. Johnson's demand for more and better statistics, according to Just, had succeeded mainly in creating a statistical deluge that proved little. Although some indicators, for example the Hamlet Evaluation Survey, were about as reliable as could be expected under the circumstances, none could predict what would happen if U.S. troops withdrew from South Vietnam. Would the Saigon government and the South Vietnamese Army alone, Just asked, be able to keep the country secure for the long term?[48]

Addressing official assertions that the Saigon government controlled some 67 percent of the South Vietnamese population, Just noted that the figure was open to considerable interpretation. The Saigon government could never control everyone who lived in so-called secure areas, nor could statistics measure such intangibles as the loyalty of the people to the regime in power. Many of the peasants living in secure areas were, indeed, refugees in flight from the violence of the war rather than loyalists fleeing the Communists. To show the unreliability of statistics, the reporter concocted a figure of his own which he said Ho Chi Minh might favor. He added one-third of the population of the cities and of all contested areas to the population of those regions that were undeniably enemy territory. The result put slightly less than 45 percent of the people of South Vietnam under Communist control. Even if the U.S. estimate of 67 percent was correct, Just added, many U.S. officials believed that the Johnson administration might yet find itself "hoist on its own statistics." What would happen "if a year from now we have, say, eighty-five percent of the Vietnamese people under Saigon control and the war has not abated and American casualties remain about the same?"[49]

Despite the criticism of the Johnson administration's claims of progress, Westmoreland's remarks appeared prominently and in detail in the press, with a number of commentators either accepting them at face value or greeting them as at least as valid an approach to the war as any other. Much of what Westmoreland had to say, indeed, had already appeared in the press, especially in a series of articles by Orr Kelly of the *Washington Star*, headlined on 7 November, "In a Military Sense the War Is Just About Won," and on the eighth, "The Enemy in

[47] Memo, Zorthian for Dep Amb Eugene Locke, 24 Nov 67, FAIM/IR; Ward Just, "President's Hard-Sell on Vietnam," *Washington Post*, 26 Nov 67; Hedrick Smith, "Optimists Vs. Skeptics," *New York Times*, 24 Nov 67.

[48] Ward Just, "The Heart-Mind Gap in Vietnam War," *Washington Post*, 19 Nov 67.

[49] Ibid.

Trouble—18 Months and No Big Victory.'' As Westmoreland's visit proceeded, Bob Considine of the *Philadelphia Inquirer* challenged Johnson's critics to ''stop griping. We're winning this lousy war. It is not, repeat not, a stalemate. The enemy has not won a substantial land battle for more than two years.'' James Reston meanwhile reviewed most of Westmoreland's themes in an article entitled ''Washington: Why Westmoreland and Bunker Are Optimistic.'' Although he believed resolution of the conflict rested with the Soviet Union and China, which would have to determine how they would respond to American successes in South Vietnam, Reston commended Westmoreland and Bunker for being effective administration spokesmen. ''They have been careful in their estimates,'' he said, ''modest in their manner, and as factual as anybody can be in reporting on such a complicated war with so many different fronts.''[50]

Before he left for Washington, Westmoreland had begun to provide the Saigon correspondents with information on enemy use of Cambodian territory. That tactic also strengthened the president's position. The State Department had long opposed publicizing the subject on grounds that it was better to tolerate the enemy's use of Cambodia than to drive the country's mercurial leader, Prince Norodom Sihanouk, into open collaboration with the Communists. Westmoreland considered the policy naive. During October, with the backing of the U.S. Mission Council in Saigon, he began to press the State Department for permission to inform the press.

On the side, after learning that Associated Press reporters George MacArthur and Horst Faas had wrangled invitations from Sihanouk to visit Cambodia's border areas, he also instructed his subordinates to brief the reporters on where to look for enemy installations.[51] Traveling to Cambodia while Westmoreland was in Washington, the two had little difficulty finding a Viet Cong base camp located some nine kilometers from the South Vietnamese border, within striking distance of Loc Ninh. The facility possessed an excellent road stretching ominously toward South Vietnam.

Learning that a news story on the camp would be forthcoming, the State Department instructed its officers to avoid an ''I told you so'' attitude in their dealings with the Cambodian government. They were to respond to questions from the press by saying only that the story spoke for itself. Recognizing nevertheless that some explanation would be necessary, the agency permitted Westmoreland to brief the press on MACV's growing conviction that the Communists were using the Cambodian port of Sihanoukville on the Gulf of Thailand as a major entrepot for war supplies.[52] Westmoreland did so on 17 November at a

[50] Orr Kelly, ''In a Military Sense, The War Is About Won,'' *Washington Star*, 7 Nov 67; Kelly, ''The Enemy in Trouble—18 Months and No Big Victory,'' *Washington Star*, 8 Nov 67; Bob Considine, ''Foe Loses Five to Our One,'' *Philadelphia Inquirer*, 17 Sep 67; James Reston, ''Washington: Why Westmoreland and Bunker Are Optimistic,'' *New York Times*, 22 Nov 67.

[51] Mission Council Action Memo 235, 16 Oct 67, sub: Actions Emerging From Mission Council Meeting of 9 October, MCAM 67 file, CMH; Msg, State 73699 to Saigon, 23 Nov 67, FAIM/IR; Westmoreland, *A Soldier Reports*, pp. 180f.

[52] Msg, State 71253 to Saigon, 18 Nov 67, FAIM/IR.

private dinner in Washington for the representatives of most of the major news organizations. He specified only that the reporters attribute the story to military sources and that they refrain from publishing it while he was in the United States. When the briefing appeared in the press shortly after MacArthur and Faas' revelation—attributed to Westmoreland despite his prohibition—it served to enhance the credibility of the Johnson administration's claims about the war and to steal the limelight from those who had been critical of official statements.[53]

The Johnson administration's public relations campaign continued into December. Shortly before he returned to Saigon, Westmoreland informed Abrams that Johnson remained keenly interested in improving the image of South Vietnam's government and army. To that end, he wanted the Military Assistance Command to do everything it could to prompt the South Vietnamese to accelerate their anticorruption efforts and to publicize South Vietnamese successes in battle. In the meantime, besides spotlighting Communist sanctuaries in Cambodia, the command was to move as quickly as possible to make body counts credible to the press and to emphasize U.S. and South Vietnamese efforts to avoid civilian casualties.[54]

To that end, during December, while an increasing number of clergymen, editors, and congressmen arrived in South Vietnam for officially sponsored tours, the Military Assistance Command attempted to demonstrate the success of its programs by opening the country's main road, Highway 1, from the Cambodian border to the Demilitarized Zone. With American troops securing the road, Vice President Ky, accompanied by newsmen, drove its length to prove how safe it had become. Then the troops withdrew, leaving those sections of the road that were of little use or too difficult to defend once more to the enemy.[55]

As the campaign continued, President Johnson stated in an interview on CBS Television that although the enemy had yet to win a single victory on the battlefield he continued to search for some way to break the will of the American people. Other officials of the administration also spoke out, among them William P. Bundy—whose explanation of why the United States was in South Vietnam appeared prominently in *U.S. News & World Report*—and General Abrams, who claimed in a *U.S. News & World Report* interview that the South Vietnamese were beginning to make "real progress." The press repeated those arguments. On 27 November Howard Handleman of *U.S. News* observed that, with the Military Assistance Command reporting 67 percent of South Vietnam's people living in government-controlled areas and with the enemy apparently incapable of scoring a significant victory, "the coin has flipped to our side in Vietnam." A month later *U.S. News* updated Handleman's report with a case study of the II Corps

[53] George MacArthur and Horst Faas, "Camp in Cambodia Linked to Viet Cong," *Washington Post*, 20 Nov 67; George C. Wilson, "Westmoreland Source of Stories on Cambodia," *Washington Post*, 25 Nov 67; Westmoreland, *A Soldier Reports*, p. 183.

[54] Msg, Westmoreland HWA 3423 to Abrams, 23 Nov 67, and Msg, Westmoreland HWA 3424 to Abrams, 23 Nov 67, both in Westmoreland Papers, CMH.

[55] Sharp and Westmoreland, *Report on the War*, p. 144.

Tactical Zone. Titled "A Report of Success in the War," the article noted that the portion of Highway 1 running through the area was safe and that the incident rate in the region was the lowest in all of South Vietnam. Meanwhile, Hanson W. Baldwin warned in a series of *New York Times* articles that much of what the Communists were doing in South Vietnam was calculated to affect public opinion in the United States.[56]

Criticism of the War Increases, December 1967

Although public opinion polls indicated that the president's popularity had risen some eleven points at the beginning of December because of Westmoreland's and Bunker's assurances that the war was going well, critics of the war remained on the offensive throughout the period. In New York City, after three days of demonstrations, 604 protesters were arrested for disrupting activities at the city's U.S. Army induction center. Angry students at California State College, Los Angeles, ejected representatives of the Dow Chemical Company, a major manufacturer of napalm, from the school's placement office. In Washington, D.C., Senator Fulbright began extensive hearings on the Gulf of Tonkin incidents, revealing for the first time that the Johnson administration had prepared a draft resolution justifying military action in South Vietnam even before the incidents occurred. Meanwhile, in the press Bernard Weintraub of the *New York Times* published a gloomy article drawn from a leaked official report circulating privately within the U.S. mission in Saigon to the effect that enemy terrorism and propaganda in South Vietnam were on the rise. Many South Vietnamese were turning against the United States, Weintraub added, because they believed their ally had become a pervasive threat to their nation's sovereignty.[57]

Criticism likewise increased in Congress, where a number of senators and congressmen were reported to have moved from support for the war to positions of doubt. One formerly prowar senator, Thruston Morton of Kentucky, had gone so far as to advertise his change of mind in a series of press conferences and Senate speeches in which he labeled U.S. policies in South Vietnam all but bankrupt. Another former supporter of the war, Congressman Thomas P. O'Neill of Massachusetts, informed reporters that, after listening to the administration's side of the story for over a year, "I've decided Rusk and McNamara and the rest of

[56] Msg, State 86286 to Saigon, 18 Dec 67, FAIM/IR; William P. Bundy, "Why U.S. Is in Vietnam: An Official Explanation," *U.S. News & World Report*, 18 Dec 67, p. 48; "A Top U.S. General Sees 'Real Progress,' " *U.S. News & World Report*, 4 Dec 67, p. 63; Howard Handleman, "The Coin Has Flipped to Our Side in Vietnam," *U.S. News & World Report*, 25 Dec 67, p. 25; Hanson W. Baldwin, "Sanctuaries Viewed as Major War Factor," *New York Times*, 28 Dec 67; Baldwin, "Vietnam Report: Foe Seeks To Sway U.S. Public," *New York Times*, 26 Dec 67.

[57] Louis Harris, "Johnson Regains Popularity," *Philadelphia Inquirer*, 4 Dec 67; "More Protests, Growing Lawlessness," *U.S. News & World Report*, 18 Dec 67, p. 6. For a summary of press reporting, Msg, State 80119 to Saigon, 6 Dec 67, and Msg, Saigon 2070 to State, 7 Dec 67, both in FAIM/IR.

them are wrong.'' Although less than 30 of 100 senators and 50 of 435 representatives had gone on record to oppose the war, there were also indications that many lawmakers were withholding judgment. As reporter Don Oberdorfer noted in an article in the *New York Times*, although a number of congressmen had become increasingly critical of the war in public, many more admitted privately that they intended to ''stay loose'' as long as possible in order to see what happened before the next year's election campaign forced them to take sides.[58]

The American news media also wavered. During January 1968 the *Boston Globe* surveyed editorial opinion among thirty-nine major U.S. metropolitan daily newspapers with a combined circulation of over twenty-two million. It found that seven—the *Charlotte Observer*, the *Cleveland Plain Dealer*, the *Detroit Free Press*, the *Kansas City Star*, the *Los Angeles Times*, the *Minneapolis Star-Tribune*, and the *Richmond Times-Dispatch*—had moved during the previous year from support for the administration's war policies to criticism. Four other papers—the *Chicago Tribune*, the *Cincinnati Enquirer*, the *New York Daily News*, and the *St. Louis Globe-Democrat*—had moved in the opposite direction, abandoning the administration's position to recommend stronger military measures and faster escalation.[59]

Recognizing that public opinion had yet to solidify, the Johnson administration pressed its public relations campaign. On 1 December Walt Rostow appeared on the ABC Television news program ''Scope'' to claim that enemy casualties were increasing in comparison to those of the United States and that the enemy's ability to move more men into the South had been limited by his need to repair the damage done by the bombing. In a speech at the AFL-CIO annual convention later in the month, Dean Rusk won resounding applause when he compared critics of the president's Vietnam policies to Adolf Hitler's storm troopers. Rusk added that the fidelity of the United States to its mutual security treaties around the world was ''the principal pillar of peace in this period of world history.'' Shortly thereafter, William P. Bundy summarized the history of the American involvement in South Vietnam in *U.S. News & World Report*, pointing out that if the United States had failed to take action when it did all of Southeast Asia would have been in jeopardy.[60]

The administration's arguments once more echoed within the news media. On 25 December *U.S. News & World Report* relayed claims by official spokesmen that the enemy was on the run in South Vietnam's II Corps Tactical Zone. Hanson Baldwin meanwhile reported that, according to nearly all the U.S. officials in Saigon from Ambassador Bunker down, the main battleground in the coming year would be in the United States. ''Intelligence appreciations are unanimous on one point,'' he said, ''that the current winter-spring offensive . . . is keyed

[58] Don Oberdorfer, ''Wobble on the War,'' *New York Times*, 17 Dec 67.

[59] Min S. Yee, ''The U.S. Press and Its Agony of Appraisal,'' *Boston Globe*, 18 Feb 68.

[60] ABC News, ''Rostow on the War,'' Radio-TV-Defense Dialog, 2 Dec 67; Helen D. Bentley, ''Vietnam Critics Scored by Rusk,'' *Baltimore Sun*, 9 Dec 67; Bundy, ''Why the U.S. Is in Vietnam: An Official Explanation.''

primarily to strengthening opposition to the war . . . and influencing American and world public opinion during a presidential election year.''[61]

The campaign continued into the new year. On 2 January 1968, Admiral Sharp dispatched a number of recommendations to strengthen MACV's public information effort. Since attempts to publicize the South Vietnamese program to eradicate corruption had served mainly to highlight the problem, he advised Westmoreland to concentrate instead on the increasing effectiveness of South Vietnamese governmental operations. In the same way, the Military Assistance Command had to exercise more care in selecting the South Vietnamese military units it recommended to the press. ''As some media representatives have stated,'' he said, ''they have heard before that the ARVN were ready to acquit themselves in combat, only to discover upon visiting the units in question that the claims would not be justified.''[62]

As work on Sharp's recommendations moved forward, Robert Komer held a news conference in Saigon on the pacification program. Although he specified carefully that the United States still had a long way to go, Komer observed that trends were nevertheless ''significantly upward.'' Advances in pacification, he said, reflected the improved performance of American and South Vietnamese troops, better organization of American and South Vietnamese officials, and better allocation of resources. Year-end reports on 12,277 hamlets showed that 67 percent of the South Vietnamese population lived in secure cities and towns or under reasonably good security conditions in the countryside—an increase over the previous year of 4.8 percent. In addition, the rice harvest was richer, more Honda motorcycles could be seen in the hamlets, and more tractors were being imported—all signs that public confidence was increasing. ''You don't start buying tractors with your piasters,'' Komer said, ''unless you expect you're going to be able to use them.''[63]

Komer's briefing marked the end of President Johnson's public relations campaign. On 27 January General Wheeler cabled Westmoreland to inform him of Johnson's satisfaction with the program and to commend General Sidle for the role he had thus far played in making it a success. Two days later the enemy launched the Tet offensive.[64]

[61] ''A Report of Success in the War,'' *U.S. News & World Report*, 25 Dec 67; Baldwin, ''Vietnam Report: Foe Seeks To Sway U.S. Public.''
[62] Msg, Sharp to Westmoreland, 5 Jan 68, Westmoreland Papers, CMH.
[63] MACOI, Text of Ambassador Komer's News Conference, 24 Jan 68, DDI Pacification 1-a file.
[64] Msg, Wheeler JCS 920 to Westmoreland, 27 Jan 68, Westmoreland Papers, CMH.

15

A Hard Blow

General Westmoreland had watched the enemy infiltrate men and materiel into the northernmost provinces of South Vietnam for months. Believing a major enemy offensive imminent, throughout December he argued vehemently, but to no avail, against plans to proclaim cease-fires over the Christmas and Tet holidays. By January 1968 he was even more alarmed. Intelligence reports indicated that an unprecedented volume of enemy supplies was moving down the Ho Chi Minh Trail and that three North Vietnamese division headquarters and seven regiments—15,000 men—had taken up station in the vicinity of the Marine outpost at Khe Sanh in Quang Tri Province. Khe Sanh could serve as an important staging point for any future American or South Vietnamese offensive into Laos to cut the trail, and loss of the base would open the populated portions of the I Corps Tactical Zone to the enemy while giving the Communists a major propaganda coup. Thus Westmoreland moved immediately to reinforce the position. Directing the bulk of U.S. B–52 raids into the region, he also began to transfer U.S. Army units northward out of such heavily populated areas as Binh Dinh Province.[1]

Unsure about his arrangements and uncertain of the marines, whose standards and tactics he believed left much to be desired, Westmoreland took several further steps. Using the infusion of Army units into the I Corps Tactical Zone as justification, he established a MACV forward command post in the region and sent his deputy, General Abrams, to act in his behalf should the need arise. Then, reasoning that the United States might find itself facing defeat if large numbers of the enemy surged across the Demilitarized Zone, he sought and received from

[1] Msg, Wheeler JCS 343 to Westmoreland, 11 Jan 68; Msg, Sharp to Wheeler, 15 Jan 68; Msg, Westmoreland MAC 547 to Wheeler, 12 Jan 68; Msg, Sharp to Wheeler, 20 Jan 68; Msg, Westmoreland MAC 686 to Sharp, 15 Jan 68; Msg, Westmoreland MAC 862 to Lt Gen Rossen et al., 19 Jan 68; Msg, Wheeler JCS 554 to Westmoreland, 19 Jan 68. All in Westmoreland Papers, CMH.

Admiral Sharp permission to begin contingency planning for the use of tactical nuclear weapons.[2]

If Westmoreland was alarmed, little of the urgency he communicated to his superiors in Washington appeared in the public statements of the Johnson administration. On 18 December General Wheeler warned in a speech before the Detroit Economic Club that the enemy might be planning an all-out offensive similar to the German drive in the Battle of the Bulge, but he coupled his remark with a reprimand to critics of the administration's war policies. The news stories that resulted played up the political aspects of his remarks, omitting or barely mentioning his warning. In the same way, President Johnson noted in a meeting with allied leaders in Australia on 21 December that he expected enemy suicide attacks and kamikaze raids in South Vietnam in the near future but failed to say anything in public. On 5 January the Joint U.S. Public Affairs Office released the text of a captured enemy notebook which read, "The central headquarters has ordered the entire army and people of South Vietnam to implement general offensive and general uprising in order to achieve a decisive victory . . . ," but few reporters or officials apparently considered the document important. One of the clearest public warnings came from General Westmoreland on 17 January, when he told the Associated Press that he expected a major enemy offensive in the I Corps Tactical Zone; but his warning was overshadowed by a comment to the effect that the enemy appeared to have temporarily run out of steam and that the ground war in South Vietnam had slipped into one of its periodic lulls. Westmoreland sounded a similar note during an interview on 22 January but again limited his comments to the I Corps Tactical Zone, which he believed would bear the brunt of the enemy's attack. In all, the few warnings of a coming offensive that appeared were so oblique or so hedged with official optimism that even the Johnson administration was unprepared for the broad extent and violence of the attack that developed.[3]

The Tet Offensive Begins

The attack on Khe Sanh that Westmoreland had predicted failed to materialize. Instead, on the evening of 30 January as the population of South Vietnam prepared to celebrate Tet, the year's most festive holiday, enemy troops, some disguised as civilians, attacked Da Nang, Nha Trang, Kontum, and a number of other cities and hamlets in the I and II Corps Tactical Zones. Westmoreland responded to queries from Washington by noting that the enemy had "displayed an attitude of almost desperation" in making the attacks and as a result had lost better

[2] Msg, Westmoreland MAC 1233 to Sharp, 26 Jan 68; Msg, Westmoreland MAC 1011 to Wheeler, 22 Jan 68; Msg, Westmoreland MAC 1164 to Sharp, 24 Jan 68; Msg, Sharp to Wheeler, 2 Feb 68. All in Westmoreland Papers, CMH.

[3] For a treatment of official statements predicting an offensive, see Braestrup, *Big Story*, 1: 60–77.

Enemy Dead on the Grounds of the U.S. Embassy, Saigon

than seven hundred killed—more than at any other time in recent memory. The situation, he said, was well in hand.[4]

Reasoning that the attacks were the prelude to the long-awaited enemy offensive, Westmoreland and Bunker prevailed upon President Thieu to cancel a 36-hour truce he had declared in honor of Tet and to recall all troops on holiday leave. Receiving word late or preoccupied with the festivities, many South Vietnamese officers and enlisted men failed to respond, leaving most of the units on duty at only 50 percent strength. President Thieu himself was so unimpressed by the warning that he left for My Tho in the Mekong Delta that afternoon to spend the holiday with his family.[5]

Although Westmoreland had placed U.S. troops on full alert several days before Tet and had deduced from the premature attacks in the I and II Corps Tactical Zones that some sort of offensive was imminent, no one anticipated the nationwide general offensive that developed. On the morning of 31 January, as late-night revelers returned home, the enemy launched simultaneous attacks on

[4] Memo, NMCC for the Chairman of the Joint Chiefs of Staff, 30 Jan 68, sub: Conversation With General Westmoreland, DDI Tet Offensive (7) file.

[5] Msg, Saigon 18268 to State, 6 Feb 68, DDI Tet Offensive (7) file; Sharp and Westmoreland, *Report on the War in Vietnam*, pp. 158f; Don Oberdorfer, *Tet!* (Garden City, N.J.: Doubleday, 1971), pp. 132f.

Westmoreland Tours the Embassy, *31 January 1968.*

five of South Vietnam's most important cities, thirty-six provincial capitals, sixty-four district capitals, and fifty hamlets. In Hue, some eight battalions of North Vietnamese and Viet Cong troops, aided by accomplices from within, penetrated the city's defenses, rapidly isolating the U.S. advisory team within its headquarters and taking virtual control of most of the city. In Saigon, eleven enemy local force battalions launched coordinated attacks on the Presidential Palace, three U.S. military billets, the South Vietnamese Joint General Staff compound, the city radio station, Tan Son Nhut Air Base, the Philippine embassy, and the newly constructed embassy of the United States.[6]

General Westmoreland was confident of his ability to repel the assault. Far from being an attempt to take control of South Vietnam's towns and cities, he told Wheeler, the attacks once more appeared to be a diversion in preparation for the long-expected offensive in the I Corps Tactical Zone. The enemy's heavy emphasis on dramatic results in populated areas seemed to indicate as well a desire to have some sort of psychological impact on world public opinion. The next morning, in a *New York Times* article by Hanson W. Baldwin, unidentified but obviously well placed U.S. sources in Washington made the same point.[7]

If Westmoreland was calm, the Saigon correspondents were aghast. Centering their attention on the fighting most accessible to them, the battle for the U.S. embassy in downtown Saigon, they turned the attacks into a *cause celebre*. At the beginning of the offensive, a nineteen-man squad of enemy sappers had breached the wall surrounding the embassy and entered the compound but failed to penetrate the chancery itself. The reporters were unable to see more than the upper floors of the building but heard a great volume of fire coming from that general direction. They thus took the word of officers at the scene and began filing mistaken reports to the effect that Communist commandos had occupied at least the lower floors of the embassy and that American troops were fighting to drive them out.[8]

[6] Msg, Saigon 18268 to State, 6 Feb 68; Msg, Westmoreland MACV 1449 to Wheeler, 31 Jan 68, Westmoreland Papers, CMH; Westmoreland and Sharp, *Report on the War in Vietnam*, pp. 158f.

[7] Msg, Westmoreland MACV 1449 to Wheeler, 31 Jan 68; Hanson W. Baldwin, "Target: Public Opinion," *New York Times*, 1 Feb 68.

[8] For a typical report, see Tom Buckley, "Foe Invades U.S. Saigon Embassy," *New York Times*, 31 Jan 68. This account of the offensive will be drawn from the MACV History, 1968, CMH files; Sharp

In contact with the embassy at all times during the attack, the State Department in Washington attempted to clarify the situation by pointing out to newsmen that the enemy had failed to penetrate the building. General Westmoreland said much the same thing at an impromptu press conference shortly after U.S. troops regained control of the embassy grounds. The news media nevertheless persisted in the error, trusting the word of military policemen at the scene over that of the commanding general. As a result, NBC News anchorman Chet Huntley told his audience that evening that enemy snipers located both in the embassy and on nearby rooftops had fired down upon American rescuers in the courtyard—the exact opposite of what had happened.[9]

General Westmoreland's news conference contributed to the reporters' misgivings. When Robert Schakne of CBS News asked how the general assessed the situation, Westmoreland implied that the enemy had suffered a great defeat. "In my opinion," he said, "this is diversionary to his main efforts which he had planned to take place in Quang Tri Province, from Laos toward Khe Sanh and across the DMZ. . . . Now yesterday the enemy exposed himself by virtue of this strategy, and he suffered great casualties." To many of the reporters present the comment seemed unreal. "How could any effort against Saigon," Peter Braestrup of the *Washington Post* later recalled, "especially downtown Saigon, be a diversion?"[10]

Gloomy news stories began to surface in the United States within hours of the attack. Orr Kelly of the *Washington Star* stressed that the United States had been caught off guard by the intensity and coordination of the offensive. CBS News correspondent Mike Wallace observed in a special television report that the raids had "demolished the myth" that allied military strength controlled South Vietnam. Free-lance reporter Sarah McClendon commented on the Washington, D.C., news program "Capital Tieline" that "the situation is very, very bad; and I think the people should realize this." Although warnings had appeared during January that an attack was imminent, she said, "just look how the American Embassy was so poorly protected." In an article entitled "Red Terror," *New York Daily News* reporter Jerry Green noted that the offensive represented a "potent propaganda victory" for the enemy, clouding a steady stream of official American optimism on the progress of the war. A *New York Times* editorial the day after the attacks began said that in combination with the enemy buildup around Khe Sanh, the raids had "undermined the optimism about the course of the war voiced in Saigon and Washington during the last few months. These are not the deeds

and Westmoreland, *Report on the War in Vietnam*, pp. 158f; RVNAF, *The Viet Cong "Tet" Offensive* (Saigon, 1969), CMH files; Colonel Hoang Ngoc Lung, *The General Offensive of 1968–1969*, Indochina Monographs (Washington, D.C.: U.S. Army Center of Military History, 1981). Two works on the role of the press have been used: Oberdorfer, *Tet!*; and Braestrup, *Big Story*.

[9] Braestrup, *Big Story*, 1:95. Westmoreland's comment was replayed on CBS Morning News, 1 Feb 68, Radio-TV-Defense Dialog.

[10] Westmoreland's comment from CBS Morning News, 1 Feb 68, Radio-TV-Defense Dialog; Braestrup, *Big Story*, 1: 124.

of an enemy whose fighting efficiency has 'progressively declined' and whose morale is 'sinking fast,' as United States military officials put it in November."[11]

Already afflicted by a crisis involving the capture by North Korean gunboats of an American electronic surveillance vessel, the USS *Pueblo*, the Johnson administration experienced, in Westmoreland's words, "great consternation" at the news from Saigon. Frequent calls and messages from Westmoreland, Bunker, and other officials in the U.S. mission attempted to restore some balance to the perceptions of official Washington, yet as Westmoreland observed in his diary, "this was more than offset by the alarming headlines and the gloom-and-doom type editorials that proceeded to propagandize the limited successes of the Viet Cong."[12] Adding to that effect was a chorus of alarmed comments from Congress, where, as *New York Times* correspondent Tom Wicker observed, the news had dealt "a hard blow." For although a number of congressmen reacted along ideological lines—opponents of the war charging that the offensive had substantiated their worst fears while supporters claimed it would prove a last desperate effort by the enemy—numerous middle-of-the-road members began to express shock and dismay. Senator John Stennis of Mississippi was blunt in his conclusions. Even if the attacks ultimately proved costly to the enemy, he told reporters, they were nevertheless humiliating to the president. "What happened?" another senator was reported to have asked; "I thought we were supposed to be winning this war."[13]

The MACV Office of Information could do little to penetrate the gloom. At the mercy of a communications system that often lagged hours behind events in the field even on a normal day, information officers issued what news they could, much of it fragmentary. On the morning of 1 February, for example, the MACV communique noted that an estimated enemy company had attacked a bridge and loading ramp in Hue City but that the marines had driven it off. The compound of the South Vietnamese 1st Division at Hue had received some small arms fire and the portion of the city north of the Perfume River was surrounded by enemy units, but no contact had occurred. That evening's release added only that two enemy mortar rounds had landed in the ammunition dump at Phu Bai, some fifteen kilometers to the south, causing light damage. In fact, by that time almost the entire city of Hue was in enemy hands.[14]

At the suggestion of President Johnson, who at the height of the fighting on 31 January had requested that Westmoreland "make a brief personal comment to the press each day . . . to convey to the American public your confidence in our capability to blunt these enemy moves," Westmoreland convened a press conference at the JUSPAO building on 1 February to bring the Saigon correspon-

[11] Orr Kelly, "U.S. Caught Off Guard by Intensity of Attacks," *Washington Star*, 31 Jan 68; Mike Wallace, "CBS News Special Report," 31 Jan 68, Radio-TV-Defense Dialog; "Capital Tieline," WRC-TV, Washington, 1 Feb 68, Radio-TV-Defense Dialog; Jerry Green, "Red Terror," *New York Daily News*, 1 Feb 68; "Bloody Path to Peace," *New York Times*, 1 Feb 68.

[12] Notes for 1 Feb 68, Westmoreland History, bk. 29, tab 1.

[13] Tom Wicker, "Viet Cong's Attacks Shock Washington," *New York Times*, 2 Feb 68.

[14] MACOI, News Releases 31-68, 1 Feb 68, and 32-68, 1 Feb 68, 334-74-593, box 66/22, WNRC.

dents up to date.[15] After outlining what had happened—again without mentioning Hue—Westmoreland gave his assessment of the enemy's strategy. He claimed that the U.S. command had foreseen attacks on South Vietnam's populated areas during the new year even though it had failed to predict an enemy initiative during the holiday itself. He then explained that the Communists had actually embarked on a three-phase campaign. The first, involving the battles of Loc Ninh and Dak To, had been designed to kill large numbers of American and South Vietnamese troops. The second, under way at that moment, concentrated on neutralizing government installations and headquarters. The third would erupt shortly. Far more violent than either of its predecessors and building upon a logistical base already in place, it would evolve in the northernmost provinces of South Vietnam and would entail an all-or-nothing effort. The enemy had already paid a dear price, Westmoreland said—5,800 men killed in the first days of the offensive—so many that it would take the units involved weeks and months to recover.[16]

Without going into as many details, General Wheeler and Secretary McNamara said much the same thing to the press in Washington. In an interview broadcast on the 1 February edition of the CBS Evening News, Wheeler observed that the enemy had lost so many men that the offensive had to be considered a failure. McNamara added the caveat that if the enemy had indeed failed militarily he might yet attempt to turn the situation to his advantage by making it appear to be a substantial psychological defeat for the United States.[17]

The Press Reacts

The attempt by official spokesmen to dispel the gloom had little effect on the news media. Official credibility had fallen so low that as late as 2 February, a day and a half after the attack on the U.S. embassy, the *New York Times* was still willing to publish a report by correspondent Tom Buckley that, according to witnesses, guerrillas had indeed penetrated the first floor of the embassy. The attacks demonstrated, Buckley said, that "after years of fighting and tens of thousands of casualties, the Viet Cong can still find thousands of men who are ready not only to strike at night and slip away but also to undertake missions in which death is the only possible outcome." A *Times* editorial the same day supported Buckley's point. Observing that the offensive was more than the diversion Westmoreland had claimed, the newspaper added that the success of the attacks threw official assertions of progress into doubt and raised serious questions about the competence of the South Vietnamese government and armed forces.[18]

[15] Msg, Wheeler JCS 8691 to Westmoreland, 31 Jan 68, Westmoreland Papers, CMH.
[16] Notes for 1 Feb 68, Westmoreland History; ASD PA Transcript, 1 Feb 68, sub: General Westmoreland Briefing, DDI Tet (1968) file.
[17] CBS Evening News, 1 Feb 68, Radio-TV-Defense Dialog.
[18] Tom Buckley, "Offensive Is Said To Pinpoint Enemy's Strengths," *New York Times*, 2 Feb 68; "More Than Just a Diversion," *New York Times*, 2 Feb 68.

Other journals were equally harsh. Observing editorially that the war was at a turning point and that peace was the only legitimate goal the United States could have in South Vietnam, the *Christian Science Monitor* published an article on 3 February in which Beverly Deepe declared that the United States for the first time faced the possibility of military defeat. The *Wall Street Journal* was also alarmed. Despite official statements that the enemy had failed, the newspaper said, there was "something . . . awfully wrong" in South Vietnam. "The South Vietnamese government, with all the vast aid of the U.S., has revealed its inability to provide security for large masses of people in countryside and city." The offensive had raised in starkest fashion "not only the question of weakness in Saigon but of whether the U.S. effort is reaching a point of diminishing returns."[19]

A somber year-end report to Congress approved by Robert McNamara before the offensive but released on 2 February added to the misgivings of the press. Warning of a further buildup of enemy forces in the South, McNamara spoke frankly of the slow pace of pacification and of the fact that enemy forces remained a formidable threat to U.S. ends. Whatever the contribution of the United States to the struggle, he said, "We cannot provide the South Vietnamese with the will to survive as an independent nation . . . or with the ability and self-discipline a people must have to govern themselves." As soon as the report appeared, the news media took up its themes. The *Christian Science Monitor* labeled it "thought provoking" and repeated McNamara's contention that victory was ultimately up to the South Vietnamese. Stressing that the report was a far cry from earlier "roseate readings," *Newsweek* termed the document "McNamara's swan song" and underscored the secretary's admission that progress in South Vietnam had been uneven. Noting a caution McNamara had inserted to the effect that some of MACV's statistics were based on "judgment factors," the magazine added that the phrase might well be translated "guesswork."[20]

The MACV Office of Information attempted to clarify the situation for the press on 3 February by calling upon the director of MACV's Command Operations Center, Brig. Gen. John Chaisson, USMC, to brief reporters. Predicting erroneously that Hue would be cleared "in the next day or so," Chaisson admitted that the attacks had been a surprise. "We were confident that something would happen around the Tet period," he said, but "our intelligence . . . never unfolded to me any panorama of attacks such as happened this week." Chaisson credited the enemy with a "very successful offensive, in its initial phases," and surprising audacity. "We have been faced this past week," he said, "with a real battle. There's no sense in ducking it. There's no sense in hiding it."[21]

At a briefing the next day the chief of MACV's Intelligence Division, Brig. Gen.

[19] "War's Turning Point," *Christian Science Monitor*, 3 Feb 68; Beverly Deepe, "Blitz Erodes U.S. Position in Vietnam," *Christian Science Monitor*, 3 Feb 68; "Vietnam, The American Dilemma," *Wall Street Journal*, 6 Feb 68.

[20] Hedrick Smith, "McNamara Wary on Trend of War," *New York Times*, 2 Feb 68; "A Thought Provoking Farewell," *Christian Science Monitor*, 5 Feb 68; "The Pentagon: Swan Song," *Newsweek*, 12 Feb 68, p. 40.

[21] Excerpts from Saigon Briefing by Gen Chaisson, ASD PA, 3 Feb 68, DDI Tet Offensive (7) file.

Phillip B. Davidson, Jr., took a different position. Although candid in admitting that the enemy had yet to commit the majority of his troops to battle and that a second wave of attacks was still possible, Davidson attempted to explain away Chaisson's assertion that the Military Assistance Command had been surprised. Implying that the command had recognized the enemy's ability to attack at Tet and had expected some sort of offensive all along, Davidson said that Chaisson had only admitted that he *personally* had been surprised.[22]

President Johnson also spoke out. At a White House Medal of Honor ceremony on 1 February, he said that the enemy would fail again and again because Americans would never yield. Continuing his remarks the next day at an unannounced news conference, he claimed that he was choosing his words carefully to deny the enemy any opportunity to twist them into an affirmation that the offensive had been some sort of Communist psychological victory. In fact, he said, the offensive had failed. The Viet Cong had lost 10,000 men while only 249 Americans and 500 South Vietnamese had fallen. Although the attacks had caused disruptions and would require adjustments, he saw no reason to change his estimate that "we have made progress."[23]

The press disputed much of what the president and his spokesmen said. Taking special notice of the contradictions apparent among the various versions of events, reporters in general suspected MACV's estimate of ten thousand enemy deaths. They also questioned whether the attacks had been as well anticipated as Johnson had implied. Cynthia Parsons of the *Christian Science Monitor* contrasted Chaisson's briefing with the one given by Davidson, observing that one general had called the offensive a surprise while the other had all but denied it. In the same way, the president had stated that the situation was under control and that the enemy had suffered severe casualties, yet Davidson and other officials in South Vietnam admitted that the enemy had still to commit the majority of his forces to combat. In a 19 February editorial entitled "Misled, In Every Sense," the *New Republic,* never a friend of the administration's war policies, was even more critical. The magazine asked rhetorically why the South Vietnamese had been so unprepared if the Military Assistance Command had been certain the enemy would attack: "Or is forewarned not forearmed in this weird war?" A year before, President Johnson had said that the enemy was losing his grip on South Vietnam. With Tet, that prophecy seemed as broken as the policy it served. "We are," the magazine avowed, quoting *New York Times* columnist James Reston, "the flies that captured the flypaper."[24]

General Westmoreland was himself wary at first of the statistics his command

[22] Excerpts from Saigon Briefing by Brig Gen Phillip B. Davidson, Jr., ASD PA, 4 Feb 68, DDI Tet (1968) file.

[23] Max Frankel, "President Foresees Khe Sanh Victory," *New York Times,* 2 Feb 68; Frankel, "Warning Is Given, President Terms U.S. Ready for a Push By Enemy at Khe Sanh," *New York Times,* 3 Feb 68; "Transcript of the President's News Conference," *New York Times,* 3 Feb 68.

[24] Lee Lescaze, "Allied Figures on Casualties Are Thrown Into Question," *Washington Post,* 3 Feb 68; Cynthia Parsons, "Saigon Briefings Puzzle Reporter," *Christian Science Monitor,* 9 Feb 68; "Misled, In Every Sense," *New Republic,* 17 Feb 68, p. 7.

was releasing. He instructed the MACV Office of Information to screen measures of progress thoroughly before releasing them to the press and ordered the MACV inspector general to monitor carefully the gathering of enemy casualty figures. As the offensive progressed, he nevertheless came to believe that the enemy had orders to do or die and that the resulting all-out effort had made MACV's body count eminently respectable. Defending his command's statistics at a 25 February news conference, he emphasized that "we seldom know the number of killed in action resulting from B–52, tactical air, and artillery strikes. . . . We never know how many die from their wounds. . . . I am convinced that these unknowns more than offset the relatively small inaccuracies of our accounting system."[25]

A Prizewinning Picture

Although Westmoreland could at least attempt to explain MACV's casualty statistics, there was little he could do about another aspect of the offensive that was drawing the attention of the press—apparent violations of the laws of war. On the morning of 2 February, a photograph by Associated Press photographer Eddie Adams, which caught the chief of South Vietnam's National Police, Brig. Gen. Nguyen Ngoc Loan, in the act of executing a newly captured Viet Cong officer, appeared in almost every important newspaper in the United States. A film of the incident by NBC News cameraman Vo Suu played on the Huntley-Brinkley Report that evening, edited just enough to eliminate the blood spurting from the man's head as he lay dead at Loan's feet.[26]

The Associated Press filed a brief story to accompany Adams' picture, noting that General Loan had told newsmen at the scene that the Viet Cong had "killed many Americans and many of my people." Limited by constraints of time, NBC allowed Suu's film to speak virtually for itself. Correspondent Howard Tuckner provide only the barest narrative. "Government troops have captured the commander of the Viet Cong commando unit," Tuckner said after describing a firefight that had occurred around Saigon's An Quang Pagoda. "He was roughed up badly but refused to talk. . . . The Chief of South Vietnam's National Police, Brig. Gen. Nguyen Ngoc Loan, was waiting for him." Tuckner said nothing more. Loan fired. The man fell. The picture faded to black. After a three-second pause, NBC went directly to a station break and a commercial announcement.[27]

[25] Transcript, Westmoreland News Conference, 25 Feb 68, DDI Tet Offensive (7) file. See also Memo, MACV CofS 68–30, 3 Feb 68, sub: CIIB Meeting, Westmoreland History, bk. 29, tab 17; Msg, Westmoreland MACV 1754 to Wheeler, 7 Feb 68, Westmoreland Papers, CMH.

[26] "Grim and Ghastly Picture," *New York Daily News*, 3 Feb 68; George A. Bailey and Lawrence W. Lichty, "Rough Justice on a Saigon Street: A Gatekeeper Study of NBC's Tet Execution Film," *Journalism Quarterly*, Summer 1972; Braestrup, *Big Story*, 1: 460f; Oberdorfer, *Tet!*, pp. 161f.

[27] Bailey and Lichty, "Rough Justice on a Saigon Street," p. 274; Unidentified AP Report, 1 Feb 68, DDI Tet (7) file; Huntley-Brinkley Report, 2 Feb 68, Radio-TV-Defense Dialog.

Loan Executes Viet Cong Officer

Although NBC and other television networks replayed Suu's film several times, it never received much attention. Television could provide at most a fleeting if shocking impression of what had happened. Adams' photograph, on the other hand, appeared again and again, winning for the photographer a Pulitzer Prize and a host of other awards.

Many of the journals that commented on Adams' picture attempted to balance it with some allusion to enemy atrocities. The *New York Times* published it over a picture of a South Vietnamese officer holding the body of his murdered child. The *New York Daily News* observed that however ghastly Loan's act, the Communists were attempting to kill as many Americans and South Vietnamese as they could. The *Chicago Daily News* commented that "there is not much point now in going queasy over a picture of one man shooting another, there is worse to come." The *Chicago Tribune* charged that the antiwar movement was quick to exaggerate U.S. and South Vietnamese atrocities but fell silent when the Viet Cong murdered families and deliberately obliterated villages.[28]

A reaction nevertheless set in almost as soon as the picture appeared, with

[28] *New York Times*, 2 Feb 68, p. 1; "Grim and Ghastly Picture"; "A Strong Stomach Helps," *Chicago Daily News*, 7 Feb 68; "The Protesters Are Silent," *Chicago Tribune*, 8 Feb 68.

Congressman Henry S. Reuss of Wisconsin leading the way. Reuss told General Wheeler that he was well aware of Communist atrocities but that despicable acts by the enemy could never "justify or excuse actions by United States or allied forces which sink to this level. Murder or torture of prisoners is horrible and un-American."[29] Shortly thereafter, the Associated Press circulated a report from its Saigon bureau that during the previous week no quarter had been asked and none given. If the enemy maintained lists of men to be killed, it said, government troops appeared just as willing to execute enemy prisoners, often with the approval of their American advisers. "If I had my way," one U.S. Army sergeant was reported to have said, "we would execute on the spot every Viet Cong and Viet Cong suspect we catch."[30]

General Wheeler responded to the criticism by directing attention to the enemy's atrocities. In a widely publicized letter to Reuss, he suggested that if Loan's act was despicable, it had nevertheless occurred "in a flash of outrage rather than 'in cold blood.' " He termed the picture of the South Vietnamese officer carrying his murdered child a "sickening indictment of our enemy's real nature" and contended that "by any decent-minded measurement," the conduct of the South Vietnamese Army was more scrupulous than that of the Communists.[31] Vice President Ky went on record with the same defense. On 5 February he told newsmen that he had given orders forbidding the mistreatment of prisoners but that the enemy was not only killing South Vietnam's soldiers, he was murdering their families. "I know the foreign press makes a lot of noise about this death," he said emotionally, "but when you see your friends die it is hard to control your reactions."[32]

The U.S. mission in Saigon cautioned the South Vietnamese on their treatment of captives, even prevailing upon them to remove several execution posts they had erected in the city's central market place, but stories of American and South Vietnamese atrocities continued to circulate. One of the most telling appeared on the morning of 19 February, when the *New York Times*, the *Washington Post*, and many other American newspapers published Associated Press photographs of a South Vietnamese Marine private shooting an enemy captive despite orders to the contrary from an officer. An accompanying article reported that an unidentified American adviser had told newsmen at the time that "we usually kill the seriously wounded Viet Cong for two reasons. One is that the hospitals are so full of our own soldiers and civilians there is no room for the enemy. The second is that when you've seen five-year-old girls with their eyes blindfolded, their arms tied behind their backs, and bullets in their brains, you look for revenge. I saw two little girls that dead [*sic*] yesterday. One hour ago I shot a Viet Cong."[33]

[29] Telg, Henry S. Reuss to Gen Wheeler, 2 Feb 68, DDI Tet Offensive (7) file.

[30] [AP], "Viet Cong, Allied Atrocities Reflect Bitterness of War," *Baltimore Sun*, 4 Feb 68.

[31] Ltr, Wheeler to Henry S. Reuss, 3 Feb 68, DDI Tet Offensive (7) file; [AP], "Wheeler Deplores Execution," *New York Times*, 6 Feb 68.

[32] Tom Buckley, "Ky Says Regime Will Arm Public," *New York Times*, 6 Feb 68.

[33] [Agence France Presse], "U.S. Cautioning Saigon on Captives' Treatment," *New York Times*, 5 Feb 68. See also Msg, Joint State/Defense 118474 to Saigon, 21 Feb 67, CMH files.

The State and Defense Departments immediately instructed the Military Assistance Command to redouble its effort to keep the actions of U.S. personnel in conformity with the laws of war. The two agencies also asked the U.S. embassy once more to inform the South Vietnamese government that the interests of both nations required closer control of troops in the field. The message closed with a warning that if the adviser was telling the truth rather than merely trying to impress a gullible newsman, a serious violation of law had occurred, one that could implicate U.S. commanders should they fail properly to investigate.[34]

Although the command and the U.S. mission complied, there was, in fact, little either could do. Without witnesses willing to come forward, legal action against offenders was almost impossible. As for bringing the South Vietnamese government into line, the United States lacked the colonial powers the French had exercised and so could only advise.

In the end, the publicity surrounding atrocities by U.S. and South Vietnamese troops during Tet probably had little effect on American public opinion. An estimated twenty million Americans watched the television account of the execution by General Loan, yet NBC received only ninety letters of protest from viewers. Fifty-six accused the network of bad taste. The rest objected because it had screened the film at a time when children were watching. Few alluded to the Vietnam War itself, apparently accepting the film as an accurate representation of what war was all about.[35]

By the end of the first week in February, reports on the status of the Tet offensive were pouring into MACV's headquarters in Saigon. Although heavy fighting continued in Saigon and at Phu Bai and Khe Sanh and although enemy units retained possession of much of Hue, all indications were that the Communists had failed to achieve most of their major objectives. Word from Kontum Province in the II Corps Tactical Zone, for example, revealed that enemy forces had hidden large quantities of weapons and ammunition in the homes of sympathizers and had lived among the people undetected for days but had never come close to instigating the sort of popular uprising against the Saigon government they had sought. Although the military commander of Kontum City, Lt. Col. Nguyen Tran Luat, had spent most of the time strengthening the defenses of his own home and demolishing the homes of his political rivals, South Vietnamese forces in the region had stood and fought rather than surrendering immediately as the enemy had expected. Much the same thing happened in the IV Corps Tactical Zone, where the people hid the enemy out of fear but nevertheless refused to turn against the government. South Vietnamese military units in the region likewise held firm—despite the fact that the corps commander hid in his mansion behind a screen of tanks and one division commander took the precaution of wearing civilian clothing beneath his uniform.[36]

[34] Msg, Joint State/Defense 118474 to Saigon, 21 Feb 67.

[35] Bailey and Lichty, "Rough Justice on a Saigon Street."

[36] Msg, Westmoreland MAC 1614 to Wheeler, 4 Feb 68, sub: Appraisal of Situation After Five Days, Westmoreland Papers, CMH. Msg, Saigon 18405 to State, 7 Feb 68, sub: The Situation in Kontum,

If the enemy's offensive miscarried, it nevertheless dealt South Vietnam a severe blow, bringing the war to the country's cities, producing large numbers of refugees, and generating great devastation in some areas. South Vietnamese troops looted portions of Can Tho, My Tho, and Chau Duc in the delta, and U.S. advisers complained that in repelling the enemy attack, American and South Vietnamese forces had unnecessarily ravaged large portions of such important towns as Can Tho and Ben Tre. So extensive was the damage in some places that U.S. officials became concerned lest the morale of the South Vietnamese population suffer irreparable harm. Although they believed that gratitude would supplant resentment if restitution was quick and effective, they had little confidence in the South Vietnamese bureaucracy's willingness to take on the task. Local officials were already making excuses. Nothing could be done about the refugees for the time being, those functionaries said, because government offices were understaffed, too many stores were closed, and the defeat of the enemy obviously came first.[37]

General Westmoreland recognized the problem. On 2 February he told his staff that the South Vietnamese appeared to be immobilized by shock and that the situation could only deteriorate if the United States failed to galvanize them into action. With the concurrence of Ambassador Bunker, he created a joint U.S.–South Vietnamese working group code named Operation Recovery to coordinate the rebuilding process. Ostensibly under the direction of Vice President Ky, the organization would take its driving force from Komer and his deputy, Maj. Gen. George Forsythe.[38]

Believing that the attacks and the enemy's extensive propaganda had produced a strongly negative effect on South Vietnamese and American public opinion, President Johnson lent his weight to Westmoreland's effort. He instructed Ambassador Bunker to inform the South Vietnamese government that it needed to move decisively to eradicate the deep-seated complaints which had made the enemy's military and political efforts so easy. The time for caution and deliberation was past, he said. The government of South Vietnam had to move urgently to eliminate corruption, strengthen the armed forces, reconstitute the intelligence services, and replace corrupt and inept officials.[39]

Although Operation Recovery succeeded over the long run, mobilizing youth groups to assist in the cleanup, distributing money and building materials to more than one million displaced persons, and alleviating food shortages, it rapidly lost momentum. Money ran out; the South Vietnamese military began drafting the officials responsible for the recovery; and minor functionaries at the province level applied the usual red tape. In addition, after only ten days, President Thieu

and Msg, Saigon 18584 to State, 8 Feb 68, sub: The Delta After the Tet Offensive, both in DDI Tet Offensive (7) file.

[37] Msg, Westmoreland MAC 1614 to Wheeler, 4 Feb 68, Westmoreland Papers, CMH; Msg, Saigon 18584 to State, 8 Feb 68.

[38] Notes for 2 Feb 68, Westmoreland History, bk. 29, tab 1.

[39] Memo, Walt W. Rostow for Secretary Rusk, 3 Feb 68, and Draft Msg, Johnson to Bunker, 3 Feb 68, both in FAIM/IR.

became concerned about the prominence Ky was attaining in the program and decided to take the vice president's place despite the fact that he was already heavily burdened with affairs of state and could spare little time for the project.[40]

The destruction and disorganization prevailing in South Vietnam received wide coverage in the American news media, where reporters and commentators emphasized the violence of the offensive and the suffering both sides were inflicting on South Vietnam's civilians. "At what point do you turn your heavy guns and fighter bombers on your own city?" Peter Arnett asked on 7 February, after visiting the city of Ben Tre in the Mekong Delta during a MACV-sponsored tour of battle areas. "When does the infliction of civilian casualties become irrelevant as long as the enemy is destroyed?" Arnett reported that the South Vietnamese Corps Command had refused to authorize air strikes and artillery fire upon Ben Tre until the total destruction of defending forces appeared imminent, but he offset that statement with a comment by an anonymous U.S. Air Force major that "it became necessary to destroy the town to save it." Featured by rewrite men in New York, the phrase rapidly became one of the war's most serviceable cliches. The *New York Times* reprinted it almost as soon as it appeared. *Time* did the same, adding that the bulk of the destruction in South Vietnam had occurred during U.S. and South Vietnamese counterattacks. The statement became a favorite of the antiwar movement, which resurrected it again and again over the years whenever a speaker needed a quotation to point up the supposed absurdity of the war.[41]

The Countryside: More or Less Secure?

Reasoning that the effort to win popular support depended largely upon the peasantry's confidence in government protection, reporters were also quick to assert that the offensive had dealt a severe blow to the pacification program. Ward Just concluded in a 5 February article for the *Washington Post* that the program was dead. A week later, Lee Lescaze of the *Post* termed pacification "one of the casualties of the offensive." A few reporters, among them Bernard Weintraub of the *New York Times*, noted cautiously that the impact of the offensive on pacification remained unclear because no one was sure what had happened in the countryside, yet most tended to agree with Charles Mohr, also of the *Times*. By withdrawing into the cities, Mohr said, South Vietnam's rural pacification teams had left the peasantry exposed to the enemy. *Newsweek*'s judgment was even harsher. Observing that Communist troops had been able to traverse supposedly pacified territories without being betrayed by the people, the magazine con-

[40] Rand Corporation Interv with Robert Komer, 7 May 70, D (1)–20104–ARPA, CMH files; Msg, Saigon 39547 to State, 5 Oct 68, sub: Final Report on Project Recovery, CMH files; Hunt, Pacification: Managing the "Other War."

[41] Arnett's report is quoted in its entirety in Braestrup, *Big Story*, 1: 254; "Survivors Hunt Dead of Ben Tre," *New York Times*, 8 Feb 68; "The War: Picking Up the Pieces," *Time*, 16 Feb 68, p. 34.

cluded that the offensive had made a mockery of Komer's claim that 67 percent of South Vietnam's population lived in secure areas.[42]

The "Southeast Asia Program Report" for February 1968, issued by the Office of the Assistant Secretary of Defense for Systems Analysis, agreed with those conclusions. On the basis of preliminary reports from the field, it said that the offensive had apparently "killed the Revolutionary Development Program as currently conceived" and that "to a large extent the Viet Cong now control the countryside."[43]

The U.S. mission in Saigon knew little about the Systems Analysis report until the end of March but attempted almost immediately to counter the stories that were appearing in the press. On 5 February Komer told Robert Schakne of CBS News that although the figures in the Hamlet Evaluation Survey would surely fall in coming weeks, longer-term considerations were far more important. "Are we going to end up with this country more secure or less secure?" he asked. "I'm not so sure it's going to be a lot less secure." Appearing on the 18 February edition of the CBS News program "Face the Nation," Ambassador Bunker added that if U.S. pacification teams had withdrawn from the countryside to protect the cities, the Viet Cong had also pulled their forces into the cities to mount the attacks—in effect, weakening Communist positions in the countryside. "Tet is the equivalent to Christmas, New Year's, Washington's Birthday, the Fourth of July and Mother's Day all rolled into one," Bunker continued. "I believe it has not been violated in over a thousand years." By breaking that tradition, he said, the enemy had undoubtedly forfeited the respect of many South Vietnamese.[44]

On 24 February, Komer held a background briefing for the press in Saigon at which he stated that although the pacification program had suffered a setback, the problem seemed much smaller than initially feared because the enemy had decided to bypass secure areas in order to keep the campaign against the cities secret. Far from abandoning the countryside, the South Vietnamese government had withdrawn only eighteen of fifty-one battalions supporting pacification and reported that up to 80 percent of the Regional and Popular Forces had remained on duty at their posts. As a result, only thirteen provinces had been seriously affected, the rest suffering slight or moderate damage. Komer claimed that the chief problems stemming from the offensive were psychological: the defensive attitude of pacification workers and popular fear of the Viet Cong. All would respond, he said, to efforts by the South Vietnamese government to reassert its authority in the countryside, a process well under way. The government was already providing for some 470,000 refugees and had allocated 174 million piasters to badly stricken provinces. Although six months might elapse before pacification regained momentum in some areas, the program was hardly as dead as some

[42] Ward Just, "Guerrillas Wreck Pacification Plan," *Washington Post*, 4 Feb 68; Lee Lescaze, "U.S. and Vietnam: Test in Battle," *Washington Post*, 11 Feb 68; Charles Mohr, "Pacification Program Is Almost at Standstill," *New York Times*, 14 Feb 68; "The VC's Week of Terror," *Newsweek*, 12 Feb 68, p. 30.

[43] Msg, Komer to Alan Enthoven, ASD (SA), 27 Mar 67, Cable file (1968), DepCORDS Papers, CMH.

[44] Robert Schakne, CBS Evening News, 5 Feb 68; "Bunker Sees Gains," *New York Times*, 19 Feb 68.

newsmen had suggested. Komer made the same point in a classified cable to the Office of Systems Analysis. The agency's criticisms were the product of preliminary reports, he said, by their very nature incomplete, unsystematic, and overly alarmed.[45]

Komer's contention was lost upon the press. Concentrating upon the negatives in the report, the *New York Times* headlined its account of the news conference, "U.S. Admits Blow to Pacification." Written by correspondent Bernard Weintraub, the article underscored Komer's concession that there were problems, only as an aside mentioning that the enemy had never attempted to control the countryside and that the withdrawal of pacification teams had been smaller than previously thought. The *Washington Post* at first paid little attention to the briefing, printing only a short Reuters account of what had transpired, but the next day it published an analysis by Murrey Marder and Chalmers Roberts which claimed that, whatever the official protestations, a substantial proportion of the protection afforded the peasantry had been withdrawn to defend the cities. Although it later retracted the allegation that pacification was dead, the Office of Systems Analysis also continued to emphasize negatives, prompting remonstrances from Komer as late as September that the agency was relying upon obsolete data with little application to current circumstances.[46]

The Battle of Hue

As allied forces regained the upper hand elsewhere in South Vietnam, the battle for Hue became a major attraction to the press. Although the MACV Office of Information was still issuing optimistic bulletins on events in the city as late as 8 February, a number of newsmen were by then on the scene and could report their observations. They emphasized that the fighting had been bitter and the enemy tenacious.[47]

Resembling the house-to-house combat that had often occurred during World War II, the battle for Hue was relatively easy for newsmen to cover. Although South Vietnamese troops were doing most of the fighting and suffering more casualties than the U.S. marines by a ratio of four-to-one, the reporters once more accompanied American units, where they felt comfortable and where they knew

[45] Msg, Saigon to State, 24 Feb 68, Cable-Chron-State file (Jan–May 68), DepCORDS Papers; Msg, Komer to Enthoven, 27 Mar 67, Cable file (1968), DepCORDS Papers. Both in CMH.

[46] Bernard Weintraub, "U.S. Admits Blow to Pacification," *New York Times*, 25 Feb 68; [Reuters], "Vacuum Remains in South Vietnam's Rural Areas," *Washington Post*, 25 Feb 68; Murrey Marder and Chalmers Roberts, "Reds Offensive Leaves U.S. With Maze of Uncertainties," *Washington Post*, 26 Feb 68, Braestrup, *Big Story*, 1: 557f; Msg, Komer MAC 12697 to Thomas Thayer, OASD SA, 19 Sep 68, Cable file, DepCORDS Papers, CMH.

[47] For an extensive analysis of the performance of the press during the battle, see Braestrup, *Big Story*, 1: 265–336. For an example of the optimism, see MACOI News Release 39–68, 8 Feb 68, MACV News Release files, 534-74-593, box 6, WNRC. It states, "U.S. marine units reported easier going in Hue yesterday as they continued to press the attack."

A Woman Mourns Her Husband Killed by the Viet Cong in Hue

their editors would want them to be. As a result, they exaggerated the American role in the battle, describing the marines fighting "inch by bloody inch" and "foot by blood soaked foot" to free the city while saying little about the South Vietnamese. Bill Brannigan of ABC News and David Greenway of *Time* did describe some of the early South Vietnamese fighting in the Citadel at Hue, but much more characteristic of the newsmen's attitude toward the South Vietnamese armed forces were reports by David Barrington of NBC News and others that South Vietnamese troops had looted portions of the city.[48]

That the press spent little time reporting on South Vietnamese units was nevertheless probably just as well from MACV's standpoint, for if many indeed fought well others did poorly. On one occasion during the battle, General Abrams complained to the chief of the South Vietnamese Joint General Staff, General Vien, that three battalions of South Vietnamese marines—according to Abrams, "the strongest force in the Citadel, whether U.S., Vietnamese, or enemy"—had moved forward less than one-half a city block in three days. "At this time," Abrams said, "I feel compelled to report to you that they have failed to perform as most

[48] Lung, *The General Offensives* of 1968–1969, p. 85; Braestrup, *Big Story*, 1: 317; David Burrington, NBC News, "The Today Show," 28 Feb 68, Radio-TV-Defense Dialog.

Bones Exhumed From Mass Grave Near Hue Laid Out for Identification

of the Vietnamese armed forces have performed. . . . In this time of great need . . . if the marines cannot [rise to the occasion] . . . they have forfeited their right to be a part of your armed forces.''[49]

As the battle for Hue continued, enemy forces within the city began systematically to execute province officials, policemen, technicians, student leaders, and anyone else who might pose a threat to Communist aims, in the end consigning more than 4,000 persons to mass graves in and around the city. First word of the executions appeared on 11 February, when the Associated Press circulated a statement by the mayor of Hue, Lt. Col. Pham Van Khoa, that some 300 civilians had been murdered. Both the *Washington Post* and the *New York Times* carried the story, but the rest of the press was slow to follow. Khoa had a reputation among newsmen as an incompetent, and there was more verifiable news at hand. Unable to confirm the rumor, the Military Assistance Command issued a communique on the subject only on 9 March, some two weeks after the battle had ended. By then the report seemed stale, just one more attempt by the U.S. command to publicize enemy atrocities. U.S. and South Vietnamese damage to Hue thus

[49] Msg, Abrams PHB 154 to Westmoreland, 23 Feb 68, sub: Information Copy of a Letter From Abrams to Vien, Westmoreland Papers, CMH.

received wide play in the press while the enemy's depredations, unseen, went largely unreported.[50]

The battle for Hue, to many senior officers the fiercest and bloodiest engagement of the offensive, ended on 24 February. The enemy, by his own count, suffered 1,042 killed and several times that number wounded while U.S. and South Vietnamese forces lost 266 killed and more than 1,200 wounded. No reliable tally of civilian casualties exists, beyond the more than 4,000 executed during the enemy occupation.[51]

Khe Sanh

Although the battle for Hue received extensive news coverage, the press, the Johnson administration, and the U.S. command in Saigon paid far more attention to another battle developing over the same period. Shortly after the Tet offensive had begun, General Westmoreland had predicted that the enemy's main effort would come in the I Corps Tactical Zone. Turning northward, U.S. officials and the press had looked beyond Hue, where the enemy was at least on the defensive, to Khe Sanh, where some 6,000 marines appeared to be surrounded by enemy forces and where the main thrust predicted by Westmoreland seemed imminent. There was always the possibility that Khe Sanh was a diversion, but as General Westmoreland told Admiral Sharp, the enemy had put too many preparations into effect for that to seem likely. "He still intends to make Khe Sanh another Dien Bien Phu," Westmoreland said, "and . . . to seize all of the two northern provinces."[52]

That possibility weighed heavily upon President Johnson. With criticism of the war rising throughout the United States and with his own political fortunes at stake, Johnson could ill afford the virtual annihilation of a major force such as the French had suffered. He therefore sought assurances from the Joint Chiefs of Staff and Westmoreland that the marines would be able to withstand any enemy assault, going so far on one occasion as to require each of the Joint Chiefs to sign a statement to that effect.[53]

Westmoreland and Wheeler attempted to reassure the president. When Johnson on 3 February requested Westmoreland's views on MACV's ability to reinforce the base, Westmoreland replied that the situation was far different from

[50] U.S. Embassy, Saigon, List of Civilians Massacred by the Communists During "Tet Mau Thuan" in the Thua Thien Province and Hue City [probably 1969], DDI Hue Massacres file; Braestrup, *Big Story*, 1: 280f; Lung, *The General Offensives of 1968–1969*, pp. 80f; "Reds Said To Execute 300 in Hue," *Washington Post*, 12 Feb 68; "Hue's Mayor Says Foe Executed 300," *New York Times*, 12 Feb 68; "How Tactics of Terror Are Used Against the South Vietnamese," *National Observer*, 19 May 69; U.S. Mission Press Release 47-68, 9 Mar 68, DDI Hue Massacres file.

[51] Lung, *The General Offensives of 1968–1969*, pp. 84f.

[52] Msg, Westmoreland MAC 1901 to Sharp, 10 Feb 68, Westmoreland Papers, CMH.

[53] "Johnson Said To Get Pledge on Khe Sanh," *New York Times*, 5 Feb 68.

Aerial View of Khe Sanh

the one that had prevailed at Dien Bien Phu, where the French had been virtually cut off from all outside help. The United States, he said, had ample tactical air power and artillery and could resupply the base entirely by air with highly accurate low-level parachute drops and helicopter runs if the enemy somehow managed to close the base's runway. In addition, B–52 bombers could pound enemy positions at will.[54]

Johnson remained dissatisfied. Aware that the U.S. command in South Vietnam was preparing contingency plans for the use of nuclear weapons, he told General Wheeler on 3 February that he was concerned the enemy might force him to drop an atomic bomb at Khe Sanh—a decision he had no wish to make. General Westmoreland assured the president that since his command had received authority to use an extremely effective antipersonnel weapon known as COFRAM, there should be no need for nuclear weapons at Khe Sanh. Unwilling nevertheless to rule out any option, Westmoreland told General Wheeler that if a major invasion of South Vietnam developed across the Demilitarized Zone, the United

[54] Msg, Wheeler JCS 1147 to Westmoreland, 1 Feb 68, and Msg, Westmoreland MAC 1586 to Wheeler, 3 Feb 68, both in Westmoreland Papers, CMH. CSM 2941, Wheeler for the President, 3 Feb 68, sub: Khe Sanh, FAIM/IR.

States should be prepared to use whatever weapons it needed to repel the attack, whether chemical or nuclear.[55]

Then on 5 February an aide of Senator Fulbright received an anonymous telephone call. The caller suggested that Fulbright's Senate Committee on Foreign Relations might inquire into why one of the United States' leading experts on tactical nuclear weapons, Professor Richard L. Garwin of Columbia University, had recently traveled to South Vietnam along with several other scientists. The committee took up the question in a closed session where, according to later newspaper accounts, several senators expressed concern that the president might feel compelled to use nuclear weapons if Khe Sanh were in danger of falling. A candidate for the Democratic Party's presidential nomination and a leading critic of the war, Senator Eugene J. McCarthy of Minnesota, made the issue public shortly thereafter, announcing at an 8 February news conference in Boston that the military had apparently already requested tactical nuclear weapons for use in South Vietnam.[56]

Both the White House and the Defense Department denied McCarthy's allegation, labeling it false and unfair speculation. Although in a conversation with several senators later reported in the press, General Wheeler refused to say that the use of nuclear weapons had been excluded as an option if Khe Sanh were in danger of being overrun, Secretary Rusk in a letter to Fulbright underscored a statement from the White House that the president "has considered no decision of this nature." On 16 February the president himself spoke out, avowing emotionally that, "so far as I am aware, they [the secretaries of state and defense and the Joint Chiefs of Staff] have at no time ever considered or made a recommendation in any respect to the employment of nuclear weapons."[57]

By 12 February criticism was beginning to rise both in the United States and around the world. In Washington, an array of senators and congressmen voiced their concern that the Johnson administration might resort to nuclear weapons. Prime Minister Harold Wilson of Great Britain meanwhile commented on the CBS News program "Face the Nation" that it would be "lunacy . . . sheer lunacy" for the United States to use nuclear weapons in South Vietnam. Prime Minister Lester Pearson of Canada agreed. Although there was no evidence to support the rumors, he said, any employment of nuclear weapons in South Vietnam would be madness.[58]

The Johnson administration quietly yielded to the pressure. On 12 February

[55] Msg, Wheeler JCS 1272 to Westmoreland, 3 Feb 68, Westmoreland Papers, CMH; Msg, Westmoreland MACV 1586 to Wheeler, 3 Feb 68. See also Westmoreland, *A Soldier Reports*, p. 338.

[56] John W. Finney, "Johnson Denies Atom Use in Vietnam Is Considered," *New York Times*, 17 Feb 68; [AP], "White House Disputes McCarthy on Atom Arms," *New York Times*, 10 Feb 68; "Viet Nukes Requested, McCarthy Says," *Washington Post*, 9 Feb 68.

[57] George Wilson, "No A Arms Requested for Vietnam, U.S. Says," *Washington Post*, 10 Feb 68; John W. Finney, "Wheeler Doubts Khe Sanh Will Need Atom Weapons," *New York Times*, 15 Feb 68; Robert C. Albright, "Fulbright and Rusk Clash on Atom Talk," *Washington Post*, 16 Feb 68; Finney, "Johnson Denies Atom Use in Vietnam Is Considered," *New York Times*, 17 Feb 68.

[58] Warren Unna, "A-Arm Use Called Lunacy by Wilson," *Washington Post*, 12 Feb 68; "Pearson on A-Arms," *Washington Post*, 13 Feb 68.

Admiral Sharp ordered Westmoreland to discontinue contingency planning for the use of nuclear weapons and to lock up all written materials generated by the project. The U.S. command was to debrief everyone even remotely involved in the study, advising each person that there could be no disclosure of the contents or existence of the plan. "Security of this action and prior actions," Sharp observed, "must be air tight."[59]

As the controversy over nuclear weapons continued in the United States, the situation at Khe Sanh grew ominous. Shortly after midnight on 7 February, Communist forces attacked a Special Forces camp at Lang Vei, some eight kilometers southwest of the base. Employing tanks successfully for the first time in the war, the enemy forced the U.S. command to evacuate the camp, in the process killing nearly 200 of its 500 native defenders and 10 of 24 American advisers. The event seemed to many to indicate that the push against Khe Sanh had finally begun.[60]

The attack never came. Instead, Communist gunners continued to hammer Khe Sanh with artillery and rocket fire. Although the enemy launched a series of bitter assaults upon the Marine outposts in the hills surrounding Khe Sanh, the only serious ground attacks upon the base itself occurred on 29 February, when enemy units stormed the positions of the 37th South Vietnamese Ranger Battalion. Detected in advance by sensors, the attackers fell victim to massed American artillery fire, radar-guided fighter bombers, and B–52 strikes.[61]

Although the marines were apparently never in any danger of annihilation, their inability to strike back and the constant casualties from high-explosive bombardment—125 killed and 812 wounded between 1 January and 25 February—inevitably took a toll on their morale. On 24 February, for example, a frustrated Marine patrol ventured against orders some 400 meters beyond American lines into enemy positions. In the melee that followed and the subsequent rescue attempt, two platoons lost 1 killed, 12 wounded, and 25 missing, according to preliminary estimates. All of the missing were later counted as dead.[62]

Although the siege of Khe Sanh never evolved into a major enemy ground assault, it still proved irresistible to the American news media. Westmoreland's statements suggesting the imminence of an enemy offensive in the northern portion of South Vietnam were in part responsible. So was the Johnson administration's obvious concern, relayed to the press through leaks as well as official statements, that the battle might somehow evolve into a debacle. After the enemy's withdrawal from Hue, the battle was also the only large-unit combat

[59] Msg, Sharp to Westmoreland, 12 Feb 68, and Msg, Westmoreland to Gen Cushman, CG, III MAF, 12 Feb 68, both in Westmoreland Papers, CMH.

[60] John A. Cash, "Battle of Lang Vei, 7 February 1968," in *Seven Firefights in Vietnam*, Vietnam Studies (Washington, D.C.: Office of the Chief of Military History, Government Printing Office, 1971), pp. 109–38.

[61] Lieutenant General Willard Pearson, *The War in the Northern Provinces, 1966–1968*, Vietnam Studies (Washington, D.C.: Department of the Army, Government Printing Office, 1975), pp. 74–78.

[62] Msg, Cushman to Abrams, 25 Feb 68, and Msg, Westmoreland MACV 2018 to Wheeler, 12 Feb 68, both in Westmoreland Papers, CMH. See also Robert Pisor, *The End of the Line: The Siege of Khe Sanh* (New York: W. W. Norton & Co., 1982), p. 226.

story left in South Vietnam. Featuring 6,000 U.S. troops surrounded and under fire, it possessed just the sort of action and drama editors and reporters had always deemed attractive to American audiences.[63]

Articles comparing the French defeat at Dien Bien Phu with the siege of Khe Sanh figured prominently in news coverage of the event, partly because the two battles indeed resembled one another superficially and partly because officials were obviously preoccupied with the parallel. During a background briefing at the Pentagon on 1 February, for example, a general officer obligingly sketched the similarities between the two sieges. Four days later, General Wheeler himself told newsmen that the United States had no intention of sustaining a defeat such as Dien Bien Phu at Khe Sanh. The Military Assistance Command attempted to counteract the pessimism appearing in the press by noting in briefings that American air power and artillery made any comparison of the two battles academic, yet the effort met with little success. Observing that U.S. firepower at Khe Sanh had thus far failed to silence enemy mortars and antiaircraft fire, Charles Mohr of the *New York Times* speculated that the Communist delay in attacking the base resembled the slow strangulation that had befallen French forces fourteen years earlier. The enemy, for his part, was quick to exploit the pessimism, broadcasting assurances to the world shortly after the fall of Lang Vei that the United States faced a defeat at Khe Sanh as humiliating as French losses at Dien Bien Phu.[64]

Although during February and March Khe Sanh was the subject of 38 percent of all Associated Press stories on Vietnam filed from outside of Saigon and figured in 18 percent of all photographs of the war published in the *New York Times* and the *Washington Post,* television news far outstripped the print media in reporting the battle. Twenty-five percent of the reports from Vietnam that appeared on evening news programs during those two months featured the siege. Indeed, CBS News devoted a full 50 percent of its film coverage of the war to the subject.[65]

Television newsmen were constrained by the need to film scenes of combat but found no fighting within camera range. In the same way, they saw the damage to U.S. positions caused by enemy rockets and mortars but were unable to witness firsthand the effect of U.S. firepower on the Communists. The reporters thus gave the enemy more credit than he perhaps deserved, contributing to the air of impending disaster surrounding the event. ''This is one place where the Americans cannot claim they have the initiative,'' CBS correspondent Murray Fromson intoned on 14 February. ''Here the North Vietnamese decide who lives and who dies . . . which planes land and which ones don't, and sooner or later they will make the move that will seal the fate of Khe Sanh.''[66]

[63] For a thorough analysis of press coverage of the battle, see Braestrup, *Big Story,* vol. 1, ch. 9.

[64] Ward Just, ''U.S. Voices Confidence Raids Were Expected,'' *Washington Post,* 1 Feb 63; Huntley-Brinkley Report, 5 Feb 68, Radio-TV-Defense Dialog; Charles Mohr, ''Khe Sanh and Dien Bien Phu: A Comparison,'' *New York Times,* 8 Mar 68; [AFP], "Hanoi Warns U.S. of Dien Bien Phu at Khe Sanh,'' *New York Times,* 11 Feb 68.

[65] Braestrup, *Big Story,* 1:338.

[66] Murray Fromson, CBS Evening News, 14 Feb 68, Radio-TV-Defense Dialog.

On 10 February the enemy shot down a Marine C-130 cargo plane loaded with fuel. Six men died when the aircraft exploded. Later a smaller C-123 went down, carrying forty-eight persons to their deaths. Two other C-123s also crashed during the battle, as did a number of Marine helicopters used to resupply the U.S. outposts on the hills around the base. Well within expectations for a battle the size of Khe Sanh, those losses figured prominently in television news accounts, where reporters emphasized them to build dramatic tension. As a result, television viewers could hardly escape the impression that a tangle of wrecked aircraft lined the runway at Khe Sanh. Describing a group of young marines waiting at the base's airstrip for a flight out, ABC News correspondent Don North, for example, observed that all were about eighteen years of age and that "their main aim in life here was to become nineteen—a final dash across the runway into . . . cargo planes for a flight back to the world." Along that runway, North added, "are the skeletons of cargo planes that didn't make it." On 6 March, almost a month after the crash of the C-130, Don Webster of CBS made the aircraft the subject of a report. When a plane landed at Khe Sanh, Webster said, much of the activity at the base stopped while everyone watched to see if the craft would survive. "This is all that's left of a C-130 that landed loaded with gasoline," he explained. "It burned and exploded when hit by enemy fire. Some escaped; others did not." A week later, after describing the fatal crash of the C-123, George Syvertsen of CBS attempted to generalize from the incident to all landings at the base. "From now on it's going to take even more courage," he said, for an Air Force pilot to fly into Khe Sanh.[67]

As the battle continued, the Johnson administration became increasingly concerned that the criticism mounting in the news media would turn U.S. public opinion against the war. As a result, shortly after Gallup polls began reporting that the number of Americans approving of the president's performance in office had fallen seven percentage points, President Johnson set off on an impromptu cross-country tour to muster support. During a visit to the USS *Constellation*, he warned that the enemy obviously believed the American public's will to win was vulnerable. "There comes a time," he said, "when men must make a stand. And for America that time has now come." At another stop in Gettysburg, Pennsylvania, where Johnson briefed former President Dwight D. Eisenhower on the war, presidential aides told newsmen that the enemy had been badly hurt in the fighting and that the South Vietnamese government and army might well emerge from the offensive stronger than ever before.[68]

The Military Assistance Command and the U.S. mission in Saigon supported

[67] Pearson, *The War in the Northern Provinces, 1966–1968*, p. 75; Bernard Nalty, *Air Power and the Fight for Khe Sanh* (Washington, D.C.: U.S. Department of the Air Force, 1973), p. 108; Don North, ABC Evening News, 19 Feb 68, Radio-TV-Defense Dialog; Don Webster, CBS Evening News, 6 Mar 68, Radio-TV-Defense Dialog; George Syvertsen, CBS Evening News, 14 Mar 68, Radio-TV-Defense Dialog; Braestrup, *Big Story*, 1: 380–404.

[68] "Johnson's Rating on Vietnam Drops," *New York Times*, 14 Feb 68; Max Frankel, "Johnson Confers With Eisenhower," *New York Times*, 19 Feb 68; "U.S. Aides Report Setbacks for Foe," *New York Times*, 19 Feb 68.

the president's efforts. Concerned that the press had failed to credit the South Vietnamese Army's accomplishments, the MACV Office of Information asked field units for human-interest stories it could use to publicize South Vietnamese heroism and sacrifices. Information officers also redoubled their efforts to provide the Saigon correspondents with well-substantiated accounts of what was happening before unconfirmed, damaging stories gained credence. The U.S. mission meanwhile began a search for captured enemy documents in which Communist commanders admitted that their offensive had failed. On 20 March it released a remarkable assessment written on 1 February in which enemy leaders confessed that they had missed their primary objectives and had failed to spark a general uprising.[69]

Changes in Information Policy

As those efforts continued, Admiral Sharp became increasingly alarmed at the amount of what he considered sensitive information appearing daily in the press. On 24 February he cabled Westmoreland that the Saigon correspondents, in reporting the bombardment of Khe Sanh, had grown accustomed to giving exact counts of the enemy rounds hitting the base and were also publishing the exact number of American casualties that resulted. Since those practices might provide the enemy with valuable information on the accuracy of his gunners, Sharp instructed Westmoreland to have them stopped immediately.[70]

Westmoreland shared Sharp's concern. With the number of accredited correspondents in South Vietnam at a record high of 636 at the end of February, the MACV Office of Information's facilities were overburdened. In addition, the sharpened competition among correspondents had produced a loosening of standards that might well have provided the enemy with information of value. Although he agreed with Zorthian and Sidle that it was better to announce friendly casualties than to permit the sort of press speculation that had led to exaggerations in the past, Westmoreland decided restrictions were necessary. In the future, he cabled Sharp, the press would have to generalize the number of incoming rounds and to apply the terms *light, moderate,* and *heavy* to casualties received in enemy attacks on fixed positions. "We are preparing a new, more detailed reminder to the press," he said, "of the meaning and importance of the ground rules from the standpoint of national security and will rigidly check violations of those rules. We also plan to limit press access to the key spots in I Corps North to ensure better control and to ease press transportation and housekeeping requirements for our field commanders." If those restrictions failed, Westmoreland con-

[69] Memo 68-34 (MC), MACV CofS, 14 Feb 63, sub: Mission Council Actions, Westmoreland History, bk. 29, tab 54; Press Release 57-68, U.S. Mission, Saigon, 20 Mar 68, sub: Viet Cong Headquarters Assesses Tet Offensive, DDI Tet Offensive (1968) file.

[70] Msg, Sharp to Westmoreland, 24 Feb 67, Westmoreland Papers, CMH.

cluded, the command would seriously consider some form of field press censorship.[71]

Both Sharp and Assistant Secretary of Defense for Public Affairs Phil Goulding approved Westmoreland's measures. Westmoreland then issued a memorandum for the press in which he outlined the changes he had instituted, explaining in detail those rules subject to misunderstanding. In a separate action he notified the press that because of limited facilities the number of newsmen visiting Khe Sanh at any one time would be restricted to fifteen. Twenty-five could visit Camp Carroll and Camp Evans, but only twelve would be allowed at Con Thien, ten at the Rockpile, and seven at Gio Linh, all prominent American bases in the I Corps Tactical Zone. Allocations for all other locations in the northern portion of the zone would be similar to those in effect for Khe Sanh: two reporters for each American television network, two each for the Associated Press and United Press International, one for Reuters, and a pool of four Americans to represent all other newspapers and agencies.[72]

Although jarred by the clampdown, the press accepted MACV's new regulations with little complaint. "If the enemy fires eighteen hundred rounds into Khe Sanh one night and hears from our side that nothing was seriously damaged and only, say, seventeen killed," *Newsweek* quoted a MACV briefer as saying, ". . . he might decide to spare ammo for the big push, because those rounds required many man hours and weeks on the supply trails. But if the enemy is killing more than a hundred of our men a day he might decide it's worth using up his supplies." The United States had "run a loose ship in Vietnam," the magazine concluded. "Previous ground rules . . . have often been violated, usually on the assumption that the enemy knows everything anyway."[73]

MACV's revision of the guidelines was the first indication that the Defense Department was reassessing its relations with the news media. McNamara's replacement as secretary of defense, Washington lawyer and lobbyist Clark Clifford, believed that the Tet offensive had been a great shock to the American people, one rendered all the more severe by the Johnson administration's policy of official optimism in the months preceding the event. If the administration persisted in playing down the damage the offensive had done and the enemy decided to launch a second wave of attacks, he told General Wheeler, the backlash within American public opinion would be so great that the credibility gap would become virtually unbridgeable.[74]

[71] Msg, MACV 6349 to the SECDEF, DAIN 588545, 4 Mar 68, sub: Control of the Press North of Hai Van Pass, Army Staff Communications Center files; Msg, Westmoreland MACV 2766 to Sharp, 27 Feb 68, Westmoreland Papers, CMH.

[72] Msg, Sharp to Goulding, 1 Mar 68, and, Msg, ASD PA to Sharp, 6 Mar 68, both in Westmoreland Papers, CMH; Msg, MACV 6349 to the SECDEF, DAIN 588545, 4 Mar 68, sub: Control of the Press North of Hai Van Pass.

[73] Msg, Sharp to Westmoreland, 29 Feb 68, Westmoreland Papers, CMH; "New Rules in Vietnam," *Newsweek*, 11 Mar 68, p. 37.

[74] This section is based on Msg, Wheeler JCS 2721 to Westmoreland, 8 Mar 68, Westmoreland Papers, CMH.

Clifford directed Wheeler's attention to statements by a "senior military spokesman" appearing in the 7 March issue of the *New York Times* as an illustration of the sort of misguided commentary that had brought about the problem. In general, the spokesman had said, the Military Assistance Command was less concerned than at any time in the previous five weeks about renewed attacks against South Vietnam's cities because the enemy was hurt and worn out. "But I do give him a capability in I Corps North, where he has large forces near Hue. In my opinion, Hue is the next objective."

Convinced that a more conservative approach was the best way to avoid any backlash, Clifford told Wheeler to lay down guidelines for the Military Assistance Command on the attitudes that would have to prevail in all future dealings with the press. Official spokesmen, he said, were never to denigrate the enemy. There were to be no forecasts of allied or enemy plans and no predictions of victory. Nor were there to be assertions of difficult fighting ahead or of residual enemy assets yet to be committed. Moderation would put the United States in a strong public position, Clifford said. If reverses occurred there would be no shock, but in the case of success, the United States and South Vietnam could modestly and without overplaying claim the credit.

Wheeler passed the instructions to Westmoreland with a request that the general put them into effect without telling anyone of their source. Westmoreland acknowledged that he had been the "senior military spokesman" and that the article in question had quoted him correctly. He had made the statement to reverse the defensive attitudes of the South Vietnamese, who spoke constantly and despairingly of a second wave of attacks in the near future. He would comply with Clifford's wishes, he said, consistent with honesty and the need to maintain the confidence of his command.[75]

[75] Msg, Westmoreland MAC 3280 to Wheeler, 8 Mar 68, Westmoreland Papers, CMH.

16

A Change of Direction

By the end of February 1968, pessimism pervaded the American news media. On 23 February *Life* published a commentary, entitled "Wherever We Look, Something's Wrong," on the problems confronting the United States at home and abroad. The magazine's editors cataloged an accumulation of woes, seemingly manageable in themselves but in combination almost unsolvable. The American city, with its seething ghettos, high crime rates, and polluted air, was becoming less livable and governable every year. American youth seemed more deeply alienated from inherited values than any previous generation. The dollar, once the world's most useful but no longer the strongest currency, had fallen victim to the Johnson administration's deficit spending. As for the war, the Tet offensive and the "looming bloodbath" at Khe Sanh had cast doubt upon the usefulness of U.S. military might as an instrument of the nation's Asian policy. "Our first necessity is to face the facts," *Life*'s editors concluded. "To acknowledge our sins is good for the soul, but honest rethinking of our purpose in the world is better."[1]

CBS News anchorman Walter Cronkite was equally pessimistic. In a widely discussed documentary aired on 27 February, he summarized the conclusions he had reached during a two-week fact-finding tour of South Vietnam. "To say we are closer to victory today," he said, "is to believe in the face of the evidence . . . optimists who have been wrong in the past. To suggest we are on the edge of defeat is to yield to unreasonable pessimists. To say we are mired in stalemate seems the only realistic yet unsatisfactory conclusion. . . . It is increasingly clear to this reporter that the only rational way out would be to negotiate—not as victims, but as an honorable people who lived up to their pledge to victory and democracy and did the best they could." Later in the week, Cronkite termed

[1] "Wherever We Look, Something's Wrong," *Life*, 23 Feb 68, p. 25.

the buildup at Khe Sanh a symbol of ''administrative intransigence and military miscalculation.'' In a subsequent radio interview with Eric Sevareid he added, ''I found very few people out there who really believe . . . Khe Sanh could be held if the North Vietnamese are determined to take it.''[2]

News commentator Howard K. Smith provided one of the few counterpoints to the mood of defeat. With the charge that the American press was contributing to the ''confusion and frustration now damaging the American spirit,'' he resigned his position at ABC News because he no longer felt he was participating in ''a great age of journalism.'' Citing black activist Stokely Carmichael as an example, Smith said that the news media had gained the power to elevate to prominence individuals who had achieved nothing and who had lacked a following before the press took notice. In the same way, reporters had exaggerated the so-called credibility gap into ''one of the most distorted over-simplifications of the time.'' Although the president of the United States had to make judgments on the basis of partial information, the news media tended to allege ''calculated deception if he does not instantly provide conclusive facts and admit failure. . . . No government has ever run that way and none ever will.'' News coverage of the Vietnam War was replete with examples of bias, Smith continued. In the case of the widely circulated photograph of General Loan executing a Viet Cong prisoner, ''not even a perfunctory acknowledgment was made of the fact that such executions, en masse, are the Viet Cong way of war.'' Smith noted that his own son had been left for dead during the Battle of the Ia Drang in 1965 and had witnessed the execution by the enemy of a dozen uniformed American soldiers.[3]

Smith's rejoinder notwithstanding, the clamor for reexamination of U.S. prospects in South Vietnam continued in the press and spread to the U.S. Congress. Senator Fulbright's committee grilled outgoing Secretary of Defense McNamara on the 1964 incidents in the Gulf of Tonkin that had led to the congressional resolution authorizing the war. Committee members came close to alleging that Congress had been tricked into approving the measure. Senator Wayne Morse charged that far from being on routine patrol as the administration had claimed, the U.S. destroyers *Maddox* and *Turner Joy* had been on an intelligence-gathering mission and had actually provoked the North Vietnamese attacks. McNamara responded indignantly that Morse's allegation was ''monstrous'' and that the Defense Department had unimpeachable but highly classified evidence proving the attacks had been planned and executed by the North Vietnamese.[4]

[2] CBS News Special Report on Vietnam, 7 Feb 68, Radio-TV-Defense Dialog; ''Cronkite Takes a Stand,'' *Time*, 11 Mar 68, p. 108.

[3] Howard K. Smith, ''A Columnist's Farewell,'' *Philadelphia Bulletin*, 18 Feb 68; ''Disillusioned With Journalism,'' *Time*, 1 Mar 68, p. 42.

[4] U.S. Congress, Senate, *Congressional Record*, 90th Cong., 2d sess., 21 Feb 68, p. S–1589; 28 Feb 68, p. S–1885; 29 Feb 68, p. S–1947. See also ''Suspicions of a Moonless Night,'' *Time*, 1 Mar 68, p. 12; ''The War: More Men, More Doubts,'' *Newsweek*, 4 Mar 68, p. 19; ''The Credibility Gulf,'' *Chicago Daily News*, 27 Feb 68; ''The Guns of August 4,'' *Time*, 1 Mar 68, p. 13.

Walter Cronkite Interviews Marines During the Battle of Hue

Public Opinion, February 1968

Whatever the merits of the committee's charges and McNamara's rebuttal, neither the dispute over the Gulf of Tonkin attacks nor the pessimism of the news media had much impact on American public opinion. Responding to the Gallup poll question, "In view of developments since we entered the fighting in Vietnam, do you think the United States made a mistake sending troops to fight in Vietnam?" Americans for the most part indicated that they had suspended judgment. Forty-five percent responded "yes," the same percentage that had given that answer in December 1967; 43 percent answered "no," a drop of 3 percent from the previous poll; and 12 percent refused to venture an opinion, an increase of 3 percent. The figures indicated uncertainty about the wisdom of American involvement but hardly constituted a repudiation of either the war or the Johnson administration's policies.[5]

Indeed, the polls indicated that far from suffering a loss of morale or fighting spirit, the majority of Americans had rallied aggressively to the side of the presi-

[5] Burns Roper, "What Public Opinion Polls Said," in Braestrup, *Big Story*, 1: 681; Hazel Erskine, "The Polls: Is War a Mistake?" *Public Opinion Quarterly* 34 (Spring 1970): 135f.

dent. Whereas in January 1968 56 percent of those polled by the Gallup organization had considered themselves hawks on the war and 27 percent doves, with 17 percent voicing no opinion, by early February at the height of the offensive, 61 percent considered themselves hawks, 23 percent doves, and 16 percent held no opinion. In the same way, the number of Americans who expressed confidence in U.S. military policies in South Vietnam rose from 61 percent in December 1967 to 74 percent in February 1968. Nor was the American public, despite the pessimism of the news media, particularly concerned that the United States would lose the war because of the offensive. Of those polled by Louis Harris, 43 percent believed the United States would still win, 39 percent expected some sort of stalemate, and only 3 percent considered outright loss probable. The majority of Americans were, indeed, unwilling to leave the matter to chance. Asked whether halting the bombing of North Vietnam would improve the possibility for peace, 71 percent opted for continuing the bombing in February 1968, a rise of 8 percentage points from the previous October. The number of those favoring a halt to the bombing fell during the same period from 26 to 15 percent.[6]

If public opinion of the war held firm during the first weeks of February, public support for President Johnson faltered. Looking to the White House for leadership and willing to back whatever action Johnson took, Americans perceived little forward motion. The president made a few comments to the press shortly after the start of the offensive but for the most part left public statements on the war to his aides and staff. The impression of indecision that resulted took its toll on his popularity, raising the percentage of disapproval of his handling of the war from 47 to 63 percent by the end of February. The lack of any effort by Johnson to marshal public opinion in his favor also affected the American public's mood of aggressiveness, which likewise began to drain away. By the end of March the percentage of those expressing confidence in U.S. military policies in South Vietnam had fallen precipitously from 74 to 54 percent, and the number of Americans considering the war a stalemate had risen from 39 to 42 percent.[7]

If pessimistic news coverage of the Tet offensive had little effect on American public opinion, it served nevertheless to reinforce the doubts that had surfaced within the Johnson administration during the previous year. Presidential speechwriter Harry McPherson described the mood.

> I felt we were being put to it as hard as we ever had, and I was extremely disturbed. I would go in two or three mornings a week and study the cable book and talk to Rostow and ask him what had happened the day before, and would get from him what almost seemed hallucinatory from the point of view of what I had seen on network television the night before. . . . Well, I must say that I mistrusted what he said. . . . I assume the reason this is so, the reason I put aside my own interior access to confidential information and was more persuaded by what I saw on the tube and in the newspapers, was that like everyone else who had been deeply involved in explaining the policies of the war

[6] Roper, "What Public Opinion Polls Said," in Braestrup, *Big Story*, 1: 679f; "The War: Thin Green Line," *Time*, 23 Feb 68, p. 15.

[7] Roper, "What Public Opinion Polls Said," in Braestrup, *Big Story*, 1: 687.

and trying to understand them and render some judgment, I was fed up with the "light at the end of the tunnel" stuff. I was fed up with the optimism that seemed to flow without stopping from Saigon.[8]

The Chief of Staff of the Army, General Harold K. Johnson, recognized what was happening but could only propose solutions that had failed in the past. On 1 March he told Westmoreland that there was "a tendency toward discouragement" within portions of the Johnson administration that hardly seemed warranted. "We suffered a loss," he admitted, "there can be no doubt about that. On the other hand, I believe that the enemy loss approaches catastrophic proportions and can become catastrophic with just a little bit of a push from us. Of course, this will mean reviving a flagging confidence and spirit in certain quarters here at home." General Johnson believed that mutual recriminations would be of no help but that "a campaign of praise and support" for the South Vietnamese armed forces might do some good. "I do not propose to obscure shortcomings," he said, "but at this stage emphasis should be on successes and not on failures." The press, he concluded, would be of no help in the effort. "It will be an uphill fight all the way."[9]

Although Secretary of Defense Clifford's instructions to play down both good and bad aspects of the war took precedence over Johnson's advice, the Military Assistance Command attempted to comply. Toward the end of March the command mounted an operation code-named QUYET THANG in the III Corps Tactical Zone at least in part to highlight the South Vietnamese Army's role in the fighting. Prompted by the MACV Office of Information, wire service reporters and several television network news teams followed the operation closely, filing stories daily. Yet only a few of those reports ever were used. When General Wheeler complained to General Abrams that the Saigon correspondents had failed to cover the operation, Abrams responded that the problem appeared to be in the United States, where editors and network news directors had for some reason decided QUYET THANG was of little importance.[10]

Doubts Rise About the Marines

In fact, so many spectacular news stories on the war were appearing in the United States that a relatively obscure operation in the III Corps Tactical Zone had little chance to compete. During most of March and into April, for example, American newspapers ran a series of articles alleging that the establishment of the MACV forward command post in the I Corps had been the result of Marine

[8] McPherson is quoted in Herbert Y. Schandler, *The Unmaking of a President* (Princeton, N.J.: Princeton University Press, 1977), p. 81.

[9] Msg, Harold K. Johnson WDC 3166 to Westmoreland, 1 Mar 68, Westmoreland Papers, CMH.

[10] Msg, Sharp to Wheeler, 10 Mar 68; Msg, Wheeler JCS 3239 to Abrams, 22 Mar 68; and Msg, Abrams MACV 4067 to Wheeler, 25 Mar 68, all in Westmoreland Papers, CMH.

incompetence. In a widely reprinted dispatch *Los Angeles Times* correspondent William Tuohy wrote that although the marines had built their reputation on dash, enterprise, style, and imagination, the top Marine leadership in South Vietnam had been singularly unimpressive. During the Battle of Hue, he said, Marine battalions had been understrength and poorly supplied; the Marine chain of command had been confused; and there had been little coordination between Marine and South Vietnamese units. Looting, as a result, had been common. The assignment of General Abrams to the I Corps zone, Tuohy concluded, reflected Westmoreland's dissatisfaction with the marines' performance and the U.S. Army's growing dominance over fighting in the region.[11]

The article drew a vehement response from General Wheeler. He told Westmoreland that although he knew that the 636 reporters in South Vietnam were bound occasionally to indulge in "unfounded and deleterious speculation," Tuohy's story was especially reprehensible and worthy of condemnation. Westmoreland agreed. Without adverting to his 22 January message to Wheeler in which he questioned the professionalism of the marines and doubted their ability to defend Quang Tri Province in an emergency, he dismissed the article as another attempt by the press to create friction between the military services in order to generate news. He immediately convened a background briefing in Saigon to clarify command arrangements in the I Corps Tactical Zone and issued a news release stating emphatically his admiration for Lt. Gen. Robert E. Cushman, Jr., and all of the marines in South Vietnam "down to the lowest private." When the Saigon correspondents nevertheless continued to pursue the subject, General Sidle managed to stop several damaging stories in the *New York Times* and on television by convincing the reporters involved that interservice rivalry was endemic only among the lowest-ranking officers and enlisted men and that relationships at command levels were completely harmonious.[12]

Although those assurances satisfied most of the Saigon correspondents, stories on the controversy appeared regularly during the next month. On 14 March the *Washington Star* published an article by Donald Kirk on what it called "The Army-Marine Feud." In April George Wilson *of the Washington Post* revealed that the Military Assistance Command planned to use Khe Sanh as a springboard for future offensive sweeps. Wilson added that U.S. Army officers at Khe Sanh had told him the marines had been so psychologically defeated during the battle for the base that "they were seeing shadows outside [the] . . . wire and wouldn't go out to pick up their dead." Westmoreland countered that the marines had stayed inside the wire at Khe Sanh on his instructions to keep from

[11] Msg, Westmoreland MACV 2717 to Cushman, 26 Feb 68; and Msg, Wheeler JCS 2581 to Westmoreland, 5 Mar 68, both in Westmoreland Papers, CMH. William Tuohy, "Marine Leadership Under Fire in Vietnam," *Washington Post*, 3 Mar 68.

[12] Msg, Wheeler JCS 2581 to Westmoreland, 5 Mar 68; Msg, Westmoreland MAC 1011 to Wheeler, 22 Jan 68, Westmoreland Papers, CMH; Msg, MACV 6587 to OASD PA, 6 Mar 68, sub: COMUSMACV Statement to the Press, DDI DMZ Barrier file; Msg, Westmoreland MACV 5120 to Gen Haines, Acting Chief of Staff, Army, 17 Apr 68, Westmoreland Papers, CMH.

being hit by friendly fire and for a time considered disaccrediting Wilson for revealing his plans for the base, a clear infraction of the MACV guidelines. He reconsidered when he learned that the Marine commander in South Vietnam, General Cushman, had revealed those plans to the reporter in an on-the-record interview, an indiscretion on the part of the general rather than the reporter.[13]

The 200,000-Man Troop Request

As the controversy over the marines ran its course, a more damaging subject caught the attention of the press. Shortly after the start of the Tet offensive, at the prompting of General Wheeler, Westmoreland had requested the deployment of an additional U.S. Army division and one-half of a Marine division. "I am making a firm request for troops," he told Wheeler, "not because I fear defeat if I am not reinforced, but because I do not feel that I can fully grasp the initiative from the recently reinforced enemy without them. On the other hand, a setback is fully possible if I am not reinforced and it is likely that we will lose ground in other areas if I am required to make substantial reinforcement in I Corps." Westmoreland informed Admiral Sharp of his request in a separate message, adding that adequate reinforcements would permit him both to contain the enemy's offensive in the I Corps Tactical Zone and to seize the initiative in other areas.[14]

Westmoreland made the request because reinforcements were always welcome. Yet in prompting it, Wheeler had a larger object in mind. He had been concerned for some time that the buildup in South Vietnam had sapped the ability of U.S. forces to respond to crises elsewhere in the world and decided to use the Tet offensive as the occasion to reconstitute the U.S. strategic reserve. In passing Westmoreland's request to the Defense Department he therefore replaced Westmoreland's assessment that defeat was at best a remote possibility with a statement emphasizing the uncertainties confronting U.S. forces in South Vietnam. He also recommended, "as a matter of prudence," that "deployment of emergency reinforcements . . . should not be made without concomitant call up of reserves sufficient at least to replace those deployed and provide for the increased sustaining base requirements of all services."[15]

[13] Donald Kirk, "The Army-Marine Feud," *Washington Star*, 14 Mar 68; Msg, Wheeler JCS 2917 to Westmoreland, 14 Mar 68, Westmoreland Papers, CMH; Msg, Wheeler JCS 3184 to Abrams, 21 Mar 68, General Creighton W. Abrams Papers, CMH; Msg, Gen Haines WDC 5588 to Westmoreland, 17 Apr 68, Westmoreland History, bk. 31, tab 38; Memo, Sidle for Westmoreland, 17 Apr 68, sub: George Wilson and UPI Stories About Future Plans at Khe Sanh, Westmoreland History, bk. 31, tab 41; Msg, Westmoreland MACV 5120 to Haines, 17 Apr 68.

[14] Msg, Wheeler JCS 1590 to Westmoreland, 9 Feb 68; Msg, Westmoreland MAC 2018 to Wheeler, 12 Feb 68; Msg, Westmoreland MAC 1975 to Sharp, 12 Feb 68. All in Westmoreland Papers, CMH. For a complete discussion of the troop request, see Schandler, *The Unmaking of a President*, pp. 92f.

[15] JCSM 91-68, 12 Feb 68, sub: Emergency Reinforcement of COMUSMACV, in *Pentagon Papers*, 4: 541. See also Msg, Westmoreland MAC 1859 to Wheeler, 9 Feb 68, Westmoreland Papers, CMH.

Although President Johnson wanted to support his troops in the field, both he and his advisers also wished to avoid any action that might spark a public outcry or require congressional approval. He therefore decided to send only a portion of the men Westmoreland had requested, one brigade of the 82d Airborne Division and a Marine regimental landing team—with support elements, some 10,500 men. He refused to activate any reserve units.[16]

To the chagrin of the White House, the Joint Chiefs of Staff immediately pointed out that deploying one-third of the 82d Airborne Division—the only combat-ready division in the United States—would render the rest of the division combat ineffective, requiring some sort of reserve call-up to restore even a semblance of balance to U.S. capabilities. To take that step, they said, the president would have to seek legislation to extend beyond 30 June 1968 the existing authority to call reserve units to active duty.[17]

Between 23 and 25 February General Wheeler visited South Vietnam to assess the situation and to consult with Westmoreland on force requirements. After both reviewed the various contingencies that might bear upon MACV's ability to fight, they settled on a figure of 205,000 men to be deployed in three increments over the next year. There was a tacit agreement between the two that only the first increment—some 108,000 men—would embark for South Vietnam. The rest would become part of the strategic reserve in the United States, deploying to South Vietnam only if the enemy appeared on the verge of defeating the South Vietnamese or if the United States adopted a new strategy—permitting the invasion of Laos, for example—that would require more men.[18]

On his return to Washington Wheeler reported to President Johnson. He avoided any mention of his understanding with Westmoreland or of the fact that the general felt no urgent need for reinforcements of the size contemplated. Instead, he painted a dark picture of the situation in South Vietnam. The enemy's attack had nearly succeeded in a dozen places, he said, failing only after the arrival of U.S. forces, which were still bearing the brunt of the battle at Hue. At the very time when Westmoreland was preparing for a possible major enemy thrust in the I Corps Tactical Zone, he was still having to "pick up the tab" for South Vietnamese forces, especially in the region around Saigon. The consensus of responsible commanders, Wheeler said, was that 1968 would be a pivotal year. If the enemy synchronized his offensive in the I Corps with attacks around the country, Westmoreland's margin would be "paper thin." For that reason Westmoreland sought reinforcements: in all, 205,000 men.[19]

Wheeler's request drew into the open many of the concerns haunting the Johnson administration. At a meeting of the president's senior advisers McNamara

[16] Memo, Maxwell D. Taylor for the President, 10 Feb 68, sub: Further Reinforcements for Vietnam, NSC file R-6, LBJ Library; *Pentagon Papers*, 4: 542; Schandler, *The Unmaking of a President*, p. 101.

[17] *Pentagon Papers*, 4: 542; Schandler, *The Unmaking of a President*, p. 101.

[18] Schandler, *The Unmaking of a President*, pp. 109f.

[19] Memo, Gen Wheeler for the President, 27 Feb 68, sub: Military Situation and Requirements for South Vietnam, NSC files, III, Mar 70, LBJ Library.

expressed grave doubts about the military, economic, political, diplomatic, and moral consequences of a larger buildup of forces in South Vietnam. A 205,000-man call-up, he said, would cost the country $10 billion in the coming fiscal year but would still be inadequate to defeat the enemy. Both Rusk and Rostow argued that at least some reinforcements would be useful to Westmoreland, but Clifford observed that the American public believed Tet was a major setback for the United States. How, in the light of the optimistic statements of the previous year, could the administration undertake yet another military buildup, especially one with major economic implications, without also leaving the impression that the United States was, as he put it, ''pouring troops down a rat hole''? Clifford concluded that a hasty decision would be inappropriate and that the United States' whole stance in Southeast Asia needed reevaluation.[20]

Ambassador Bunker also had serious doubts. When Secretary Rusk asked him to comment on the troop request, he responded that an increase of the size contemplated might well nullify the purposes that had brought the United States into the war in the first place by destroying what was left of South Vietnamese initiative. The United States, intentionally or not, would come increasingly to play the role of a colonial power. Far more effective would be an effort to equip the South Vietnamese armed forces so that they could equal the enemy at least in firepower.[21] ''No matter what we achieve here,'' Bunker concluded, ''the American press and probably certain of our congress will never regard it as sufficient, given their tendency sometimes to demand standards of perfection which even we have not attained.''[22]

On 27 February, after the president's senior advisers had informed Johnson of their doubts, Walt Rostow recommended that a joint task force under Clifford investigate all of the options open to the United States. The next day Wheeler returned to Washington. Invited to the White House, he once more drew the darkest possible picture of the situation in South Vietnam. Caught between Wheeler's pessimism and his advisers' misgivings, Johnson accepted Rostow's suggestion. Because Clifford brought to the problem ''a new pair of eyes and a fresh outlook,'' he asked the new secretary of defense to chair a study to determine the least objectionable course.[23]

Clifford reported the results of the study and made his recommendations to the president on 4 March. Task force members had expressed concern, he said, that the addition of 205,000 men would merely move the United States further along a course that had yet to produce a viable South Vietnam living at peace. ''As we build our forces . . . [the Communists] build theirs. . . . Even if we were to meet this full request for 205,000 men, and the pattern continues as it has,

[20] Handwritten Notes of Meeting Involving Rusk, McNamara, Clifford, Rostow, William Bundy, Nicholas Katzenbach, Joseph Califano, and Harry McPherson, 27 Feb 68, NSC files, I, Mar 70, LBJ Library.

[21] Ibid.; Msg, State 12437, Rusk to Bunker, 5 Mar 68, and Msg, Saigon 21733 to State, for Rusk from Bunker, 11 Mar 68, both in FAIM/IR.

[22] Msg, Saigon 22096 to State, Bunker to Rusk, 14 Mar 68, FAIM/IR.

[23] Schandler, *The Unmaking of a President*, pp. 117f; Johnson, *The Vantage Point*, pp. 390f.

McNamara and Clifford

it is likely that by March . . . [Westmoreland] may want another 200,000 to 300,000 men with no end in sight.'' An infusion of troops of the magnitude Westmoreland and Wheeler sought would, in addition, mean significantly higher American casualties and a greater temptation to broaden the war into Laos and Cambodia. In the process, the South Vietnamese, who looked to the United States for additional assistance every time a crisis occurred, would lose even more of their ability to help themselves. ''We can no longer rely just on the field commander,'' Clifford concluded. ''He can want troops and want troops and want troops. We must look at the overall impact on us, including the situation here in the United States. We must look at our own economic stability, our other problems in the world, our other problems at home; we must consider whether or not this thing is tying us down so that we cannot do some of the other things we should be doing.'' That said, Clifford recommended sending only enough troops to cover problems arising in South Vietnam in the next two to four months, perhaps 22,000 men.[24]

Walt Rostow amplified Clifford's remarks in a memorandum for the president. The participants in the study had been unsure, he said, whether Westmoreland's

[24] MFR, 4 Mar 68, sub: Notes of the President's Meeting With His Senior Foreign Policy Advisers, NSC files, I, Mar 70, LBJ Library.

request was designed to prevent deterioration of the U.S. position in South Vietnam or to permit a vigorous U.S. offensive in the second half of the year. The thought ran deep both at State and Defense that negotiation with North Vietnam was the only way for the United States to attain its ends. The addition of more American troops would merely complicate the process, leading to "extremely ugly and determined opposition" from Fulbright and other members of Congress.[25]

For his part, Rostow favored vigorous prosecution of the war. Because of the surge in public opinion favorable to military action, he believed that the American people would rally to the president's side as soon as Johnson decided to act. Thus he counseled the president to go before the public to demand maximum support for an all-out effort to defeat the enemy. Whether the president agreed with him or not, he told Johnson on 8 March, action of some sort was necessary. "The country badly needs a presidential decision," he said, "even if only an interim decision, and a presidential speech."[26]

Rostow's advice notwithstanding, Johnson chose to wait. Aware that the dispatch of 205,000 troops was politically difficult yet unwilling to deny his commander in the field, he allowed discussion to continue in order to determine the least number of troops he needed to send. On 9 March Wheeler advised Westmoreland that 30,000 men might be available but that strong resistance had developed to any larger deployment. Westmoreland immediately accepted that number, which he would use to form seven maneuver battalions and a military police battalion. Ambassador Bunker supported the request, once again urging greater efforts to equip and reinforce the South Vietnamese so that they could fight their own war.[27]

President Johnson at first favored sending the troops. On 13 March he authorized the deployment of some 35,000 men and a limited call-up of reserves. But he reversed his decision a short time later, when he learned from the Joint Chiefs that the 35,000 men would have to be supplemented by another 13,500, needed as support for the 10,500 dispatched during February. In addition, Westmoreland chose that moment to reveal his plan for a major offensive in the I Corps Tactical Zone to relieve the garrison at Khe Sanh. That announcement led Johnson to conclude that since Westmoreland could mount so large an operation with only the troops on hand, additional deployments were hardly as necessary as he had supposed. In the end, on 28 March, Johnson decided to send only the 13,500 support forces and rejected further troop deployments.[28]

[25] Memo, Walt Rostow for the President, 6 Mar 68, sub: The Clifford Committee, NSC files, vol. 65, W. Rostow Memos to the President, LBJ Library.

[26] Memo, Walt Rostow for the President, 8 Mar 68, NSC files, S-66, LBJ Library; Schandler, *The Unmaking of a President*, pp. 177f.

[27] Msg, Wheeler JCS 2767 to Westmoreland, 9 Mar 68, and Msg, Westmoreland MAC 3385 to Wheeler, 11 Mar 68, both in Westmoreland Papers, CMH. Msg, Saigon 22096 to State, Bunker to Rusk, 14 Mar 68; Schandler, *The Unmaking of a President*, pp. 229f.

[28] Johnson, *The Vantage Point*, pp. 407f; *Pentagon Papers*, 4: 593f; Schandler, *The Unmaking of a President*, pp. 231f.

Johnson Confers With His Advisers

While the president was making his decision, the "extremely ugly and determined" opposition Rostow had predicted came into being. On the morning of 10 March, Hedrick Smith and Neil Sheehan revealed in the *New York Times* that Westmoreland had requested some 206,000 more men. Basing their story on information from a number of sources rather than on a major leak, the reporters summarized the discussions under way within the Johnson administration. They underscored the depth and scope of the debate as well as the doubts within official circles spawned by the Tet offensive. "At every level of government," the two reported, "there is a sense that the conflict, if expanded further, can no longer be called 'a limited war.' Officials acknowledge that any further American involvement carries serious implications for the civilian life of the nation—not only the call up of military reserves and enactment of a tax increase but problems with the budget, the economy, and balance of payments."[29]

Shortly after the story appeared, critics of the war in Congress decided to exploit the opportunity. "I think it would be a mistake," Senator Robert Kennedy of New York said, "for the president to take a step toward escalation of the con-

[29] Hedrick Smith and Neil Sheehan, "Westmoreland Requests More Men," *New York Times*, 10 Mar 68.

flict without having the support and understanding of the Senate and the American people." Senator Clifford Case of New Jersey added that "it is now a question as to whether or not the war is winnable without the destruction of South Vietnam and much of American might itself." Senate Majority Leader Mike Mansfield of Montana stated boldly that "we are facing today the most troublesome days in the entire history of the republic. . . . We are in the wrong place and we are fighting the wrong kind of war." Senator Frank Church of Idaho was even more graphic. The Johnson administration, he told reporters, seemed "poised to plunge still deeper into Asia, where huge populations wait to engulf us and legions of young Americans are being beckoned to their graves."[30]

On the evening of the day the article appeared, NBC News aired a special program outlining how the United States had lost the initiative to the enemy in South Vietnam. Citing Westmoreland's troop request as evidence, correspondent Frank McGee charged that only one conclusion could be drawn from the facts of the Tet offensive—that the United States was losing the war. The next morning Secretary of State Rusk appeared before the Fulbright committee to begin one and one-half days of nationally televised testimony on the war. Pressed to disclose whether the United States was considering escalation, he would state only that the administration was reexamining its policies from A to Z and that it would certainly consult with Congress about sending additional troops to South Vietnam.[31]

On the same day Rusk finished his testimony, Senator Eugene McCarthy, an avowed antiwar candidate for the Democratic Party's presidential nomination, came within a few hundred votes of defeating President Johnson in the New Hampshire Democratic primary election. Although more than half of those who had voted for McCarthy considered themselves in favor of the war, the election results appeared to signify popular support for McCarthy's antiwar stand. Louis Harris and other experts immediately warned that the vote represented public dissatisfaction with Johnson's handling of a number of issues, but a Gallup poll published the day after the election emphasized the possibility that the nation was indeed turning against further U.S. involvement in South Vietnam. According to that poll, 69 percent of all Americans, irrespective of political affiliation or opinion of the war, favored a phased withdrawal of American troops from South Vietnam as soon as enough South Vietnamese could be trained and equipped to do all of the fighting themselves.[32]

On 16 March General Wheeler summarized for Westmoreland the issues confronting the Johnson administration. Smith's and Sheehan's revelations had

[30] "Demand for a Voice," *Time*, 15 Mar 68, p. 14.

[31] Jack Gould, "U.S. Losing War in Vietnam, NBC Declares," *New York Times*, 11 Mar 68; John W. Finney, "Rusk Tells Panel of A to Z Review of Vietnam War," *New York Times*, 12 Mar 68; Finney, "Rusk Tells Panel, 'We Will Consult on Any Troop Rise,'" *New York Times*, 13 Mar 68.

[32] "Poll of Democrats Finds Many Hawks Backed McCarthy," *New York Times*, 15 Mar 68; Louis Harris, "How the Voters See the Issues," *Newsweek*, 25 Mar 68, p. 26; "69% in Poll Back a Pullout in War," *New York Times*, 13 Mar 68.

created an extremely difficult political situation, and problems with the balance of payments and the outflow of gold had greatly reduced the possibility of spending more money on the war. Citing the Gallup poll, Wheeler added, "all these things have, I judge, together with the gloom and doom generated by the Tet Offensive, affected heavily support for our war effort." Bearing Wheeler out, *Newsweek*'s chief congressional correspondent, Samuel Shaffer, noted a few days later that the troop request had brought Congress close to mutiny. "Hawks," he said, "are being converted overnight to doves and House members in particular are falling over each other to get resolutions in the hopper demanding that no more troops be sent." General Westmoreland was amazed. As far as he was concerned, the troop request had been mainly an academic exercise designed to strengthen the strategic reserve while providing for contingencies in South Vietnam. That the three increments of troops he and Wheeler had discussed totaled 205,000 men, he later noted in his memoirs, had never even crossed his mind.[33]

Compounding the Johnson administration's problems were two news stories by Neil Sheehan that appeared in the *New York Times* on 19 and 21 March. In the first, Sheehan cited "well-placed informants" who alleged that the Central Intelligence Agency had reopened the order of battle controversy supposedly settled in November 1967. According to those sources, U.S. military analysts had drastically underestimated enemy strength prior to Tet. Sheehan added, erroneously, that the forces excluded from the estimate had later played a major role in the attack on South Vietnam's cities. The reporter's second story, based on a secret year-end report on the war leaked to the *Times*, continued in the same vein. In it Sheehan observed that Westmoreland's view of the situation was so unrealistic that just before Tet the general had predicted gains for 1968 far in excess of anything the United States had achieved in South Vietnam during 1967. Contrasting Westmoreland's optimism with the events of Tet, Sheehan argued that the United States had in fact suffered "a massive failure of intelligence."[34]

Considering the leaks serious breaches of national security, Secretary Clifford ordered an investigation to determine their source. Nothing apparently came of it. Newsmen with long experience in the Pentagon noted that security investigations of alleged news leaks came in cycles and could be expected whenever a new secretary of defense took office.[35]

[33] Msg, Wheeler JCS 3024 to Westmoreland, 16 Mar 68, Westmoreland Papers, CMH; "Growing Dissent," *Newsweek*, 25 Mar 68, p. 33; Westmoreland, *A Soldier Reports*, pp. 357f.

[34] Neil Sheehan, "U.S. Undervalued Enemy Strength Before Offensive," *New York Times*, 19 Mar 68; Sheehan, "68 Gain Was Seen by Westmoreland," *New York Times*, 21 Mar 68. Sheehan's informant on the order of battle controversy was Daniel Ellsberg. See Jack Anderson, "Daniel Ellsberg: The Other Leaks," Washington Post, 28 Sep 75.

[35] "Clifford Orders Inquiry on Leaks," *New York Times*, 24 Mar 68.

Johnson Announces That He Will Not Run for Reelection

A Move Toward Peace

By the end of March, President Johnson and his advisers were urgently seeking some way to regain the political initiative. Early in the month Secretary Rusk had proposed another bombing halt since the weather was usually too poor in North Vietnam during April to make bombing profitable. Convinced that a halt would do little to move the North Vietnamese to negotiations, President Johnson hesitated before accepting the idea. On 31 March General Wheeler nevertheless cabled Westmoreland that the president had made up his mind and that a halt would shortly go into effect. During a speech to be televised that evening, he said, Johnson would announce the deployment to South Vietnam of the 13,500 additional troops and the call-up of 48,500 reservists to replenish the strategic reserve. To make that move more palatable to the American public, the president would then proclaim another initiative to achieve peace. ''Since the Tet Offensive,'' Wheeler explained, ''support of the American public and congress for the war in Southeast Asia has decreased at an accelerating rate. Many of the strongest proponents of forceful action in Vietnam have revised their positions, have

moved to neutral ground, or are wavering. If this trend continues unchecked, public support of our objectives in Southeast Asia will be too frail to sustain the effort.'' The president hoped that a unilateral move toward peace would blunt the criticism of foreign nations while arresting opposition to the war in the United States. As an afterthought, Wheeler asked Westmoreland to make every effort to keep the members of his command from criticizing the decision to the press.[36]

Although Wheeler appears to have considered the president's speech a public relations ploy, Johnson had other motives. He announced the partial bombing halt as planned but also went to great lengths to emphasize his hope that the move would lead to early negotiations. Ambassador W. Averell Harriman would be his personal representative in any official contacts that developed. Specifying neither a time limit after which the bombing would resume nor conditions the North Vietnamese would have to fulfill, he described the limited deployments he planned yet spent most of the rest of the speech detailing the accomplishments of his administration and pleading for national unity. He then electrified the nation and the world by declaring that, in order to spend all of his time in the pursuit of peace, he would not accept the nomination of his party for another term as president.[37]

Confronted by so dramatic a gesture, the North Vietnamese decided to go along rather than concede a major propaganda victory to the United States. Three days later, on 3 April, they declared their readiness to talk, engendering hope in the United States and around the world that an accommodation was in the offing and that the war would soon be at an end.[38]

[36] Msg, Wheeler JCS 3561 to Sharp, Westmoreland, 31 Mar 68, Westmoreland Papers, CMH. See also *Pentagon Papers*, 4: 594f; Schandler, *The Unmaking of a President*, pp. 183f.

[37] Address by President Lyndon Johnson, ''A New Step Toward Peace,'' in Department of State *Bulletin*, 15 April 1968, p. 481.

[38] Westmoreland, *A Soldier Reports*, p. 533; White House Press Release, Statement by North Vietnam, 3 Apr 68, in Department of State *Bulletin*, 22 April 1968, p. 513.

17

Conclusion

Most of the public affairs problems that confronted the United States in South Vietnam stemmed from the contradictions implicit in Lyndon Johnson's strategy for the war. The president was convinced that the conflict was necessary but believed that the American public and Congress lacked the will, without very careful handling, to carry it to a successful conclusion. Accordingly, he sought to move the country toward an acceptance of war, but in so doing to alienate as few Americans as possible. A policy of gradually increasing pressures against North Vietnam seemed the best approach. Besides minimizing public relations problems and preserving as much leeway as possible for his domestic agenda, it would reduce the chance of a major confrontation with North Vietnam's allies, the Soviet Union and Communist China, and might persuade Hanoi to abandon its aggression against South Vietnam before all-out war erupted. At the very least, it would introduce the American public and Congress to the war by degrees while buying time for the military to prepare a proper base of action in South Vietnam. Doing just enough to placate scattered but vocal prowar elements in Congress and the news media, it would also preserve options for the president that might disappear if the so-called hawks gained ascendancy.

Johnson had his way, but at the cost of his own credibility. By postponing some unpopular decisions while making others only after weighing how the press and public might react, he indeed hardened the American people and Congress to the necessity for military action, enabled the armed services to build up strength in South Vietnam, and kept the hawks largely at bay. Yet in the process he also peppered the public record with so many inconsistencies and circumlocutions that he prompted one commentator to observe that the record of his administration's "concealments and misleading denials . . . is almost as long as its impressive list of achievements."[1]

Once the United States had become fully committed to the war, major flaws

[1] Charles Roberts, "LBJ's Credibility Gap," *Newsweek*, 19 Dec 66.

in the administration's strategy created more public relations problems. Given the restrictions and limited goals Johnson had adopted—no extension of the ground war to North Vietnam, no invasion of Laos or Cambodia, no action that would induce Communist China to enter the war—the practical initiative rested with the enemy. He could choose when or where to fight. If American or South Vietnamese forces delivered a serious blow, he could withdraw into his sanctuaries to mend and regroup. All the while, his adherents could hide among the South Vietnamese, subverting the military and civilian bureaucracies and preparing for the day when the United States would tire and withdraw. Under the circumstances, the only viable option open to the United States was to convince the enemy that there was no hope for the Communist cause. To do that, however, the administration had first to convince the American people that South Vietnam was either worth a prolonged war of attrition or that U.S. forces could win in the end without a major sacrifice of lives and treasure.

Neither alternative was possible. For many reasons—political immaturity brought on by years of French misrule; a corrupt, entrenched bureaucracy; a lack of initiative aggravated by the "can do" impatience of the U.S. military; a basic American failure to understand the oriental mentality—the South Vietnamese were unreceptive to the sort of reforms that might have made their cause attractive to the American public. As for the enemy, with the Soviet Union and Communist China replenishing his materiel losses and with the number of young men coming of age every year in North Vietnam outpacing his battlefield casualties, he could lose every battle and still win. He had only to endure until the cost of the war for the United States increased to levels intolerable over the long term. Although there were moments of insight—Westmoreland's reflections on the "political attrition" the Johnson administration was inflicting upon itself and the military, Ward Just's article on the marines' programs in the I Corps Tactical Zone—neither the administration nor the press appears to have recognized the implications and potential consequences of the president's strategy.

As the war progressed, the frustrations endemic to the conflict nevertheless found their way into the press with disconcerting regularity. While capable of victories, reporters claimed, the South Vietnamese Army was all too ready to surrender the burden of the fighting to the United States. In the same way, they noted, American forces won on the battlefield but made little progress toward a satisfactory settlement of the war. Meanwhile, corruption remained rampant within the South Vietnamese bureaucracy, and the pacification program appeared either to make little headway or, as at Ben Suc, to be counterproductive.

The Johnson administration responded with public relations campaigns to demonstrate that the South Vietnamese were indeed effective, that pacification was working, and that American forces were making progress. The press dutifully repeated every one of the president's assertions. Yet as the Saigon correspondents continually demonstrated, each official statement of optimism about the war seemed to have a pessimistic counterpart and each statistic showing progress an equally persuasive opposite. When General Westmoreland commented at the

National Press Club in November 1967 that the enemy could no longer conduct large-unit operations near South Vietnam's cities, his statement received wide, mostly straightforward coverage in the press. Then, only two months later, the Tet offensive established that Communist forces retained the ability to attack the cities and to confound even the most astute advertising claims.

As the war progressed, information officers found themselves caught between the president's efforts to bolster support and their own judgment that the military should remain above politics. Beginning with General Wheeler's decision to disregard the advice of the Honolulu information conference that the Military Assistance Command should leave justification of the war to elected officials in Washington, they found themselves drawn progressively into politics, to the point that by late 1967 they had become as involved in "selling" the war to the American public as the political appointees they served.

Complicating the situation further was a conflict in Saigon between the American press and the military. With censorship politically impossible, the military had to make do with a system of voluntary guidelines that largely eliminated security problems but left reporters free to comment on the inconsistencies that plagued the U.S. effort. Believing that the press had in most cases supported official policies in earlier American wars, especially World War II, many members of the military expected similar support in Vietnam. When the contradictions engendered by President Johnson's strategy of limited war led instead to a more critical attitude, the military tended increasingly to blame the press for the credibility problems they experienced, accusing television news in particular of turning the American public against the war.

In so doing, critics of the press within the military paid great attention to the mistakes of the news media but little to the work of the majority of reporters, who attempted conscientiously to tell all sides of the story. They also misassessed the nature of television coverage, which, despite isolated instances to the contrary—the burning of Cam Ne, General Loan's execution of the Viet Cong officer—was most often banal and stylized. What alienated the American public, in both the Korean and Vietnam Wars, was not news coverage but casualties. Public support for each war dropped inexorably by 15 percentage points whenever total U.S. casualties increased by a factor of ten.[2]

The news media, for their part, responded in kind. Citing the clandestine bombing of Laos, the slowness with which information from the field reached Saigon, and instances of perceived dissembling, reporters accused the military of attempting to mislead the American public. Yet even as they leveled this charge, they yielded far too readily to the pressures of their profession. Competing with one another for every scrap of news, under the compulsion of deadlines at home, sacrificing depth and analysis to color, they created news where none existed—Arnett's story about the use of tear gas, Webster's report on the severing of an enemy's ear, a whole string of stories on the dire position of the marines at Khe

[2] John Mueller, *War, Presidents, and Public Opinion* (New York: Wiley, 1973).

Sanh—while failing to make the most of what legitimate news did exist. The good and bad points of the South Vietnamese Army and government, the wars in Laos and Cambodia, the policies and objectives of Hanoi and the National Liberation Front, the pacification program—all received less coverage in the press, positive or negative, than they probably should and could have. It is undeniable, however, that press reports were still often more accurate than the public statements of the administration in portraying the situation in Vietnam.

In the end, President Johnson and his advisers put too much faith in public relations. Masters of the well-placed leak, adept at manipulating the electorate, they forgot at least two common-sense rules of effective propaganda: that the truth has greater ultimate power than the most pleasing of bromides and that no amount of massaging will heal a broken limb or a fundamentally flawed strategy. Even if Zorthian, Bankson, Sidle, and the others had managed to create the sort of objectivity they sought in the press, they would have failed in their larger purpose. For as long as the president's strategy prevailed, the enemy would hold the initiative and casualties would continue, inexorably, to rise.

Bibliographical Note

The war in Vietnam is at once the best- and the worst-documented conflict in American history. Although the Washington National Records Center (WNRC) in Suitland, Maryland, contains an unprecedented volume of records on the subject, serious gaps in the chronicle of the war remain. Important record collections were abandoned to the North Vietnamese when Saigon fell to the Communists in 1975. Others were broken up and scattered in the melee that preceded the final collapse, when clerks at Tan Son Nhut Air Base loaded boxes of documents on any flight leaving the country, whatever the destination. Office managers in Washington meanwhile routinely thinned the files under their control, often without regard to the historical value of the collections they were destroying.

The records of the various agencies that dealt with the U.S. government's public relations during the Vietnam War are a case in point. The archives of the MACV Office of Information and of the Joint U.S. Public Affairs Office have all but disappeared, apparently the victims of the rush to evacuate Saigon. Also missing are the files of the Southeast Asia Desk of the Office of the Assistant Secretary of Defense for Public Affairs and the records of the Bureau of Public Affairs at the State Department.

Official Records

There are, nevertheless, many ways to approach a subject. With easy avenues of access closed, this study benefited from the bureaucracy's practice of making multiple copies of documents; in effect, the author reconstructed the official record from the documentation scattered among the archival collections of the various agencies in Washington that had some say in the government's public affairs. Many of the most important of those documents reside at the U.S. Army Center of Military History among the papers of General William C. Westmoreland. In addition to keeping a detailed diary, Westmoreland set aside and retained a wealth of back-up materials, including MACV position papers and studies, statistical summaries, memorandums, and a large number of important messages between the U.S. mission in Saigon and Washington agencies. Westmoreland's collection of backchannel messages has also been of value in providing a view of the political concerns that underlay many of the Johnson administration's public affairs policies and initiatives. Augmenting Westmoreland's papers are the records amassed by the general's military aide, Paul Miles, also on file at the Center. Although these documents are in many cases duplicates of items in the Westmoreland papers, they do contain a number of messages and memorandums unavailable anywhere else.

Westmoreland's papers provide the bones for a study of the military's relations with the news media, but they are hardly complete. A number of other sources must also figure in, especially the central files of the State Department. Although split into subjects and arranged chronologically—a circumstance that requires the researcher to plod methodically through mountains of paper to find the documents he needs—those records provide a relatively complete view of how MACV's public affairs policies meshed with those of other agencies, especially the Department of State. Record Group 951 for the early years of the war and the military operations file, Pol. 27 Viet S, for the period from 1965 were especially important. These records undoubtedly contain many of the documents that were once present in the files of the Bureau of Public Affairs.

The Directorate of Defense Information has apparently lost or destroyed the records of the Southeast Asia Desk of the Office of the Assistant Secretary of Defense for Public Affairs, but those records did exist when this study was beginning, a fact that allowed the author to survey and photocopy at least some of them. Those materials, filed with the author's working papers at the Center of Military History, augment and complement the other sources cited in the study, at times providing unique insights into the Johnson administration's public relations efforts.

Although the records of the MACV Office of Information were lost, some routine material was retired to the WNRC, mostly press release files but also a few policy documents. The Office of the MACV Historian also included a few of the agency's working papers and memorandums as back-up material for MACV's annual history. An index of those records, most on file at WNRC, is available at the Army's Office of the Adjutant General. Among the most significant of them is accession 69A702, boxes 2 to 4, 12, and 21, which contain a large number of background materials.

Among the secondary collections that contributed to this book, mention must be made of the U.S. Information Agency papers on file at WNRC, a group of records at one time so lost that even the originating agency had forgotten its whereabouts. Although short on information about operations in the field and mainly concerned with psychological warfare against the enemy, these papers contain a number of documents bearing upon relations with the press. Record Group 306, accession 72A5121, boxes 92 to 99, 226, and 228 contain administration, management, journalist, and press conference files. Accession 71A2101, boxes 67 to 80 contain considerable information on the censorship question and the U.S. mission's press center activities.

Of very great value, especially for understanding the Buddhist disturbances between 1963 and 1966, are the papers of Dr. Richard A. Gard, on file at the Center. Gard was the USIA's specialist on Buddhist affairs. He apparently kept much of the material that crossed his desk, amassing a major collection of studies and intelligence reports on Buddhist beliefs and activities. Also worthy of note are the collected records of MACV's pacification program, the DepCORDS Papers, on file at the Center and the Center's chronological file, a collection of miscellaneous studies, memorandums, and messages on possible courses of action in South Vietnam dating from 1964 to 1967. The papers of Clark Clifford at the Lyndon Baines Johnson Library in Austin, Texas, provided considerable information on the discussions within the Johnson administration that followed the Tet offensive. General Creighton Abrams' backchannel messages, on file at the Center, are likewise of importance in understanding that period and in shedding light on the order of battle controversy. So, too, are the personal papers of Thomas Thayer, who served as director of the Southeast Asia Intelligence and Force Effectiveness Division of the Southeast Asia Programs Office under the Assistant Secretary of Defense for Systems Analysis for most of the period between 1967 and 1975. Thayer's papers are on file in the Center. The author has made extensive use of the documents and histories on file at the Marine Corps History Office at the Washington Navy Yard. The Corps maintains a virtually complete record of its activities in South Vietnam, and its collections have been invaluable to this work. The author also consulted the notes and working papers of other Center historians—especially those of Vincent Demma, Richard Hunt, Jeffrey Clarke, and George MacGarrigle.

News Media Sources

This book would have been impossible to write without the newspaper, magazine, and television reports on file with the Air Force News Clipping and Analysis Service in the Pentagon. Broken into subjects and containing items on military affairs dating back to 1950, the collection draws together clippings from dozens of newspapers and magazines on every aspect of the Vietnam War covered by the news media. It is the only source for many of the early television reports on the war, all of which were excerpted verbatim for publication in the clipping service's in-house, Radio-TV-Defense Dialog. The author has also made extensive use of the *New York Times*, the *Wall Street Journal*, and the *Washington Post* and of such periodicals as *Time, Life, Newsweek,* the *New Yorker,* and *U.S. News & World Report*, all of which are on file either in the Pentagon Library or the Newspaper Reading Room of the Library of Congress.

Printed Works

Several published official histories have been of special use to this study. The most important collection is the so-called Pentagon Papers printed by the U.S.

Government Printing Office in twelve volumes under the title, *United States–Vietnam Relations, 1945–1967: A Study Prepared by the Department of Defense.* Citations in this work are taken from *The Senator Gravel Edition of the Pentagon Papers* (Boston: Beacon Press, 1971), which is generally more accessible to the reading public than the government edition. Also valuable is the MACV History, a multivolume, year-by-year record of the war compiled by the Office of the MACV Historian. This history provides a general view of the war's more technical aspects. The volume for 1968 provides an excellent picture of the operations of the MACV Office of Information during General Sidle's tenure as chief of information. Other volumes devote less space to the Office of Information but provide good coverage of MACV's position on many important disputes with the press. General Westmoreland's and Admiral Sharp's *Report on the War in Vietnam (as of 30 June 1968)* (Washington, D.C.: Government Printing Office, 1970) likewise provides essential background on the war along with a number of statistical summaries.

Transcripts or reports from three sets of congressional hearings also contributed immeasurably to this book. The first provides an important starting point for the reconstruction of official press policies in Vietnam during the early portion of the war and is officially titled: U.S. Congress, House, Eleventh Report by the Committee on Government Operations, Subcommittee on Foreign Operations and Government Information, *United States Information Problems in Vietnam,* 88th Cong., 1st sess., 1 Oct 63, H. Rpt 797. The second goes into more detail. Known informally as the Moss Report, it is cited as: U.S. Congress, Senate, Hearings Before the Committee on Foreign Relations, *News Policies in Vietnam,* 89th Cong., 2d sess. The third, dealing with the M16 controversy, is titled: U.S. Congress, House Committee on Armed Services, *Hearings Before the Special Subcommittee on the M-16 Rifle Program,* 90th Cong., 1st sess.

Interviews

Although this study draws mainly upon documentary and published sources, it relies as well on a number of interviews with the men responsible for creating MACV's information program conducted by the author or others. Wide-ranging interviews with Barry Zorthian, Rodger Bankson, and Maj. Gen. Winant Sidle clarified a number of questions unexplained by documentary sources. An interview with Col. Ralph Ropp and Lt. Col. Richard Bryan provided important information on MACV's handling of the tear gas controversy in 1965. An extremely candid interview with Col. Robert J. Coakley, the chief of the U.S. Army, Vietnam (USARV), Information Office during 1966 (CMH Interview VNIT-27 by Maj. Robert H. Van Horn, CO, 3d MHD), remains an important source for anyone attempting to piece together the structure and operations of the MACV Office of Information during the period.

Some Significant Secondary Works

A number of books and articles on press coverage of the war deserve special mention. Of extreme importance in understanding the period between 1963 and 1965 is John Mecklin's *Mission in Torment* (Garden City, N.Y.: Doubleday, 1965). David Halberstam's *The Making of a Quagmire* (New York: Random House, 1965) reads as though the author saw the State Department's classified file on the period. Although highly impressionistic, Jim Lucas' *Dateline: Vietnam* (New York: Award House, 1966) provides some insights into the problems confronting newsmen who reported the war during the early years.

Dale Minor presents a highly opinionated view of official press policies in *The Information War: How the Government and the Press Manipulate, Censor, and Distort the News* (New York: Hawthorne Books, 1970) as does Phillip Knightley in *The First Casualty* (New York: Harcourt Brace Jovanovich, 1975). Phil G. Goulding's book *Confirm or Deny* (New York: Harper & Row, 1970) supplies a more officially oriented perspective. For information on how the Army's general officers felt about the press, see Douglas Kinnard's *The War Managers* (Hanover, N.H.: The University Press of New England, 1977).

Louis Harris' *The Anguish of Change* (New York: W. W. Norton, 1973) provides some insight into public opinion of the war, as does the more puckish book on polling and pollsters by Michael Wheeler, *Lies, Damn Lies, and Statistics: The Manipulation of Public Opinion in America* (New York: Liveright, 1976). Perhaps the best book on the subject to date is John Mueller's *War, Presidents, and Public Opinion* (New York: Wiley, 1973). Mueller published a digest of his findings in "Trends in Popular Support for the Wars in Korea and Vietnam," *American Political Science Review* 65 (1971): 358. Hazel Erskine's article, "The Polls: Is War a Mistake," *Public Opinion Quarterly* 34 (Spring 1970): 134, is the main source for polling data on the so-called mistake question. One of the best polls on the war was conducted by the National Opinion Research Center of the University of Chicago. Its findings are in Sheldon Appleton, ed., *United States Foreign Policy: An Introduction With Cases* (Boston: Little, Brown & Co., 1968). Burns Roper's essay on public opinion during and after the Tet offensive is also revealing and important. See Burns Roper, "What Public Opinion Polls Said," in Peter Braestrup, *Big Story: How the American Press and TV Reported and Interpreted the Crisis of Tet in Vietnam and Washington*, 2 vols. (Boulder, Colo.: Westview Press in Cooperation with Freedom House, 1977), 1:674. Also of use are the many essays on the war in Peter Braestrup, ed., *Vietnam As History* (Washington, D.C.: Woodrow Wilson International Center for Scholars, 1984).

Peter Braestrup's monumental, two-volume history of press coverage of the Tet offensive, *Big Story*, provides a matrix for any study of the subject. Also important is Don Oberdorfer's *Tet!* (Garden City, N. J.: Doubleday, 1971). For an excellent South Vietnamese view of the offensive with important intelligence information, see Colonel Hoang Ngoc Lung, *The General Offensives of 1968–1969*, Indochina Monographs (Washington, D.C.: U.S. Army Center of Military History, 1981). George A. Bailey and Lawrence W. Lichty's study of how NBC News handled General Loan's execution of the Viet Cong officer is also enlightening. See Bailey and Lichty, "Rough Justice on a Saigon Street: A Gatekeeper Study of NBC's Tet Execution Film," *Journalism Quarterly*, Summer 1972, p. 274.

Index

☆ U.S. GOVERNMENT PRINTING OFFICE : 1989 O - 250-255 : QL 2

Made in United States
Cleveland, OH
05 November 2026

47668302R10233